Jürgen Wittmann
Werner Bergholz

# Introduction to Quality Management in the Semiconductor Industry

Volume I: General

Students' Version

# Preface

There are many excellent standard textbooks on quality management (QM), so why another one? The answer is simple: The semiconductor industry, i.e. microelectronics, photovoltaics, flat panel display and LED manufacturing and last but not least printed electronics all share the necessity for a 100% stringent quality management and quality engineering. Without strict adherence to QM principles, a microelectronics factory would come to a standstill within a few months if not weeks. The other mentioned industries are similar in the sense that sustainable operation and long-term survival of the operations is impossible without QM. A general textbook on quality management is helpful, but there is no textbook, which covers the special needs of the semiconductor industries.

So, the primary purpose of our book is to provide the "fast lane" to practical quality engineering and management in the semiconductor industries. In line with this objective, this book is meant to be more a guide to practical quality engineering and management, rather than a scientific treatise. Although it has been written for the semiconductor technology community, it goes without saying that it is useful for almost all other industrial areas, since in a way semiconductor technology (in particular microelectronics) is a good model case how 100% stringent QM can be implemented in practice.

Based on many years of practical experience of both authors in the semiconductor industry (semiconductor material and technology as well as quality management) this book provides a general overview of methods and tools in quality management in connection with and with special emphasis on semiconductor specific topics related to quality. It also contains many generic technical and business processes which have to be detailed and adjusted according to particular company, product or business segment requirements.

Volume I of this introduction to quality management in the semiconductor industry is composed of three major areas: statistical and quality engineering basics, quality management tools and methods and business excellence. It represents a general introduction to quality management and quality engineering with special emphasis on semiconductor technology

Volume II focuses more on semiconductor materials and technology specific defects such as point defects, dislocations, stacking faults and precipitates, voids and inclusions. The topics are covered in a systematics which is aligned to QM principles. For every defect type and formation mechanism the following topics will be covered:

- Why is it relevant, in other words, what are the detection of the potential performance and reliability problems caused by such defects?
- How can the problem be detected in mass production with acceptable cost?
- What is the formation mechanism and root cause of the problem?
- How can the problem be prevented?

For each topic there will be practical examples.

Experts from the semiconductor industry gave us valuable advice and suggestions for improvement. In particular, we want to thank P. Brysch, Dr. W. Gustin, Dr. K. Haidn, Dr. K. Langer and Dr. E. Müller from Infineon Technologies AG for their helpful advice.

Finally we want to encourage experts and professionals in the field of semiconductor related quality management to send us their publications to cover new knowledge in subsequent versions of this book or in a separate publication. Those articles may be detailing and/or updating already existing chapters or may be an entirely new chapter, e.g. in the area of functional safety, failure tree analysis (FTA), requirements engineering or six sigma.

# Authors

**Prof. Jürgen Wittmann** worked for fifteen years in the semiconductor and photovoltaics industry mainly in technology, materials & suppliers and quality management (production, suppliers, customers). He conducted more than seventy worldwide supplier audits. His publications focus on silicon substrate materials and quality management. Today he teaches quality management and microsystems technology at Beuth University of Applied Sciences, Berlin. He can be contacted at jwittmann@beuth-university.de.

**Prof. Werner Bergholz** has worked for 17 years for Siemens Semiconductor and Infineon Technologies AG in various positions (R&D, production, supply chain management, quality management and business excellence). He is a certified auditor for the ISO 9001 standard and an EFQM assessor. He served as Professor of Electrical Engineering at Jacobs University until his retirement in 2015, where his research topics included microelectronics, photovoltaics, LEDs and quality management and has authored more than 100 publications and contributed to several books.

He is the co-founder of the International Standards Consulting GmbH & Co KG. He serves as Assistant Secretary of the IEC Technical Committee 113 Nanoelectronics and as the European Co-Chair of the SEMI Regional European Standards Committee and the Silicon Wafer Technical Committee.

© Jürgen Wittmann
Beuth University of Applied Sciences, Berlin, Germany
Luxemburger Str. 10, 13353 Berlin, Germany

ISBN-13: 978-1718637504
ISBN-10: 1718637500

Introduction to Quality Management in the Semiconductor Industry Students' Version – Vol. 1 General
1st Ed., May 1st, 2018

This work is subject to copyright. All rights are reserved.

Printed by CreateSpace Independent Publishing, Amazon Media EU S. à.r.l., 5 Rue Plaetis, L-2338, Luxembourg in Luxembourg

# Contents

Contents ................................................................................................. v
Abbreviations ....................................................................................... xii
0. Introduction ..................................................................................... 1

**I. Mathematical, Statistical and Quality Engineering Fundamentals** ............... 7

1. Fundamental Statistical Methods ........................................................ 8

1.1. Visual Illustration of Data .................................................................. 8
    1.1.1. Error Lists ................................................................................. 8
    1.1.2. Value Beam .............................................................................. 9
    1.1.3. Histogram & Step Curve ......................................................... 10
    1.1.4. Pareto Diagram ...................................................................... 11
    1.1.5. Scatter Plot & Correlation Diagram ......................................... 12
1.2. Quantitative Description of Data ...................................................... 14
    1.2.1. Arithmetic Mean .................................................................... 14
    1.2.2. Median & Mode ...................................................................... 15
    1.2.3. Range & Mean absolute deviation (MAD) .............................. 16
    1.2.4. Variance & Standard Deviation .............................................. 18
    1.2.5. Quantiles, Quartiles & Interquartile Range (IQR) .................... 18
    1.2.6. Box- and Whisker Plots .......................................................... 20
1.3. Continuous Distributions .................................................................. 21
    1.3.1. Normal Distribution ................................................................ 21
    1.3.2. Log-normal Distribution .......................................................... 28
    1.3.3. Exponential Distribution ......................................................... 31
    1.3.4. Weibull Distribution ................................................................ 33
    1.3.5. $\chi^2$ Distribution ................................................................ 38
    1.3.6. t-Distribution (Student Distribution) ......................................... 39
1.4 Discrete Distributions ........................................................................ 40
    1.4.1. Binomial Distribution .............................................................. 40
    1.4.2. Poisson Distribution ............................................................... 41

2. Capabilities ..................................................................................... 43

2.1. Introduction ...................................................................................... 43
2.2. Machine Capability .......................................................................... 44
2.3. Process Capability ........................................................................... 47
    2.3.1. Normally Distributed Data with Upper and Lower Specification Limit ............ 48
    2.3.2. Non-Normally Distributed Data and One-Sided Specification Limits .......... 51
    2.3.3. Manual Estimation of the Process Stability and Distribution ........ 56
2.4. Gauge Capability ............................................................................. 56
    2.4.1. Introduction ............................................................................ 57
    2.4.2. Gauge Capability (type I analysis) .......................................... 58
    2.4.3. Gr&R Study (type II analysis) ................................................. 59
    2.4.4. Gr Study (type III analysis) ..................................................... 62
2.5. Business processes for capability studies ....................................... 63

3. Statistical Process Control ............................................................... 68

- 3.1. Motivation ................................................................................................................ 68
- 3.2. Quality Control Chart ............................................................................................. 70
  - 3.2.1. Introduction ...................................................................................................... 70
  - 3.2.2. Sampling ........................................................................................................... 72
  - 3.2.3. Design of Quality Control Charts ..................................................................... 73
  - 3.2.4. Shewart Control Charts for Attributive Characteristics ................................... 74
    - 3.2.4.1. p-Chart ..................................................................................................... 74
    - 3.2.4.2. np-Chart ................................................................................................... 75
    - 3.2.4.3. c-Chart ..................................................................................................... 75
    - 3.2.4.4. u-Chart ..................................................................................................... 76
  - 3.2.5. Shewart Control Charts for Variable Characteristics ....................................... 77
    - 3.2.5.1. Individual Value Chart ............................................................................. 77
    - 3.2.5.2. X Bar Chart .............................................................................................. 78
    - 3.2.5.3. X Bar Chart with Extended Limits .......................................................... 79
    - 3.2.5.4. S Chart ..................................................................................................... 80
    - 3.2.5.5. R Chart ..................................................................................................... 81
    - 3.2.5.6. Median Chart ........................................................................................... 81
  - 3.2.6. Acceptance Control Charts ............................................................................... 82
  - 3.2.7 Pearson Control Chart ....................................................................................... 83
- 3.3. Choice of Charts Based on Time Dependent Distribution Models ......................... 84
- 3.4. Instability Rules (Western-Electric-Rules) ............................................................. 91
- 3.5. Business processes .................................................................................................. 93

# 4. Fundamentals of Reliability Engineering ............................................................... 98

- 4.1. Introduction ............................................................................................................. 98
  - 4.1.1. Bath Tub Curve ................................................................................................. 98
  - 4.1.2. Failure Levels and Requirements ..................................................................... 99
  - 4.1.3. Failure Rate and FIT ....................................................................................... 100
  - 4.1.4. MTBF & MTTF .............................................................................................. 105
- 4.2. Acceleration Models for different stress factors ................................................... 106
  - 4.2.1. Approach ......................................................................................................... 106
  - 4.2.2. Arrhenius Model ............................................................................................. 108
  - 4.2.3. Eyring Model .................................................................................................. 112
  - 4.2.4. Peck Model ..................................................................................................... 113
  - 4.2.5. Coffin-Manson Model .................................................................................... 114
  - 4.2.6. Mission Profile ............................................................................................... 116
- 4.3. Reliability Stress Testing ...................................................................................... 119
  - 4.3.1. Preconditioning (PC) ...................................................................................... 119
  - 4.3.2. High Temperature Storage Life (HTSL) ........................................................ 120
  - 4.3.3. Temperature Cycling (TC) ............................................................................. 121
  - 4.3.4. Temperature Humidity Bias (THB) ............................................................... 122
  - 4.3.5. Autoclave (AC) .............................................................................................. 123
  - 4.3.6. (Unbiased) Highly Accelerated Stress Test ((U)HAST) ................................ 123
  - 4.3.7. High Temperature Operating Life (HTOL) .................................................... 124
- 4.4. End of Life Testing ............................................................................................... 124
- 4.5. Wear out mechanisms ........................................................................................... 124
  - 4.5.1. Front End of Line related failure mechanisms ............................................... 125
    - 4.5.1.1. Hot Carrier Stress ................................................................................... 125
    - 4.5.1.2. Time Dependent Dielectric Breakdown (TDDB) .................................. 127
    - 4.5.1.3. Negative Bias Temperature Instability (NBTI) ..................................... 128
    - 4.5.1.4. Electromigration .................................................................................... 130

    4.5.1.5. Stress Migration .................................................................. 131
    4.5.1.6. Metal Corrosion ................................................................... 132
    4.5.1.7. Alpha Particle Induced Soft Error ......................................... 132
    4.5.1.8. Non-Volatile Memory Reliability .......................................... 133
    4.5.1.9. Surface Inversion ................................................................. 135
  4.5.2. Assembly Process Related Failure Mechanisms ............................. 135
    4.5.2.1. Wire Bonding Reliability ....................................................... 135
    4.5.2.2. Ag Ion Migration ................................................................... 136
    4.5.1.3. Moisture Resistance ............................................................ 137
4.7. System Reliability ..................................................................................... 137
  4.7.1. Serial and Parallel Systems ............................................................ 137
  4.7.2. Bridge Structure .............................................................................. 140
  4.7.3. Different Failure Modes ................................................................... 141

## II. Quality Management: Tools and Operations ............................................. 142

## 5. FMEA .............................................................................................................. 143

5.1. Introduction ............................................................................................... 143
  5.1.1. History ............................................................................................. 143
  5.1.2. Motivation ........................................................................................ 144
  5.1.3. FMEA as Part of the Company Know-How ..................................... 146
  5.1.4. Product- and Process-FMEA ........................................................... 147
5.2. FMEA Execution ....................................................................................... 148
  5.2.1. Preparation ...................................................................................... 148
    5.2.1.1. Responsibilities .................................................................... 148
    5.2.1.2. Documents ........................................................................... 149
    5.2.1.3. Steps .................................................................................... 149
  5.2.2. General Approach ........................................................................... 150
  5.2.3. Example: Product FMEA ................................................................. 153
  5.2.4. Example: Process FMEA ................................................................. 156
  5.2.5. The FMEA Form .............................................................................. 159
5.3. Risk Priority Number ................................................................................. 161
  5.3.1. Occurrence ..................................................................................... 162
  5.3.2. Severity ........................................................................................... 163
  5.3.3. Detection ......................................................................................... 163
  5.3.4. Action Limits ................................................................................... 164
5.4 Weaknesses of the Method ....................................................................... 167
  5.4.1 Absolute Value of Risk Priority Numbers ......................................... 167
  5.4.2 Scaling of Risk Priority Numbers ..................................................... 167
5.5. Business Process ..................................................................................... 168

## 6. Process and Product Development & Qualification ................................. 175

6.1. Introduction ............................................................................................... 175
6.2. Requirements of the relevant standards .................................................. 176
  6.2.1. Planning .......................................................................................... 177
  6.2.2. Inputs .............................................................................................. 177
  6.2.3. Development Control Process ........................................................ 178
  6.2.4. Outputs ........................................................................................... 178
  6.2.5. Change Management during Development .................................... 179
6.3. Generic product development process ..................................................... 179
  6.3.1. Market and customer requirements ................................................ 179

  6.3.2. Approach ................................................................................. 180
  6.3.3. Phases and milestones ............................................................... 182
  6.3.4. Phases and deliverables .............................................................. 184
 6.4. Qualification & Market Introduction ................................................. 187
  6.4.1. Qualification Standards ............................................................. 187
  6.4.2. Family Concept .......................................................................... 189
  6.4.3. Customer Sampling & Market Introduction (non automotive) ...... 192
  6.4.4. Market Introduction Automotive: Safe Launch ........................... 195
  6.4.5. Robust Validation" ..................................................................... 195

## 7. Supplier Quality Management ................................................................. 199

 7.1. Requirements of the standards· ........................................................... 199
  7.1.1. General ....................................................................................... 199
  7.1.2. Control of external supply .......................................................... 200
  7.1.3. Information ................................................................................ 200
 7.2. Supplier Management Process ............................................................. 201
  7.2.1. Supplier Strategy ........................................................................ 201
  7.2.2. Supplier Selection & Qualification ............................................. 205
  7.2.3. Supplier Evaluation .................................................................... 212
  7.2.4. Supplier Development ................................................................ 217
  7.2.5. Supplier Integration .................................................................... 219
 7.3. Supplier Audit ..................................................................................... 219
  7.3.1. Definition and Motivation ........................................................... 219
  7.3.2. Supplier Audit Process ............................................................... 220
  7.3.3. Quantification of the Audit Result .............................................. 231
  7.3.4. Questionnaires ............................................................................ 234
 7.4. Quality Assurance Agreement ............................................................. 239
  7.4.1. Motivation and Purpose .............................................................. 239
  7.4.2. Contents of a QAA ..................................................................... 239
 7.5. Technical Material Management .......................................................... 241
  7.5.1. Material Specification ................................................................. 241
  7.5.2. Introduction of a New Material ................................................... 244
  7.5.3. Material Evaluation .................................................................... 245

## 8. Production Quality Management .............................................................. 247

 8.1. Requirements of the Standards· ........................................................... 247
  8.1.1. Measuring .................................................................................. 247
   8.1.1.1. General ............................................................................... 247
   8.1.1.2. Traceability ........................................................................ 247
  8.1.2. Production .................................................................................. 247
   8.1.2.1. Control of Production ........................................................ 248
   8.1.2.2. Marking and Back Tracing ................................................. 248
   8.1.2.3. External Property ............................................................... 248
   8.1.2.4. Preservation of Conformity ................................................ 249
   8.1.2.5. Post Delivery ..................................................................... 249
   8.1.2.6. Changes ............................................................................. 249
  8.1.3. Product Release .......................................................................... 249
  8.1.4. Nonconformances ....................................................................... 249
 8.2. Process and Product Control ................................................................ 250
  8.2.1 Clean Room ................................................................................. 250
   8.2.1.1 Particles ............................................................................... 250

  8.2.1.2. Clean Room Management .................................................................. 251
 8.2.2. Inline Process Control ........................................................................... 253
 8.2.3. Advanced Process Control (APC) ......................................................... 257
 8.2.4. Equipment Maintenance ........................................................................ 258
 8.2.5. Electrical Parameter Control ................................................................. 259
 8.2.6. Yield ...................................................................................................... 260
8.3. Reliability in Production ............................................................................... 263
 8.3.1. Wafer Level Reliability ........................................................................ 263
 8.5.2. Burn In ................................................................................................. 265
8.4. Deviation Management ................................................................................ 266
 8.4.1. Identification and Risk Assessment ...................................................... 267
 8.4.2. Analysis and Disposition ....................................................................... 268
 8.4.3. Preventive Actions and Closure ............................................................ 272
8.5. Process Change Management ....................................................................... 272
 8.5.1. Proposal for a Change ........................................................................... 273
 8.5.2. Change Assessment by a Review Board ............................................... 273
 8.5.3. Classification ........................................................................................ 274
 8.5.4. Qualification, Approval and Implementation ....................................... 275
 8.5.5. Customer Notification ........................................................................... 279
8.6. Control of Measurement Equipment ............................................................. 279
 8.6.1. Calibration Traceability ........................................................................ 279
 8.6.2. Calibration Interval ............................................................................... 280
 8.6.3. Calibration Record ................................................................................ 281
 8.6.4. Measurement System Analysis ............................................................. 282
8.7. Traceability of the Product ............................................................................ 282

## 9. Customer Quality Management .................................................................. 283

9.1. Requirements of the Standards ..................................................................... 284
 9.1.1. Customer Focus .................................................................................... 284
 9.1.2. Determination of Product Requirements .............................................. 284
 9.1.3. Customer Communication .................................................................... 284
 9.1.4. Customer Property ................................................................................ 285
 9.1.5. Customer Satisfaction ........................................................................... 285
9.2. Customer & Product Requirements .............................................................. 286
 9.2.1. Specification ......................................................................................... 286
 9.2.2. Quality Function Deployment (QFD) ................................................... 287
 9.2.3. Customer Survey .................................................................................. 291
9.3. Planned Changes ........................................................................................... 292
9.4. Complaint Management ................................................................................ 298
 9.4.1. General Process .................................................................................... 298
 9.4.2. 8D: Eight Disciplines ............................................................................ 299
 9.4.3. Root Cause Analysis ............................................................................. 301
  9.4.3.1. Fishbone / Ishikawa Diagram ......................................................... 302
  9.4.3.2. 5-Why ............................................................................................ 302
9.5. Product Recall ............................................................................................... 306
9.6. PPAP .............................................................................................................. 309
 9.6.1. PPAP Requirements .............................................................................. 310
 9.6.2. Customer Communication .................................................................... 315
 9.6.3. PPAP Process ........................................................................................ 315

## 10. Sub dpm Quality ......................................................................................... 316

10.1. Introduction ................................................................................................. 316
10.2. Automotive Semiconductor Customer Expectation ................................... 317
10.3. Sub dpm Quality Measures in the Semiconductor Industry ....................... 318
    10.3.1. Design Phase ...................................................................................... 320
        10.3.1.1. Quality Function Deployment (see 9.2.2. QFD) ........................ 321
        10.3.1.2. Design FMEA (see chapter "FMEA") ...................................... 321
        10.3.1.3. Redundancy ................................................................................. 321
        10.3.1.5. Built in Self Test (BIST) ............................................................. 324
        10.3.1.6. Design for Testability ................................................................. 325
        10.3.1.7. Design for Manufacturability (DfM) ......................................... 326
        10.3.1.8. Design for Reliability ................................................................. 329
        10.3.1.9. Simulation ................................................................................... 331
        10.3.1.10. Design for Analysis .................................................................. 331
        10.3.1.11. Characterization ....................................................................... 332
    10.3.2. Manufacturing ..................................................................................... 332
        10.3.2.1. Process and Product FMEA ........................................................ 332
        10.3.2.2. Statistical Analysis of Variance .................................................. 332
        10.3.2.3. Control Plan ................................................................................ 333
        10.3.2.4. Statistical Process Control .......................................................... 333
        10.3.2.5. Wafer Level Failure Mechanism Monitoring ............................. 334
    10.3.3. Test ...................................................................................................... 334
        10.3.3.1. Good Die and Bad Bin in a Bad Cluster .................................... 334
        10.3.3.2. Part average testing" .................................................................. 335
        10.3.3.3. Statistical Bin Yield Analysis .................................................... 336
        10.3.3.4. Screens ........................................................................................ 337
    10.3.4. Application .......................................................................................... 338
        10.3.4.1. Part Qualification & Robust Validation ..................................... 338
        10.3.4.2. Part Derating .............................................................................. 338
        10.3.4.3. System Engineering .................................................................... 339

## III. Business Excellence ....................................................................................340

## 11. Process Management ....................................................................................341

11.1 Introduction ................................................................................................. 341
    11.1.1. Basic Definitions ................................................................................. 341
    11.1.2. Motivation for Process Management .................................................. 344
    11.1.3. Modelling Techniques & Process Description ................................... 346
11.2 Process Management .................................................................................. 349
    11.2.1. Process IS-Analysis & Documentation ............................................... 352
    11.2.2. Performance Indicators ........................................................................ 352
    11.2.3 Target Setting ....................................................................................... 353
    11.2.4. Process Improvement .......................................................................... 355
        11.2.4.1. Continuous Process Optimization .............................................. 355
        11.2.4.2. Re-Engineering .......................................................................... 357
    11.2.6. Business Process ................................................................................. 361
11.3. Process Assessment ..................................................................................... 361
    11.3.1. Maturity Models .................................................................................. 361
    11.3.2. PPA: Pragmatic Process Assessment .................................................. 365
    11.3.3. Business Process: PPA ........................................................................ 373

## 12. Improvement, Excellence and Self Assessment ..........................................376

12.1. Evolution of the quest for quality .............................................................. 376
12.2. Continuous Improvement ............................................................................ 377
   12.2.1. Deming Cycle ...................................................................................... 378
   12.2.2. Continuous Improvement Circles ........................................................ 379
   12.2.3. Employee Improvement Suggestion Scheme ...................................... 380
   12.2.4. Improvement Initiatives ....................................................................... 382
12.3. Six Sigma ..................................................................................................... 383
   12.3.1. History and Definition .......................................................................... 383
   12.3.2. Six Sigma Organization ....................................................................... 385
   12.3.3. Method ................................................................................................. 386
12.4. Business Excellence & Quality Awards ..................................................... 390
   12.4.1. Introduction to TQM & Business Excellence ...................................... 390
   12.4.2. Deming Prize ........................................................................................ 391
      12.4.2.1. Introduction and History .............................................................. 391
      12.4.2.2. Assessment Criteria of the Deming Prize .................................... 392
   12.4.3. EFQM & EFQM Excellence Award ..................................................... 393
      12.4.3.1. Introduction and History .............................................................. 393
      12.4.3.2. Eight basic concepts of excellence ............................................... 393
      12.4.3.3. Nine Elements ............................................................................... 395
      12.4.3.4. RADAR ......................................................................................... 400
      12.4.3.5. Levels of Excellence ..................................................................... 403
   12.4.4. Malcolm Baldrige National Award ...................................................... 403
      12.4.4.1. Introduction and History .............................................................. 403
      12.4.4.2. Categories and Items .................................................................... 404
      12.4.4.3. Scoring ........................................................................................... 409
12.5. Self Assessment ........................................................................................... 410
   12.5.1. Introduction .......................................................................................... 410
   12.5.2. Execution .............................................................................................. 411
   12.5.3. Examples .............................................................................................. 413
      12.5.3.1. DIN EN ISO9004:2009 ................................................................. 413
      12.5.3.2. Standard Independent Approach .................................................. 414

## 13. Addendum: Statistical Tables ..................................................... 416

## 14. List of References ............................................................................ 420

14.1. Books / Reports / Thesis / Lectures ............................................................ 420
14.2. Journals & Conferences ............................................................................... 423
14.3. Company internal ........................................................................................ 429
14.4. International and Consortia Standards ....................................................... 429
14.5. Internet ......................................................................................................... 431

# Abbreviations

| | |
|---|---|
| AC | Autoclave |
| ADLI | Approach, Deployment, Learning, Integration |
| AEC | Automotive Electronics Council |
| AF | Acceleration Factor |
| Ag | Silver |
| Al | Aluminum |
| AltPSM | Alternating Phase Shift Mask |
| AMR | Anistropic Magneto Resistance |
| AOQ | Average Outgoing Quality |
| APC | Advanced Process Control |
| ATPG | Automatic Test Pattern Generation |
| Au | Gold |
| AV | Appraiser Variation |
| AQL | Accepted Quality Level |
| BE | Business Excellence |
| BIST | Built in Self Test |
| $B_{LCL}$ | statistical constant, see addendum |
| $B_{LWL}$ | statistical constant, see addendum |
| BPM | Business Process Management |
| BPR | Business Process Re-engineering |
| $B_{UCL}$ | statistical constant, see addendum |
| $B_{UWL}$ | statistical constant, see addendum |
| b2b, b-to-b | Business to Business |
| CAD | Computer Aided Design |
| CEO | Chief Executive Officer |
| $c_g$ | Gauge Capability Index |
| $c_{gk}$ | Critical Gauge Capability Index |
| CI | Continuous Improvement |
| CIC | Continuous Improvement Circle |
| $c_m$ | Machine Capability Index |
| $c_{mk}$ | Cricital Machine Capability Index |
| CMMI | Capability Maturity Model Integration |
| CMP | Chemical Mechanical Polishing |
| $c_p$ | Process Capability index |
| $c_{ph}$ | Cycles per Hour |
| $c_{pk}$ | Critical Process Capability index |
| CPU | Central Processing Unit |
| CTQ | Critical to Quality |
| Cu | Copper |
| DfM | Design for Manufacturability |
| DfT | Design for Testability |
| DIN | Deutsches Institut für Normung |
| $D_{LCL}$ | statistical constant, see addendum |
| $D_{LWL}$ | statistical constant, see addendum |
| DMAIC | Define, Measure, Analyze, Improve, Control |
| dpm | Defects per Million |
| dpmo | Defects per Million Opportunities |
| DRAM | Dynamic Random Access Memory |
| $D_{UCL}$ | statistical constant, see addendum |
| $D_{UWL}$ | statistical constant, see addendum |
| $E_A$ | Activation Energy |
| EBIT | Earnings Before Interest and Taxes |
| ECU | Electronic Control Unit |
| EEM | Electrical / Electronic Modules |
| EFQM | European Foundation for Quality Management |
| EIPS | Employee Improvement Proposal System |
| EN | Europäische Norm (European Standard) |

| | | |
|---|---|---|
| ERP | | Enterprise Resource Planning |
| ESD | | Electrostatic Discharge |
| EU | | European Union |
| EV | | Machine Variation |
| eV | | Electron Volt |
| FDC | | Fault Detection and Classification |
| FET | | Field Effect Transistor |
| FIB | | Focused Ion Beam |
| FIT | | Failure in Time |
| FMEA | | Failure Mode and Effects Analysis |
| FPGA | | Field Programmable Gate Array |
| FTA | | Fault Tree Analysis |
| fWLR | | fast Wafer Level Reliability |
| GE | | General Electric |
| GD&T | | Geometric Dimensioning and Tolerancing |
| GM | | General Motors |
| GMR | | Giant Magneto Resistance |
| Gr | | Gauge Repeatability |
| Gr&R | | Gauge Repeatability and Reproducability |
| HAST | | Highly Accelerated Stress Test |
| HR | | Human Resources |
| HTGB | | High Temperature Gate Bias |
| HTOL | | High Temperature Operating Life |
| HTRB | | High Temperature Reverse Bias |
| HTSL | | High Temperature Storage Life |
| HVST | | High Voltage Stress Test |
| HW | | hard ware |
| IC | | Integrated Circuit |
| $I_{DD}$ | | Supply Current |
| $I_{DDQ}$ | | Supply Current in Quiescent State |
| IEC | | International Electrotechnical Commission |
| IGBT | | Insulated Gate Bipolar Transistor |
| IQR | | Interquartile Range |
| ISO | | International Organization for Standardization |
| IT | | Information Technology |
| JEDEC | | Joint Electron Device Engineering Council |
| JESD | | JEDEC Standard |
| JIT | | Just in Time |
| JUSE | | Union of Japanese Scientists and Engineers |
| k | | Boltzmann's constant |
| KCC | | Key Control Characteristics |
| $k_g$ | | Customer specific Constant for Gauge r & R |
| $k_{gk}$ | | Customer specific Constant for Gauge r & R |
| KPI | | Key Performance Indicator |
| LBIST | | Logic Built in Self Test |
| LDD | | Lightly Doped Drain |
| LED | | Light Emitting Diode |
| $L_{eff}$ | | Channel Length |
| LeTCI | | Level, Trend, Comparison, Integration |
| LSL | | Lower Specification Limit |
| LTL | | Lower Tolerance Limit |
| LWL | | Lower Warning Limit |
| MADµ | | Mean Absolute Deviation (from mean value) |
| $MAD_{Median}$ | | Mean Absolute Deviation (from Median value) |
| MBIST | | Memory Built in Self Test |
| MIL | | Military Standard |
| MOS | | Metal Oxide Semiconductor |
| MR | | Median Rank |
| MRB | | Material Review Board |
| MS | | MicroSoft |
| MSA | | Measurement System Analysis |

| | |
|---|---|
| MTBF | Mean Time Between Fail |
| MTTF | Mean Time To Fail |
| MTTR | Mean Time To Repair |
| MV | Megavolt |
| NASA | National Aeronautics and Space Administration |
| NBTI | Negative Bias Temperature Instability |
| NBTS | Negative Bias Temperature Stress |
| NDA | Non Disclosure Agreement |
| NIST | National Institute of Standards and Technology |
| NMIP | National Metrology Institute of Japan |
| NMOS | n-type Metal Oxide Semiconductor |
| NNR | Nearest Neighbour Residual |
| NPL | National Physical Laboratory |
| OAI | Off Axis Illumination |
| OSHA | Occupational Safety and Health Administration |
| OPC | Optical Proximity Correction |
| PAT | Part Average Testing |
| PC | Preconditioning |
| PC | Personal Computer |
| PCSM | Product Special Characteristics Matrix |
| PCB | Printed Circuit Board |
| PCB | Process Change Board |
| PCN | Process Change Notification |
| PCRB | Process Change Review Board |
| PDCA | Plan, Do, Check, Act |
| PEM | Power Electronic System |
| PEM-Model | Process and Enterprise Maturity-Model |
| PMOA | p-type Metal Oxide Semiconductor |
| POR | Process of Records |
| PPA | Pragmatic Process Assessment |
| PPAP | Production Part Approval Process |
| PSM | Phase Shift Mask |
| PTB | Physikalisch Technische Bundesanstalt |
| ppm | Parts per Million |
| psia | Pounds Per Square Inch |
| PSW | Part Warrant Submission |
| PV | Process Variation |
| QA | Quality Assurance |
| QAA | Quality Assurance Agreement |
| QC | Quality Control |
| QCC | Quality Control Chart |
| QFD | Quality Function Deployment |
| QM | Quality Management |
| RADAR | Results, Approach, Deployment, Assessment, Refinement |
| RH | Relative Humidity |
| RIF | Robustness Indicator Figure |
| RMA | Return of Material |
| RoHS | Restriction of certain Hazardous Substances |
| ROI | Return of Invest |
| RPN | Risk Priority Number |
| R&D | Research and Development |
| R2R | Run-to-Run |
| SBL | Statistical Bin Limit |
| SEI | Software Engineering Institute of the Carnegie Melon University |
| SEM | Scanning Electron Microscope |
| Si | Silicon |
| SIMS | Secondary Ion Mass Spectroscopy |
| $SiO_2$ | Silicondioxide |
| SiP | System in Package |
| SMART | Specifically, Measurable, Actively Influencable, Relistic, Timed |
| SoC | System on Chip |

| | | |
|---|---|---|
| SOP | | Standard Operating Procedure |
| SPC | | Statistical Process Control |
| SPICE | | Software Process Improvement and Capability dEtermination |
| SRAF | | Sub Resolution Assist Feature |
| SW | | soft ware |
| SYL | | Statistical Yield Limit |
| $T_a$ | | Ambient Temperature |
| TC | | Temperature Cycling |
| TDDB | | Time Dependent Dielectric Breakdown |
| THB | | Temperature Humidity Bias |
| $T_j$ | | Junction Temperature |
| TMR | | Tunnel Magneto Resistance |
| TPS | | Toyota Production System |
| TQC | | Total Quality Control |
| TQM | | Total Quality Management |
| UHAST | | Unbiased Highly Accelerated Stress Test |
| US | | United States (of America) |
| USL | | Upper Specification Limit |
| UTL | | Upper Tolerance Limit |
| UWL | | Upper Warning Limit |
| VAR | | Variance |
| VDA | | Verband der Automobilindustrie |
| VIA | | Vertical Interconnect Access |
| VLVT | | Very Low Voltage Test |
| VoC | | Voice of the Customer |
| WLR | | Wafer Level Reliability |
| $x_r$ | | reference value of the standard |
| $\overline{X}_{Diff}$ | | Operator to Operator Variation |
| ZF | | Zahnradfabrik Friedrichshafen |

# 0. Introduction

Quality is understood as meeting customer needs and lack of deficiencies for the satisfaction of the customer.

> " 'Quality' means those features of products which meet customer needs and thereby provide customer satisfaction". [1]

> " 'Quality' means freedom from deficiencies" [2]

In order to achieve these objectives active management of quality is mandatory. According to J. Juran[3] "managing for quality makes extensive use of three '...' managerial processes":

- Quality planning
- Quality control
- Quality improvement"

The purpose of this book is to provide Quality Management methods and approaches to reach customer satisfaction in the field of semiconductor technology applications by quality planning, control and improvement.

<u>Semiconductor market and market requirements</u>
Semiconductor technology applications are split into six major segments (computer, consumer, communications, automotive, industrial/medical, and government/military) with significantly different product quality requirements regarding operation time, operating temperature and so forth.

For instance[4]

---

[1] Juran, J. (1999), section 2.1.
[2] Juran, J. (1999), section 2.2.
[3] Juran, J. (1999), section. 2.5.
[4] Tiederle, V. (2008)

➢ The overall operating time ranges from 2 up to 25 years (e.g. mobile phones vs. telecommunication or automotive)
➢ The weekly operating time ranges from 10 hours up to 168 hours (i.e. 100% usage)
➢ The operating temperature is assumed to be -40°C up to 85°C and may be much higher in an automotive environment e.g. close to the engine

In addition to this set of different requirements, over time the requirements are getting more demanding and increase in numbers, too. For example for the automotive semiconductor segment the following additional drivers of requirements were identified[1]. However, those are not limited to automotive semiconductors.

➢ more extreme operating conditions, i.e. high temperature above 140°C: the product is exposed to higher and higher temperatures, in particular in automotive applications with high expectations regarding reliability[2]: within twenty years the ambient temperature silicon based devices have been exposed to roughly 50 Kelvin higher operating temperatures in automotive applications.
➢ more extreme operating conditions in general, e.g. high humidity and vibration
➢ higher system power density

At the same time further shrinking of feature size and use of new materials (e.g. to be heat resistant) may lead to an acceleration or change of the dominant wear out mechanisms (e.g. regarding electro migration).

The growth rate of the number of ICs (integrated circuits) sold has been stable at 9% in average per year resulting in an exponential curve. The worldwide IC consumption per capita increased from six ICs in 1991 up to 28 ICs in 2013[3]. The highest annual growth rate of 10.8% (2013 – 2018) is observed in the automotive semiconductor segment[4]. In case of automotive the reason behind is an increase in infotainment, safety systems as well as driver assistance systems.

In general, however, with an increasing number of devices per system, i.e. a car, the fail rate of a single device has to decrease in order to keep the overall fail rate of the system at least stable.

Example[5]:

---

[1] Arthur D Little (2015)
[2] Von Tils (2006)
[3] Schaefer, U. (2008)
[4] Fangaria, P. (2014)
[5] Tiederly, V. (2008)

Assume forty ECUs (Electronic Control Units) in a car with a fail rate of one fail in $10^9$ hours and an operating time of 1200 hours per year. Assume in addition three hundred devices per ECU on average then the expected fail rate is 7.2% in five years!

Hence there is a wide range of requirements and an increasing pressure to improve the quality performance of the devices. Microelectronics is spreading into many applications with an increasing diversity of requirements. This is also expected to drive further tightening of quality standards.

Technology and Product
While facing increasing requirements regarding operating conditions and pressure to improve IC quality and reliability, at the same time feature sizes shrink at a dramatic pace and the number of transistors on a die is increasing fast. Also the chip area increases over time, which makes a chip more susceptible to defects.

> Moore's Law
> The number of components integrated into a semiconductor circuit doubles each year. [1]

In other words more and more components are placed on a single die, which drives the complexity of the IC and increases likelihood of a fail (fig. 0.1). In order to achieve this increase in performance, the minimum feature size on a chip has to be reduced constantly and significantly (fig. 0.2), e.g. by introducing new technologies, new materials, new manufacturing equipment and so forth.

Overall the complexity of semiconductor devices is increasing, too. Design, technology and production systems have to deal with complex procedures and products (fig. 0.3). Obviously high complex systems are more prone to failure or defect than simple systems.

---

[1] Mack, C. A. (2011)

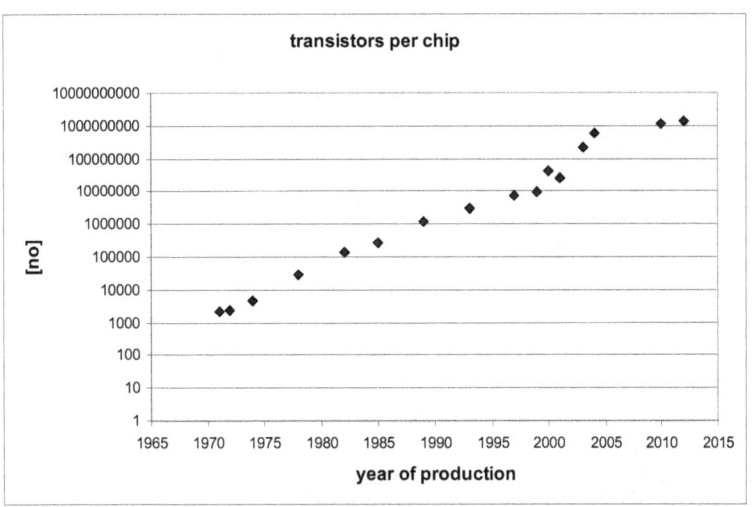

Fig. 0.1 Increasing number of transistors on an INTEL CPU[1,2]

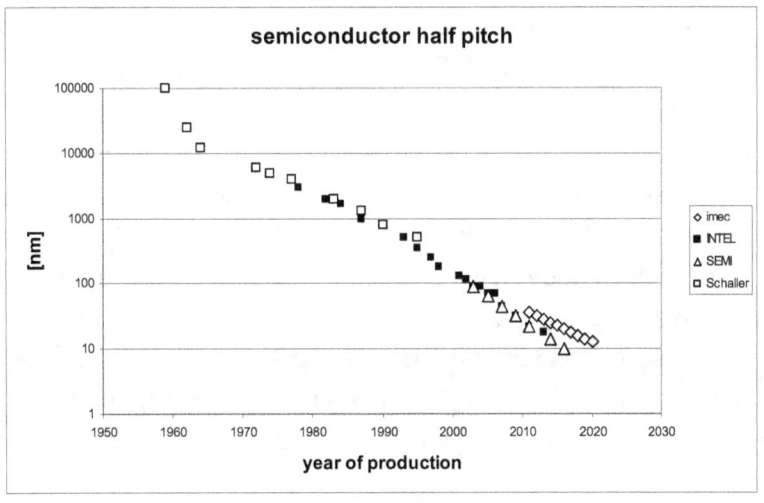

Fig. 0.2 Feature size reduction of semiconductor devices[3,4,5] and DRAMs in particular[6,7]

---

[1] INTEL (2) (2015)
[2] INTEL (3) (2015)
[3] Schaller, R. (2004)
[4] INTEL (2015)
[5] SEMI (2014)
[6] Allan, A. (2002)
[7] De Schepper, P. (2014)

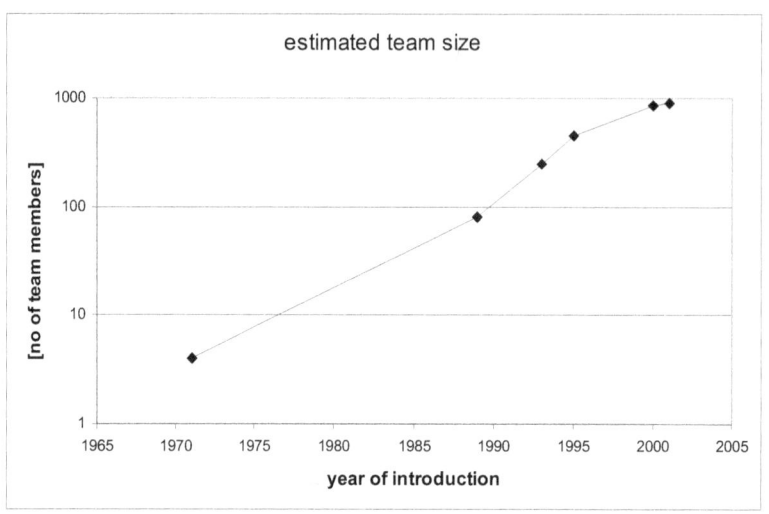

Fig. 0.3 Complexity of product development expressed in number of engineers working in the development team[1]

Production Process

On the production side the increase of complexity can e.g. be measured by counting number of mask levels or number of non-litho steps per critical mask level. In case of DRAM manufacturing, for example, the number of mask levels increased by 35% from the 30 nm technology node down to 20 nm technology node. At the same time the number of non-litho steps per critical mask level increased by 110%[2]

Manufacturing of semiconductor devices, in particular front end production, i.e. chip manufacturing, is highly challenging in regards to process and quality control & management. The semiconductor production process is mainly characterized by the following features:

- High complexity of the manufacturing process (technology)
- Clean room environment required & cleanness of materials is a must, hence special equipment e.g. for contamination analysis is required
- Continuous feature size reduction with new or modified technologies, equipment and materials
- The semiconductor content of many products is increasing, e.g. in case of automobiles the semiconductor content is growing at an estimated 8.1%[3].

---

[1] Schaller, R. (2004), p. 69
[2] Byers, E. (2015)
[3] Hartsell, M. (2008)

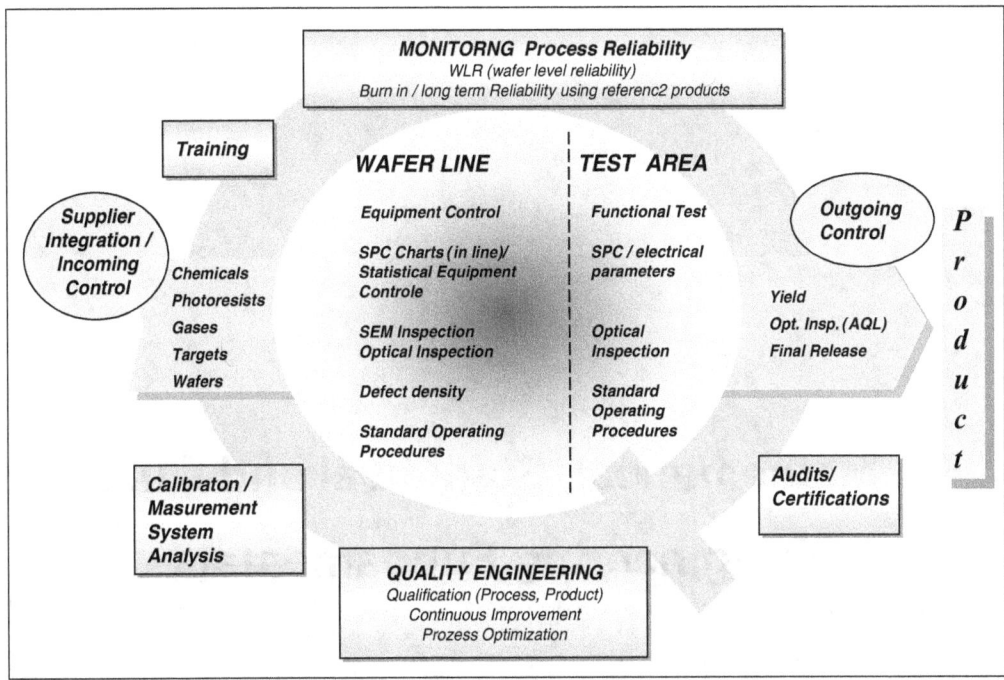

Fig. 0.4 Quality related activities in production using a semiconductor manufacturing environment[1]

Thus the semiconductor production is under significant pressure to deliver high quality products. Fig. 0.4 provides an overview of quality related activities in a semiconductor fab. Many of those activities are also relevant for any other production environment, e.g. training of workers, incoming control, outgoing control.

However, this section also introduces semiconductor specific activities like wafer level reliability, defect density control, SEM (scanning electron microscopy) inspection and many others.

---

[1] Geleng (1999) and Bergholz, W. (2008)

# I. Mathematical, Statistical and Quality Engineering Fundamentals

# 1. Fundamental Statistical Methods

In many companies today's quality management philosophy is to achieve zero defects in production as well as in product development and other areas. In order to achieve this or at least to get close to this target, quality management methods need to be in place within the company which support this goal.

A low number of failures or defects, stable process results close to the target value or rejection of bad batches require knowledge and use of the relevant quality engineering methods.

This chapter contains fundamental statistical concepts as well as quality tools to meet these requirements.

## 1.1. Visual Illustration of Data

It is well known fact that visualization of data facilitates the analysis and interpretation of the empirical findings. In the following the most important standard methods are described.

### 1.1.1. Error Lists

Error lists in form of tally sheets represent a very simple way to illustrate which and how many failures occur in a period of time (fig. 1.1). Each error found is documented in the list by adding a tally at the respective failure type.

Setting up an error list requires
- The definition of the type of the defects and/or defect classes (e.g. red and green discoloration is summarized in "discoloration")
- The definition of the respective time periods
- Master data for clear identification of product or process (e.g. name, ident number, product number)
- Place and method of inspection
- Definition of the sampling rate
- To summarize all of this in a controlled document including date and name of inspector

> „Controlling documents per ISO means tracking documents to ensure that employees who refer to them always have the correct document available. ISO document control procedures specify measures you can use to control quality assurance documents and satisfy the requirement to withdraw issued documents when they are obsolete and replace them as revisions are approved." [1]

---

[1] Markgraf, B. (2014)

| Error list | | | | | | | |
|---|---|---|---|---|---|---|---|
| Product ID: 3A 412 751  Name: 1G DRAM | | | Place of inspection: outgoing control  Kind of inspection: visual inspection | | | | |
| Nr. | type of failure | number of failures | | | | | |
| | | cw5 2013 | | | | | |
| | | Feb 5th | Feb 6th | Feb 7th | Feb 8th | Feb 9th | Feb 10th | Feb 11th |
| 1 | Scratch | ⅢⅢI | ⅢⅢIII | ⅢⅢ | II | ⅢⅢ I | ⅢⅢII | ⅢⅢ I |
| 2 | Missing line | III | I | I | III | II | I | |
| 3 | Corrosion | III | III | ⅢⅢII | I | II | | II |
| 4 | Discoloration | ⅢⅢ | | III | | | ⅢⅢ ⅢⅢI | |
| 5 | Others | ⅢⅢ ⅢⅢ | ⅢⅢ ⅢⅢII | ⅢⅢ ⅢⅢ | III | ⅢⅢ ⅢⅢ | ⅢⅢ ⅢⅢ | ⅢⅢ ⅢⅢ |

Fig. 1.1 Error list

The error list combines several advantages like quick and simple compilation of the format of the list and a very simple way to edit it. In addition, this method, assuming the number of different failure types is not too big, enables a quick identification of failure trends and focus areas.

### 1.1.2. Value Beam

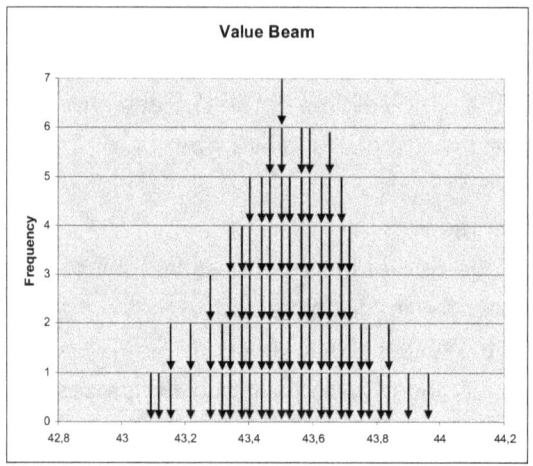

Fig. 1.2 Value beam

Value beams show the frequency distribution of a characteristic or of an incident based on individual values[1]. The x-axis represents the range of values and the y-axis the quantity of each value. The class width along the x-axis is basically defined by the resolution of the measurement equipment. Each single value is considered which prevents potential loss of information due to data classification.

---

[1] Dietrich, E., Schulze, A. (2014), p. 81

## 1.1.3. Histogram & Step Curve

Histograms are bar-diagrams showing a frequency distribution of a characteristic or of an incident etc. Data is summarized in a compact manner which allows a quick evaluation of the important parameters of a set of data, e.g. location of the data compared to the target value or the specification limits, rough estimate of the form of distribution. The advantage of histograms over value beams is a clearer appearance.
As a rule of thumb the number of classes k equals the square root of the number of data points n[1]:

$$k \approx \sqrt{n}$$

For large n (n ≥ 1000) the number of classes k can be determined by[2]:

$$k \approx 10 \cdot \lg(n)$$

The class width b can generally be calculated e.g. using[3]

$$b = \frac{(x_{max} - x_{min})}{k}$$

or[4]

$$b = \frac{(x_{max} - x_{min})}{(1 + 3{,}32 \cdot \lg(n))}$$

Class widths and class limits have a significant impact on the appearance of a histogram and may lead to wrong conclusions, if the data assessment is not executed complying to a consistent procedure. The disadvantage of histograms over value beams is a minor loss of information by classifying the data.

In addition to the histogram the step curve can be used, too, to present the same data in a different form (fig. 1.4). In this case the data is stepwise cumulated.
Often, the cumulated frequency is illustrated using percentage numbers (cumulative frequency plot).

---

[1] Hedderich, S., Sachs, L. (2012), p. 101
[2] Hedderich, J., Sachs, L. (2012), p. 101
[3] Dietrich, E., Schulze, A. (2014), p.82
[4] Hedderich, J., Sachs, L. (2012), p.101

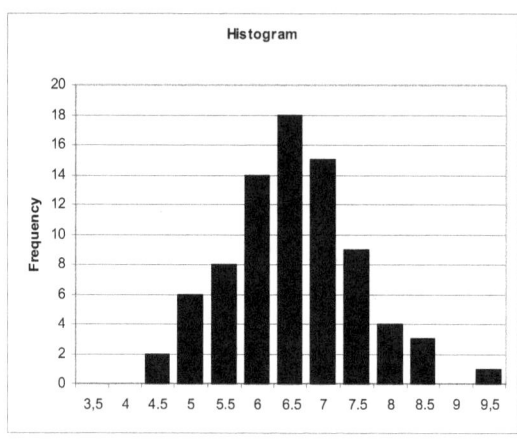

Fig. 1.3 Example for a histogram

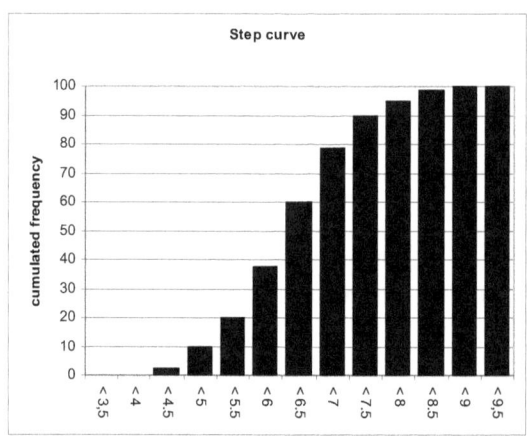

Fig. 1.4 Example of a step curve / cumulative frequency plot

### 1.1.4. Pareto Diagram

The pareto diagram is again a bar diagram showing the frequency of a characteristic or of problems. The bars are arranged sequentially descending in frequency from the left to the right (fig. 1.5).

The purpose of arranging data in this way is to quickly visualize and address the top 3 or top 5 topics representing ~ 80% of the problems. Hence, it is a priorization chart.

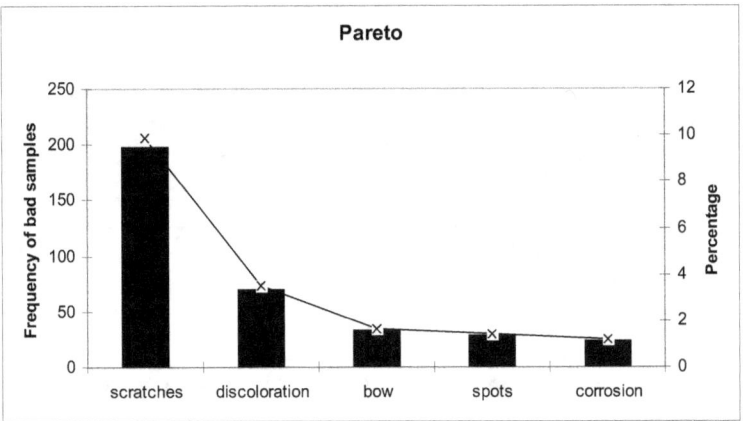

Fig. 1.5 Example for a pareto diagram

### 1.1.5. Scatter Plot & Correlation Diagram

A scatter diagram is used to visually display and compare two sets of related quantitative, or numerical, data[1]. It's a graphical illustration of the relationship between two sets of data, in particular it may show a potential correlation between those sets of data (fig. 1.6). It's then also called correlation diagram.

For the linear regression[2] the best fit line is described by

$$y = a + bx = \bar{y} + b \cdot (x - \bar{x})$$

with:
$\bar{x}, \bar{y}$ : mean values
$b$ : slope of the fitted line by linear regression

a and the slope of the fit line b is estimated using the ordinary least square method (Carl Friedrich Gauss, 1777 – 1855) to be by the following formulas:

$$a = \bar{y} - b \cdot \bar{x} \qquad b = \frac{n \sum_{i=1}^{n} x_i y_i - \sum_{i=1}^{n} x_i \sum_{i=1}^{n} y_i}{n \sum_{i=1}^{n} x_i^2 - (\sum_{i=1}^{n} x_i)^2}$$

---

[1] Wagner, T., Statistics Handbook (2008)
[2] Hedderich, J., Sachs, L. (2012), p. 116

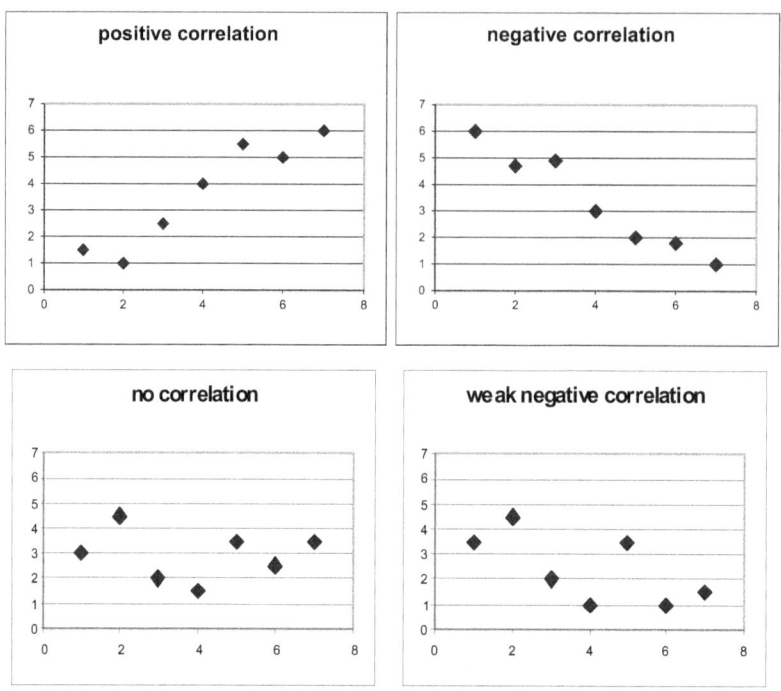

Fig. 1.6 Examples for correlations of data

Using the mean values (see 1.2.1. Arithmetic Mean) b can be described in the following way, too[1]

$$b = \frac{\sum_{i=1}^{n}(x_i - \bar{x})(y_i - \bar{y})}{\sum_{i=1}^{n}(x_i - \bar{x})^2}$$

K. Pearson[2] used a quantitative method to classify the strength of a correlation between two sets of data. The strength of the correlation is described using the correlation coefficient R which is found to be in the interval $[-1;1]$[3].

The correlation is

- very strong for  $0.87 \leq Abs(R) \leq 0.99$
- strong for  $0.71 \leq Abs(R) \leq 0.86$
- medium for  $0.50 \leq Abs(R) \leq 0.70$

---

[1] Kappes, S. (2014)
[2] Pearson, K. (1895)
[3] Klotz, S. (2014)

> weak for          Abs(R) ≤ 0.50

> R = 0:   there is no linear correlation
>
> R = -1:  there is a perfect negative linear correlation
>          the bigger X, the smaller Y and vice versa
>
> R = 1:   there is a perfect positive linear correlation
>          the bigger X, the bigger Y and vice versa

The empirical correlation coefficient R is calculated according the following equation[1]:

$$R = \frac{\sum_{i=1}^{n}(x_i - \bar{x}) \cdot (y_i - \bar{y})}{\sqrt{\sum_{i=1}^{n}(x_i - \bar{x})^2} \cdot \sqrt{\sum_{i=1}^{n}(y_i - \bar{y})^2}}$$

The square of the correlation coefficient $R^2$, the coefficient of determination, is interpreted as the part of the spread of a variable caused by another variable in percent. If, for instance, R = 0.6, then $R^2$ is 0.36, i.e. 36% of the spread of the Y-values can be explained linearly by the spread of the X-values.

| Correlation Coefficient R | Coefficient of Determination $R^2$ |
|---|---|
| 0.87 | 0.75 |
| 0.71 | 0.50 |
| 0.5 | 0.25 |

Table 1.1 Correlation coefficient R vs. Coefficient of determination $R^2$

## 1.2. Quantitative Description of Data

### 1.2.1. Arithmetic Mean

The arithmetic mean is the sum of all observed values divided by the number of observations[2]. It is also called average, average value or mean:

---

[1] Hedderich, J., Sachs, L. (2012), p. 109
[2] Hedderich, J., Sachs, L. (2012), p. 78

$$\bar{X} = \frac{X_1 + X_2 + X_3 + ... + X_n}{n} = \frac{1}{n}\sum_{i=1}^{n} X_i$$

Example:
Territory of
| | | |
|---|---|---|
| Canada | : | 9.984.670 km² |
| USA | : | 9.826.675 km² |
| Mexiko | : | 1.972.550 km² |
| Overall territory | : | 21.783.895 km² |
| Mean | : | 7.261.298 km² |

The mean includes all data points of a distribution.

**Outliers have a significant impact on the value of the mean.**

In addition to the normal arithmetic mean also the weighted mean is often used. In this case each value is weighted according the importance or the meaning of the value.

$$\bar{X} = \frac{w_1 X_1 + w_2 X_2 + w_3 X_3 + ... + w_n X_n}{w_1 + w_2 + w_3 + ... + w_n} = \frac{\sum_{i=1}^{n} w_i X_i}{\sum_{i=1}^{n} w_i}$$

with   $X_i$ : individual value; $w_i$: weighting factor

Example
A group of people of 10 persons has an average income of $ 1.000.000 per year. Another group of people of 1 million persons has an average income of $ 35.000 per year. The weighted mean is calculated to be

$$\bar{X} = \frac{10 \cdot 1.000.000 + 1.000.000 \cdot 35.000}{1.000.010} = 35.099,65$$

## 1.2.2. Median & Mode

The <u>median</u> is the value with the same number of smaller and larger values within a distribution. When all n values in a distribution are sorted by size the median is the[1]

---

[1] Gellert, W. et al. (1974)

$[(n+1)/2]^{th}$ value,            if n is odd

and the mean value of the

$(n/2)^{th}$ and the $[(n+1)/2]^{th}$ value,      if n is even

Hence the median $\tilde{X}$ is defined as:

$$\boxed{\tilde{X} = X_{(n+1)/2}} \qquad \text{if n is odd}$$

$$\boxed{\tilde{X} = \frac{1}{2}(X_{n/2} + X_{(n+1)/2})} \qquad \text{if n is even}$$

Example:
The following sequence of numbers is given:     1 / 7 / 1.5 / 0.5 / 1 / 0 / 2 / 0 / 3 / 4
Sorting by magnitude leads to:     0 / 0 / 0.5 / 1 / **1** / **1.5** / 2 / 3 / 4 / 7
Median: (1+1.5) / 2 = 1.25

> Outliers have no impact on the value of the median.

The <u>mode</u> is the mostly observed data point within a distribution of data. A distribution can have more than one mode, in case two or more values are observed with the same rate of occurrence.

Example:
The following sequence of numbers is given:     0 / 0 / 0 / 0.5 / 1 / 1.5 / 2 / 2 / 3 / 4
Mode: 0; As shown in the example the mode may be at the edge of a distribution, too.

### 1.2.3. Range & Mean absolute deviation (MAD)

The <u>range</u> of a distribution is the maximum value minus the minimum value of this distribution.

> The range of a distribution is very sensitive to extreme values.

Example:
Test of airbag inflation time in quality control (sec.): 0.70 / 0.80 / 0.85 / 0.95 / 0.80
In this case the spread is 0.25 sec.

The <u>mean absolute deviation</u> $MAD_\mu$ describes the average deviation of a value from the mean value of the distribution.

$$MAD_\mu = \frac{\sum_{i=1}^{i=n} |x_i - \mu|}{n}$$

with:
n : number of data points
$x_i$ : individual value
$\mu$ : average value

All data points are included in the calculation of the MAD.

**The MAD is less sensitive to extreme values than the range.**

Example:
Two experiments are executed leading to the identical mean value $\mu = 10$. Both experiments comprise five measurement points.

Experiment 1
Measured values         : 5.0 / 7.0 / 11.0 / 12.0 / 15.0
Deviations from mean    : 5   3   1   2   5
Mean absolute deviation : 3.2

Experiment 2
Measured values         : 9.0 / 9.5 / 10.0 / 9.5 / 12.0
Deviations from mean    : 1   0.5   0   0.5   2
Mean absolute deviation : 0.8

The <u>mean absolute deviation</u>[1] $MAD_{Median}$ describes the average deviation of a value from the median value of the distribution.

$$MAD_{Median} = \frac{\sum_{i=1}^{i=n} |x_i - med|}{n}$$

with:
n : number of data points
$x_i$ : individual value
med : median value of the distribution

---
[1] Hedderich, J., Sachs, L. (2012), p. 71

## 1.2.4. Variance & Standard Deviation[1]

The variance is the average squared deviation from the mean value of the distribution.

$$VAR = \frac{\sum_{i=1}^{i=n}(x_i - \mu)^2}{n-1}$$

Example:
Using the same data like in the previous section two distributions with an identical mean value lead to different variance values:

Experiment 1:
Measured values      : 5.0 / 7.0 / 11.0 / 12.0 / 15.0
Squared deviations   : 25   9   1   4   25
Variance             : 16

Experiment 2:
Measured values      : 9.0 / 9.5 / 10.0 / 9.5 / 12.0
Squared deviations   : 1   0.25   0   0.25   4
Variance             : 1.375

The standard deviation σ is the positive square root of the variance.

$$\sigma = +\sqrt{VAR}$$

Experiment 1 and experiment 2 results in standard deviations of 4 and 1.173, respectively. The standard deviation is the primary figure to characterize the statistical spread of a normal distribution.

## 1.2.5. Quantiles, Quartiles & Interquartile Range (IQR)

Quantiles
A quantile, also known as fractile, is the fraction of values which are smaller than the quantile value. A α-quantile defines the data point which separates α*100% or smaller values from (1-α)*100% or larger values of the distribution.

The calculation[2] of quantiles is done according to these rules:

---
[1] Hedderich, J., Sachs, L. (2012), p. 80
[2] Hedderich, J., Sachs, L. (2012), p. 71

(1) Arrange the data with n data points in an ascending sequence of numbers
(2) Calculate the position number k of the respective quantile:

$$k = \text{int}[n \cdot \alpha] + 1 \quad \text{if } n*\alpha \text{ not integer}$$

$$k = n \cdot \alpha \quad \text{if } n*\alpha \text{ integer}$$

(3) Assign the respective value to the position

$$x_\alpha = x(k) \quad \text{if } n*\alpha \text{ not integer}$$

$$x_\alpha = [x(k) + x(k+1)]/2 \quad \text{if } n*\alpha \text{ integer}$$

Example: 33% quantile

3 / 5 / 1 / 2 / 6 / 2.1 / 8 / 9 / 11 / 1.9 / 1.4 / 0.5 / 0 / 4

(1)  0 / 0.5 / 1 / 1.4 / 1.9 / 2 / 2.1 / 3 / 4 / 5 / 6 / 8 / 9 / 11

(2)  k = int(14*0,33)+1 = int(4,62)+1 = 5

(3)  $x_{33\%}$ = 1.9

## Quartiles

Three values fragment a frequency distribution into four parts. The central value is the median. The other two values are the lower (1st quartile: $Q_1$) and upper (3rd quartile: $Q_3$) quartile, i.e. 25% of the values of the distribution are smaller than $Q_1$ and 75% of the values of the distribution are smaller than $Q_3$.

> The first quartile $Q_1$ is the value which stands at the end of the 1st quarter of data points.
>
> The third quartile $Q_3$ is the value which stands at the end of the 3rd quarter of data points.

The calculation of the quartiles is according to the calculation of $Q_1$ and $Q_3$ is performed in the following way

(1) Arrange the data with n data points in an ascending sequence
(2) Calculate the position number k of the respective quartile:

$$k = \text{int}[(n+1) \cdot 0.25] \quad \text{for } Q_1$$

$$m = \text{int}[(n+1) \cdot 0.75] \quad \text{for } Q_3$$

(3) Assign the respective value to the position

$$Q_1 = x(k) \quad \text{for } Q_1$$

$$Q_3 = x(m) \quad \text{for } Q_3$$

Inter quartile range (IQR)
The inter quartile range is another figure to describe the spread of a distribution. It is simply the distance between the upper and lower quartile.

$$IQR = Q_3 - Q_1$$

## 1.2.6. Box- and Whisker Plots

A box- and whisker plot, briefly a box plot, summarizes range and skewness of a distribution in one diagram (fig. 1.7). It shows extreme values or high and low percentiles, the median and upper and lower quartile in one diagram.
The box plot provides a quick overview if a distribution is symmetric or skewed. Various distributions can be compared very easily.
The box between lower and upper quartile counts for 50% of the distribution. The extreme values are either the maximum and minimum values. In case the end of the whiskers represent percentiles (e.g. 1%, 99%), then single outliers are shown as single dots below and/or above the whisker's end. The distance between the quartile and the respective extreme value is illustrated by the whiskers.
The box plot is used independently of the shape of the distribution.

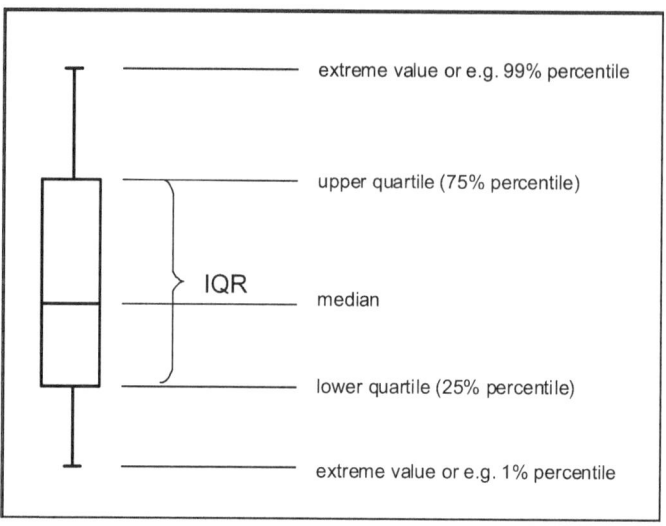

Fig. 1.7 Illustration of a box- and whisker plot

## 1.3. Continuous Distributions

"A continuous probability distribution[1] is a probability distribution which possesses a probability density function", i.e. a variable can take on any particular value between two specified values.

### 1.3.1. Normal Distribution

The normal distribution, also known as the Gauss distribution, is one of the most frequently used distributions to describe the distribution of a characteristic within a population or the distribution of a measurement value in case of many independent small incidents. The normal distribution is symmetric. Mean value, median and mode are identical.

Due to the enormous practical importance of the normal distribution function, this section is elaborated in more detail than the sections for the practically relevant other distribution functions.

Mathematical description[2]
Probability density function (PDF)

$$f(x) = \frac{1}{\sqrt{2\pi}} \frac{1}{\sigma} \exp\left(\frac{(x-\mu)^2}{-2\sigma^2}\right)$$

---

[1] Continuous probability distribution, www.princeton.edu (2015)
[2] Hedderich, J., Sachs, L. (2012), p. 238

with  μ    : mean value
      σ    : standard deviation

The normal distribution is transformed to the so called standard normal distribution[1] (fig. 1.8) by the introduction of u:

$$u = \frac{x - \mu}{\sigma}$$

Hence:

$$f(u) = \frac{1}{\sqrt{2\pi}} \exp\left(\frac{-u^2}{2}\right)$$

The respective cumulative distribution function (CDF) is the integral of the probability density function

$$F(u) = \frac{1}{\sqrt{2\pi}} \int_{-\infty}^{u} e^{-\frac{v^2}{2}} dv$$

with the following properties:

$F(-\infty) = 0$
$F(0) = 0{,}5$
$F(\infty) = 1$

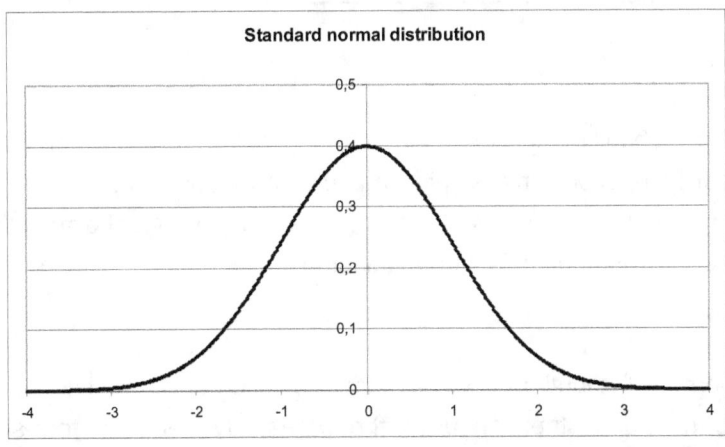

Fig. 1.8 Probability density function of the standard normal distribution (created using office software)

---

[1] Weiß, M., Österreicher, F. (2014)

## Quantitative Characteristics

Unfortunately, the distribution of data is sometimes not perfectly composed according the symmetric normal distribution. In particular skewness and kurtosis are additional characteristics describing distributions (not only normal distributions)[1].

Skewness describes the asymmetry of a distribution and can be calculated:

$$\sqrt{b_1} = \frac{\frac{1}{n}\sum_{i=1}^{n}(x_i - \mu)^3}{\left(\sqrt{\frac{1}{n}\sum_{i=1}^{n}(x_i - \mu)^2}\right)^3}$$

For normal distributions: $\sqrt{b_1} = 0$

Skewness < 0 means the distribution is bent to the right hand side (negative skewness), whereas for skewness > 0 the distribution is bent to the left hand side (positive skewness)

Kurtosis $b_2$ describes the shape of the probability density function.

$$b_2 = \frac{\frac{1}{n}\sum_{i=1}^{n}(x_i - \mu)^4}{\left(\sqrt{\frac{1}{n}\sum_{i=1}^{n}(x_i - \mu)^2}\right)^2} - 3$$

For normal distributions: $b_2 = 0$

Kurtosis < 0 means the probability function is more semi-circular shaped, whereas for kurtosis > 0 the probability function is more spiky than the normal distribution.
If kurtosis and/or skewness are too far away from 0 instead of the normal distribution, frequently e.g. a log-normal distribution is a good assumption.

## Probability plot

Before assuming a normal distribution, a verification should be performed. A simple method to verify the normal distribution uses the probability net in a four-step-approach[2,3]:

---

[1] Wagner, T., Statistics Handbook (2008)
[2] Weiß, M., Österreicher, F. (2014)
[3] Müller, U. (2007)

(1) Arrange the data according to the density distribution in ascending order and defined classes
(2) Assign a percentage number to each class of data
(3) Cumulate the percentage numbers
(4) Plot the data of the cumulative distribution function in a logarithmic graph

➔ If the plot is approximately a straight line, then the distribution is normal.

Example (adopted from M. Weiß and F. Österreicher[1]):
The following set of data shows the duration of pregnancies (in days) which were collected in 1978 in one of the County General Hospitals in the US.

| 251 | 264 | 234 | 283 | 226 | 244 | 269 | 241 | 276 | 274 |
|---|---|---|---|---|---|---|---|---|---|
| 263 | 243 | 254 | 276 | 241 | 232 | 260 | 248 | 284 | 253 |
| 265 | 235 | 259 | 279 | 256 | 256 | 254 | 256 | 250 | 269 |
| 240 | 261 | 263 | 262 | 259 | 230 | 268 | 284 | 259 | 261 |
| 268 | 268 | 264 | 271 | 263 | 259 | 294 | 259 | 263 | 278 |
| 267 | 293 | 247 | 244 | 250 | 266 | 286 | 263 | 274 | 253 |
| 281 | 286 | 266 | 249 | 255 | 233 | 245 | 266 | 265 | 264 |

Step 1:
Arranging the data means defining classes of data. In case it's not obvious like in this example, the number of classes k is $k = \sqrt{n}$ (see 1.1.3.). Dividing the range (maximum value – minimum value) by k yields the required width of one class.

```
22 | 6
23 | 0 2 3 4 5
24 | 0 1 1 3 4 4 5 7 8 9
25 | 0 0 1 3 3 4 4 5 6 6 6 9 9 9 9 9
26 | 0 1 1 2 3 3 3 3 4 4 4 5 5 6 6 6 7 8 8 8 9 9
27 | 1 4 4 6 6 8 9
28 | 1 3 4 4 6 6
29 | 3 4
```

Fig. 1.9 Manual illustration of the density function

Step 2 & 3:

---

[1] Weiß, M., Österreicher, F. (2014), p. 34

| classes | frequency | cumulated frequency | cumulated percentage |
|---------|-----------|---------------------|----------------------|
| 22 | 1 | 1 | 1,43 |
| 23 | 5 | 6 | 8,57 |
| 24 | 10 | 16 | 22,86 |
| 25 | 16 | 32 | 45,71 |
| 26 | 23 | 55 | 78,57 |
| 27 | 7 | 62 | 88,57 |
| 28 | 6 | 68 | 97,14 |
| 29 | 2 | 70 | 100,00 |

Table 1.2 Cumulative frequency distribution in percent

Step 4:
Plot the data on the probability plot (fig. 1.10). The probability plot shows an approximate straight line. Thus the distribution can be assumed to be normally distributed.

The probability plot is very useful to illustrate the graphical test for normal distribution during the lectures. Alternatively, nowadays statistical software uses the Quantile-Quantile-Plots[1].

Test for outlier[2]
"An outlier is an observation that appears to deviate markedly from other observations in the sample."[3] Assuming a distribution with at least n = 10 single values (better n ≥ 25) a particular value is considered to be an outlier[4], if it is outside

$$\boxed{\bar{x} \pm 4s}$$

Both s and mean value have to be calculated without the suspected outlying value!

Consequences of the central limit theorem
A random variable, which can be perceived as the sum of a big number of independent summands, each of them contributing only negligibly, can be considered normally distributed.[5]

---

[1] Hedderich, J., Sachs, L. (2011), p. 414
[2] Thümmel, A. (2010)
[3] Engineering Statistics Handbook (2014)
[4] Hedderich, J., Sachs, L. (2011), p. 428
[5] Bronstein, I.N., Semendjajew K.A., (1987), p. 677

> The mean values of sufficiently large samples are approximately normally distributed

Hence, mean values of independent events, e.g. product characteristics measurement results, are normally distributed, even if the individual values are not normally distributed.

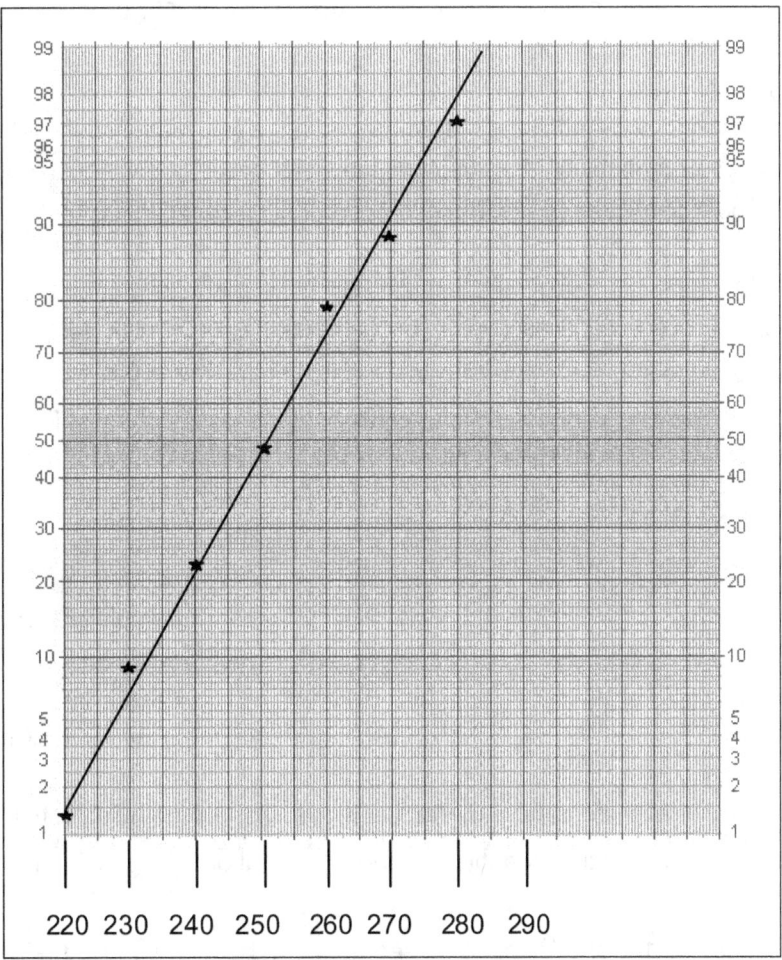

Fig. 1.10 Probability plot of duration of pregnancy, template e.g. from
http://de.wikipedia.org/wiki/Wahrscheinlichkeitsnetz

Thus all rules and equations for normal distribution can be used. Mean values of normal distributions are normally distributed, too.

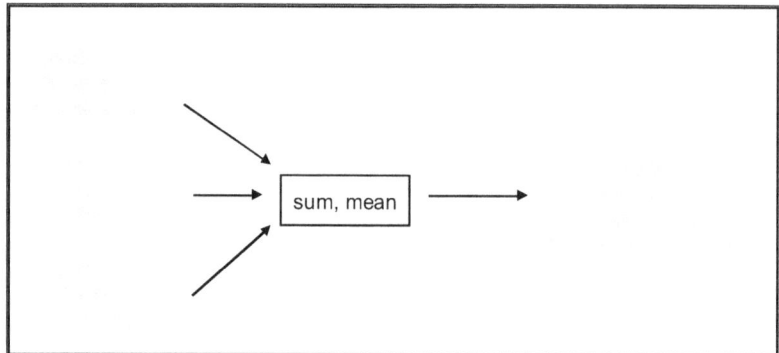

Fig. 1.11 mean values of non-normal distributions are normally distributed

Confidence Interval

The confidence interval indicates the range where the real mean value of the population lies with a given probability. The actual mean value of the population is not known. For instance, the mean value of a sample may be 6.3, the mean value of the population is not 6.3.

Hence an interval with lower and upper limits (μl, μu) can be defined, where the real mean value is assumed to be with a defined probability 1-α (with the premise that the mean value of the sample is approximately matching the mean value of the population). α is the probability that the mean of the population is not within the confidence interval. For two-sided distributions the calculation of the limits is performed[1]:

$$\mu_u = \bar{x} + t_{n-1, 1-\alpha/2} \frac{s}{\sqrt{n}}$$

$$\mu_l = \bar{x} - t_{n-1, 1-\alpha/2} \frac{s}{\sqrt{n}}$$

with
$\bar{x}$ : sample mean
s : sample standard deviation
n : sample size
$t_{n-1, 1-\alpha/2}$ : t-distribution quantile
(see 1.3.6. t-Distribution)

The confidence level for finding the actual mean of the population within the defined range is correlated to the quantiles of the t-distribution $t_{n-1, 1-\alpha/2}$. Typical values are confidence levels 1-α of 95% or 99% (see table 13.1). $\frac{s}{\sqrt{n}}$ is called "standard error".

$1-\alpha/2$ stands for the fact, that the quantile is subtracted from the upper end of the distribution as well as from the lower end of the distribution.

---

[1] Meyna, A., Pauli, B. (2010), p. 488

Example:
Assuming the following measured values, the 95% and 99% confidence interval, respectively, shall be calculated: 85 / 74 / 95 / 86 / 76 / 76 / 85 / 84 / 65 / 83
The mean value $\bar{x}$ = 80.44, the respective sample standard deviation s = 8,9597867.

From table 13.1:

for (1-α) = 95%:   $t_{n-1,1-0.05/2} = t_{17-1,0.975} = t_{16,0.975}$ = 2.262
for (1-α) = 99%:   $t_{n-1,1-0.01/2} = t_{17-1,0.995} = t_{16,0.995}$ = 3.25

Hence there is a 95% probability that the population mean value is found to be in [74.0354; 86.8534] interval and there is a 99% probability that the population mean value is found to be in the [71.2361; 89.6527] interval. Higher probabilities lead to larger intervals.

Use cases
Typical use cases for the normal distribution are measurement values of industrial manufacturing processes. From a mathematical point of view, the normal distribution is actually not applicable to describe product or component lifetimes. However, in practice, assuming the standard deviation of a distribution to be much smaller than its mean value the mathematical error may be neglected[1].

## 1.3.2. Log-normal Distribution

The density function of the log-normal distribution is positively right skewed with only positive values. It is limited by a natural limit on the left side of the distribution. A random variable X is said to be log-normally distributed if log(X) is normally distributed[2].

Mathematical description
Probability density function:

$$f(x) = \frac{1}{\sqrt{2\pi}} \frac{1}{\sigma} \frac{1}{x} \exp\left(-\frac{(\ln x - \mu)^2}{2\sigma^2}\right)$$

with
x > 0
μ   : mean value
σ   : standard deviation

The respective cumulative distribution function is the integral of the probability density function

---

[1] Meyna, A., Pauli, B. (2010), p. 78
[2] Limpert, E., et al. (2001)

$$F(x) = \frac{1}{\sqrt{2\pi}} \frac{1}{\sigma} \int_0^x \frac{1}{t} \exp\left(-\frac{(\ln t - \mu)^2}{2\sigma^2}\right) dt$$

Quantitative Characteristics

Different form the normal distribution μ is a parameter of scale and σ is a parameter of form and they are estimated as follows[1,2]

$$\mu = \frac{1}{n} \sum_{i=1}^{n} \ln(x_i)$$

$$\sigma = \sqrt{\frac{1}{n} \sum_{i=1}^{n} (\ln x_i - \mu)^2}$$

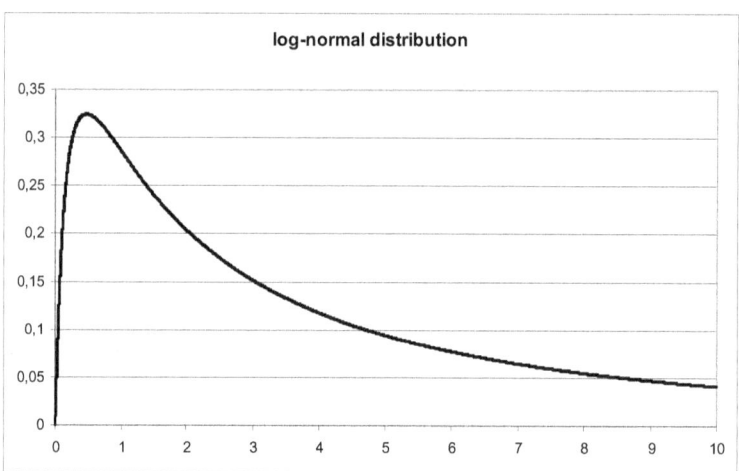

Fig. 1.12 Log-normal distribution (μ = 1.4; σ = 1.5) [0.001;10] (generated using office software)

The expected value E(x), variance, skewness and kurtosis are calculated as follows[3]:

---

[1] Scheid, S. (2001)
[2] Hedderich, J., Sachs, L. (2012), p. 251
[3] Wagner, T., „Statistics Handbook" (2008)

$$E(x) = \exp\left(\mu + \frac{\sigma^2}{2}\right)$$

$$VAR = \exp(2\mu + \sigma^2) \cdot (\exp\sigma^2 - 1)$$

$$Skewness(x) = (\exp(\sigma^2) + 2) \cdot \sqrt{\exp(\sigma^2) - 1}$$

$$Kurtosis(x) = \exp(4\sigma^2) + 2\exp(3\sigma^2) + 3\exp(2\sigma^2) - 3$$

Finally the mode and the median are:

$$Mode = \exp(\mu - \sigma^2)$$

$$Median = \exp\mu$$

Probability plot
A log-normal distribution with original scale (fig. 1.12) appears approximately like a normal distribution when applying a logarithmic scale for the independent variable (fig. 1.13).
Plotted on a probability paper with two logarithmic scales the log-normal distribution appears to be a straight line.

Test for outlier
If the data in question is approximately log-normally distributed the logarithm of the data is approximately normally distributed. Now the outlier test described in the previous section can be used.

3-parameter log-normal function
Adding an additional parameter λ describing the location of the distribution makes a 2-parameter log-normal density function a 3-parameter log-normal density function[1].
If Y = ln(X-λ) is normally distributed, then X is called a 3-parameter log-normal random variable.

$$f(x) = \frac{1}{\sqrt{2\pi}} \frac{1}{\sigma} \frac{1}{(x-\lambda)} \exp\left(\frac{(\ln(x-\lambda) - \mu)^2}{-2\sigma^2}\right)$$

---

[1] Scheid, S. (2001)

with : x > λ

It describes a log-normal distribution which is shifted by λ.

Use cases

The lognormal distribution can be used to describe maintenance times, mileage and material wear out in mechanical engineering. In the area of semiconductor manufacturing it can be used to describe defect density in the production line, electrical product parameters and e.g. electromigration data.

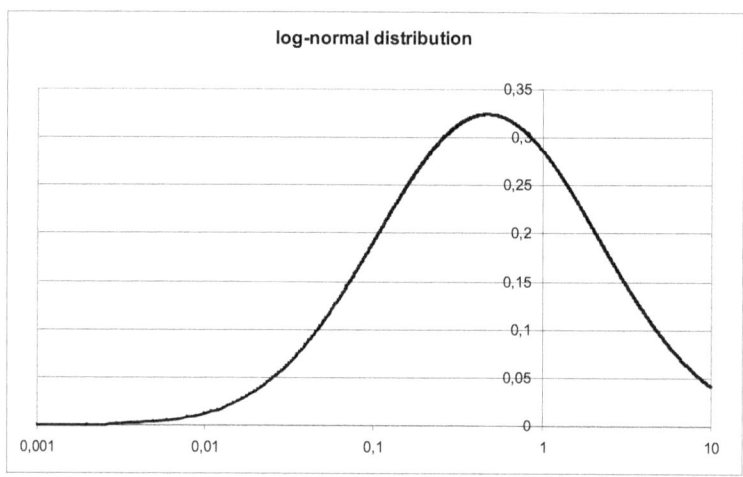

Fig. 1.13 Log-normal distribution with logarithmic scale [0.001;10] (created using office software)

### 1.3.3. Exponential Distribution

In reliability engineering the exponential distribution[1] is often used to describe the failure rate of a product, e.g. of electronic devices, over its lifetime. It is easy to show, that with exponential distributions the fail rate λ is constant over time. In other words the remaining life time does not depend on the elapsed life time.

Mathematical description[2]

Probability density function with variable t (time):

$$f(t) = \lambda \cdot e^{-\lambda \cdot t}$$

for t ≥ 0

$$f(t) = 0$$

---

[1] Meyna, A., Pauli, B. (2010), p. 58
[2] Meyna, A., Pauli, B. (2010), p. 58

for t < 0

The respective cumulative distribution function is the integral of the probability density function

$$F(t) = 1 - e^{-\lambda \cdot t}$$ for t ≥ 0

$$F(t) = 0$$ for t < 0

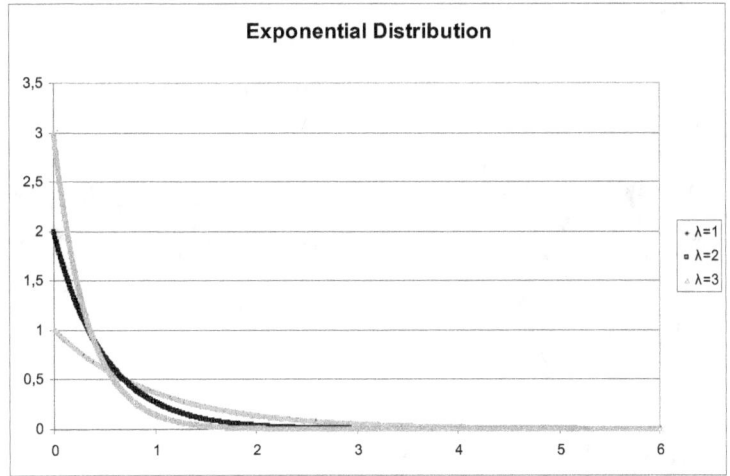

Fig. 1.14 Exponential density distribution function (with negative exponent) for several λ (created using office software)

The expected value μ and variance are calculated as follows[1]

$$\mu = \frac{1}{\lambda}$$

$$VAR = \sigma^2 = \frac{1}{\lambda^2}$$

The expected value μ equals the "mean time between fails" (MTBF)

---

[1] Hedderich, J., Sachs, L. (2012), p. 256

Use cases

Typical examples for exponentially distributed values are life times of products and components or operational times of machines in between two downtimes.

### 1.3.4. Weibull Distribution

In reliability engineering the Weibull distribution[1] is often used, too, to describe the failure rate of a product, e.g. of electronic devices, over its lifetime. The Weibull function is the generalized form of the exponential function.

Due to the fact that the Weibull distribution is mainly used to describe product life times, x is replaced by t in the following equations.

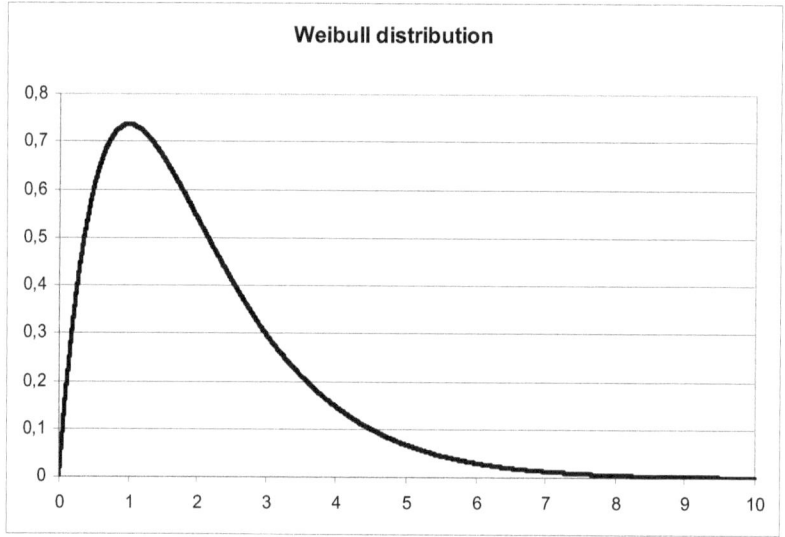

Fig. 1.15 Weibull density distribution function ($\beta$ = 2; $\alpha$ = 1, $t_0$ = 0) (created using office software)

Mathematical description[2]

Probability density function

$$f(t) = \alpha \cdot \beta \cdot t^{\beta-1} \cdot e^{-\alpha \cdot t^{\beta}} \quad \text{for } t \geq 0$$

with  $\alpha$ : characteristic parameter
      $\beta$ : shape parameter (Weibull slope)

---

[1] Weibull, W. (1951)
[2] Meyna, A., Pauli, B. (2010), p. 63

The respective cumulative distribution function is the integral of the probability density function

$$F(t) = 1 - e^{-\alpha \cdot t^\beta} \qquad t \geq 0$$

$$F(t) = 0 \qquad t < 0$$

With β = 1 the Weibull distribution is the exponential distribution.

Inserting the **characteristic life time** $\eta = \alpha^{-\frac{1}{\beta}}$ the cumulative distribution function becomes:

$$F(t) = 1 - e^{-\left(\frac{t-t_0}{\eta}\right)^\beta} \qquad t \geq t_0$$

η characterizes a 63.2% fail rate: $F(\eta) = 1 - e^{-1} \approx 0{,}632$

From there the conditional failure rate h(t) can be calculated to be:

$$h(t) = \alpha \cdot \beta \cdot t^{\beta-1}$$

The conditional failure rate is the probability of failure under the condition that the product has so far survived. Mathematically, $h = f/R$, where R = 1 − F. R is called the reliability function.

The expected value μ and variance are calculated as follows[1]

$$\mu = \eta \cdot \Gamma\left(\frac{1}{\beta} + 1\right) \qquad VAR = \sigma^2 = \alpha^{-\frac{2}{\beta}} \cdot \left[\Gamma\left(\frac{2}{\beta} + 1\right) - \Gamma^2\left(\frac{1}{\beta} + 1\right)\right]$$

The gamma function used here is described by the following equation:

$$\Gamma(x) = \int_0^\infty t^{x-1} \cdot e^{-t} dt$$

---

[1] Hedderich, J., Sachs, L (2012), p. 257

Weibull plot[1]

In reliability engineering F(t) represents the failure probability and can be transformed into a linear equation using logarithm and double logarithm for x and y and the reliability function R(t):

$$x = \ln(t)$$

$$y = \ln\left[\ln\left(\frac{1}{R(t)}\right)\right]$$

This results in the linear equation $y = a + bx$ with b = β and a = ln(α) with a straight line in the Weibull plot (fig. 1.16). This approach yields three important pieces of information about the distribution:

(1) it represents a graphical test of the distribution: if single points (roughly) hit the straight line, then the distribution is a Weibull distribution
(2) b represents the Weibull slope
(3) the constant term a represents the vertical position of the straight line, and thus provides information about the characteristic life time

Application to the Weibull plot[2]

After confirming that the distribution is Weibull distributed several steps have to be executed in order to determine the relevant parameters.

Step 1: Median ranks
First of all the median ranks[3] MR are calculated for the Weibull plot:

$$\boxed{MR = \frac{i - 0.3}{N + 0.4}}$$ with

i : failure order number
N : total sample size

The median rank is the position of the median value in the list of values.

Step 2: Data point arrangement
After fail of all samples the fail times are arranged in an ascending sequence according to the median ranks

---

[1] Mayna, A., Pauli, B. (2010), p. 525
[2] Linß, G. (2005)
[3] Antonitsin, A. (2009)

Step 3: Plot
The median ranks are then plotted against the ordered failure times (in a log₁₀ scale).

Step 4: Fitting
A straight line R is fitted through the bulk of the points, i.e. the extreme values are neglected. The intersection of R with a horizontal line at 63.2% determines the estimated value for the characteristic lifetime on the x-axis.

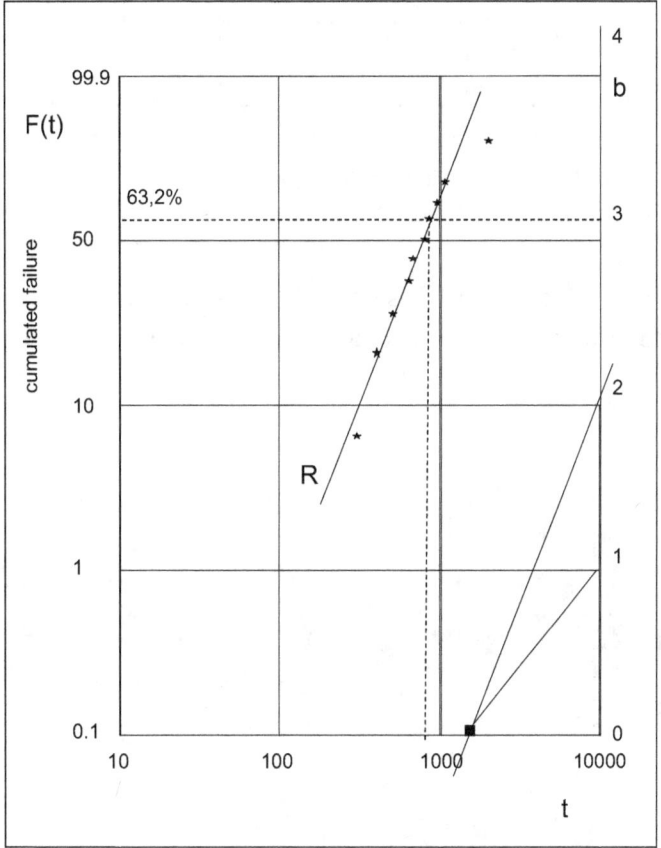

Fig. 1.16 Weibull plot

Step 5: Parallel translation
The resulting straight line R is parallel translated in a way that this line crosses the pole on the Weibull plot which results in an intersection of this line at the right hand y-axis. From there the slope b can be determined.

Example (adopted from Warwick Manufacturing Group[1]):
During a test the following times of failures were measured: 410, 1050, 825, 300, 660, 900, 500, 1200, 750 and 600 hours.

| failure number | failure time | cumulative % failed F(t) |
|---|---|---|
| 1 | 300 | 6,7 |
| 2 | 410 | 16,2 |
| 3 | 500 | 25,9 |
| 4 | 600 | 35,5 |
| 5 | 660 | 45,2 |
| 6 | 750 | 54,8 |
| 7 | 825 | 64,5 |
| 8 | 900 | 74,1 |
| 9 | 1050 | 83,8 |
| 10 | 1200 | 93,3 |

Table 1.3: Data preparation for the Weibull plot

Use cases

The use cases of the Weibull Distribution is clearly to model product lifetimes, in particular gate oxide dielectric breakdown or fracture mechanics.

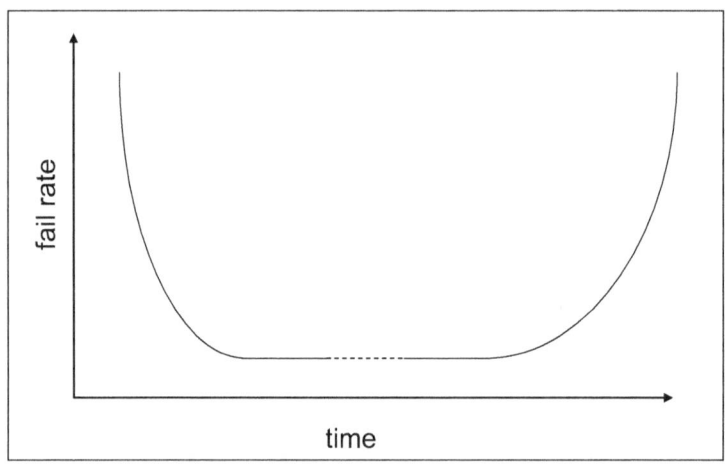

Fig. 1.18 Bath tub curve

Frequently the fail rate of products is described using the bath tub curve (fig. 1.18). Three sections (falling, stable and rising fail rate) can be described with the Weibull function by inserting different Weibull parameters in the equations and by shifting the starting point of the Weibull function. Decreasing fail rates are modelled with β < 1 (=

---

[1] Warwick Manufacturing Group (2007), p. 23

early fails), stable fail rates (time of use) are achieved with β = 1 (= random fails) and finally increasing fail rates are described using β > 1 (= wear out).
The bath tub curve thus is the sum of three Weibull distributions.

### 1.3.5. χ² Distribution

The chi-square (χ²) distribution is a right positively skewed distribution. It is typically used as a test distribution. A χ² distribution is characterized by the sum of the squares of v independent and normally distributed variables:

$$\chi_v^2 = Z_1^2 + Z_2^2 + \ldots + Z_v^2$$

v describes the degree of freedom, i.e. the number of independent variables Z (also known as df = degree of freedom).
The degree of freedom (v = df) is defined as the number of observations n, i.e. of samples, minus the number of estimates a based on the observations.

$$v = df = n - a$$

Mathematical description[1]
Probability density function with variable t (for integration) and the independent variable x:

$$f(x) = \frac{1}{2^{v/2} \cdot \Gamma\left(\frac{v}{2}\right)} \cdot x^{\frac{v}{2}-1} \cdot e^{\frac{-x}{2}}$$

with :

$$\Gamma(x) = \int_0^\infty t^{x-1} \cdot e^{-t} dt$$

The respective cumulative distribution function is[2]:

$$F(x) = \frac{1}{\Gamma\left(\frac{n}{2}\right)} \cdot \int_0^x \left(\frac{x}{2}\right)^{\frac{n}{2}-1} e^{\frac{-x}{2}}$$

The expected value μ and variance are calculated as follows[3]

---
[1] Hedderich, J., Sachs, L. (2012), p. 265
[2] Walk, C. (2007)
[3] Hedderich, J., Sachs, L. (2012), p. 268

$$\mu = E(\chi_v^2) = v$$

$$VAR(\chi_v^2) = 2v$$

Use cases

The χ² distribution is used for the χ² test and for the estimation of variances, e.g. test if two characteristics are independent of each other or if two samples are from the same population.

## 1.3.6. t-Distribution (Student Distribution)

If the variance of the population is unknown and a characteristic is normally distributed, then the t-distribution is used instead of the normal distribution. For large sample sizes it is a normal distribution[1] and is calculated as follows[2]:

$$t = \frac{\bar{x} - \mu}{s/\sqrt{n}}$$

with

- x : standard normal distributed random variable
- s : standard deviation (sample)
- n : sample size

Thereby the $\chi^2$ distribution is the sum of independent, normally distributed random variables $x_i$ with $\bar{x} = 0$ and $s = 1$.

$$\chi^2_{df=n} = x_1^2 + x_2^2 + x_3^2 + \ldots + x_n^2$$

The degree of freedom n is the number of parameters which are free, i.e. not determined by other parameters, in this case:

$$df = n - 1$$

Typically the quantiles of the t-distribution are used for the analysis of standard distributions. The confidence interval for the mean value of the distribution with an estimated standard deviation is:

$$\bar{x} \pm t_{df, 1-\alpha/2} \cdot s/\sqrt{n}$$

with

- s : sample variance
- n : number of observations

---

[1] Ludwig-Mayerhofer, W. (2015)
[2] Dietrich, E., Schulze, A. (2012), p. 144

The quantiles of the t-distribution are taken from table 13.1.

Example:
Estimate the t-quantile for the 95% confidence interval for the mean value µ for large n.

$$t_{\infty,1-0.05/2} = t_{\infty,0.975} = 1.96$$

## 1.4 Discrete Distributions

Typically, the number of discrete random variables is a finite number. So the discrete variables and the respective probabilities can, in principle, be listed in a table.
The mean value µ and variance VAR are defined to be[1]:

$$\mu = \sum_{i=1}^{n} x_i \cdot P(x_i)$$

with:
- $x_i$ : value of a variable
- $P(x_i)$ : respective probability

$$VAR = \sigma^2 = \sum_{i=1}^{n} (x_i - \mu)^2 \cdot P(x_i)$$

### 1.4.1. Binomial Distribution

The starting point for looking at the binomial distribution is the so called Bernoulli experiment.

> A Bernoulli trial is an experiment that can have two outcomes:
> True or False

The binomial variable is asking how many events k are successful when assuming n independent Bernoulli experiments and a success probability p.
This probability is defined to be:

$$p(k) = \frac{n!}{k! \cdot (n-k)!} p^k (1-p)^{n-k}$$

with: k = 0, 1, 2, 3, … n

---
[1] Barr, T. (2010)

The success probability does not change.

Example:
In outgoing inspection, the rate of defective chips in a batch of 14 is 20%. What is the probability of at least one defective die.

Calculation:   p(at least one defective) = 1-p(zero defective)

$$n = 14; \quad p = 0.2, k = 0$$

$$p(0) = \frac{14!}{0! \cdot (14-0)!} 0.2^0 (1-0.2)^{14-0} = 1 \cdot 1 \cdot 0.04398$$

p(>0) = 1-0.04398 = 95,6%

Mean and variance are defined to be:

$$\mu = n \cdot p$$

$$VAR = n \cdot p \cdot (1-p)$$

For p → 0 and n → ∞ with E(x) = µ = np = const. the Binomial distribution becomes the Poisson distribution. Therefore for large n (n ≥ 80) and small probabilities p (p ≤ 0.1) the Binomial distribution can be approximated by the Poisson distribution.
For np(1-p) → ∞ the Binomial distribution can be approximated by the normal distribution.

Use cases
The binomial distribution is used to determine defects and yields in the semiconductor manufacturing process or in quality control.

## 1.4.2. Poisson Distribution

Assuming the number of observations n becomes infinite and the probability p of an event to happen is very small, then, with n*p is neither infinite nor zero, the binomial distribution approaches the poisson distribution.
With x representing the number of events within a defined time or within a defined volume etc. under the assumption that a particular mean value can be expected[1] and assuming that those events are random and independent from each other the Poisson density distribution then defines the probability of an event as:

---

[1] Reusch, W. (2002)

$$p(x) = \frac{\lambda^x \cdot e^{-\lambda}}{x!}$$

with

$\lambda$ : expected value

$x$ : 0, 1, 3, 4, …. (number of events)

The respective distribution function F(x):

$$F(x) = \sum_{x_i < x} \frac{\lambda^{x_i}}{x_i!} \exp(-\lambda)$$

The expected value $\lambda = \mu$ (i.e. the mean value) is:

$$\mu = \sum_{x=0}^{\infty} x \cdot p(x) = n \cdot p$$

Expected value $\lambda$ and variance VAR are identical:

$$\mu = VAR = \sigma^2$$

Example:
Incoming inspection detects 2 defects per week. So the defect frequency is described by a Poisson distribution with $\mu = 2$. What is the probability for having less than 4 defects in 2 weeks?

Calculation:

P(≤3) = P(0) + P(1) + P(2) + P(3) =

$$= \frac{4^0 \cdot e^{-4}}{0!} + \frac{4^1 \cdot e^{-4}}{1!} + \frac{4^2 \cdot e^{-4}}{2!} + \frac{4^3 \cdot e^{-4}}{3!}$$

= 0.018318 + 0.073264 + 0.146528 + 0.195367

= 0.4335 = 43.35%

### Use cases

The Poisson distribution is used to describe rare events, e.g. defects in the semiconductor process. It is also has been used to model yield in semiconductor manufacturing[1].

---

[1] Xu, Q. (2012)

# 2. Capabilities

In section 1, the mathematical foundations have been explained, which are needed for the characterization of the process capability. Managing a process in a way that the natural process variation is stable and known is the prerequisite for a production process that will turn out predictably product that fulfils the specifications for performance and durability, i.e. high quality products.

In the beginnings quality assurance meant to perform production and eventually sort out bad products in the functional test. This is a waste of resources and does not lead to improvement of the production processes. Hence today's quality management approach in production is to keep production equipment, production processes and the respective measurement tools stable in order to yield only small variations well within defined tolerances. In addition to having stable processes, this approach also leads to continuous improvement, built into the process itself, since it is stable and the natural process variation is known, process changes can be assessed as to whether the process variation is decreased (i.e. the process improved) or is increased, which constitutes a deterioration of the process.

Despite the fact that nowadays capability studies are typically performed supported by statistics software, it is useful to have a basic understanding of the fundamentals, which is why they are described in this section.

## 2.1. Introduction

Capability studies are performed in order to demonstrate that a piece of equipment, a process[1] or a measurement tool is able to meet the requirements for the final product, i.e. whether or not the product can be produced within the upper and lower specification limits for that product. The way capability studies are performed is similar for equipment and process capabilities. Depending on the product status or the production process status different kinds of capability studies have to be performed.

Gauge capability studies to characterize a measurement method follow a different approach in the way they are performed as well as if they are performed, e.g. a type II measurement system analysis is not making sense if the type I measurement system analysis was not successful (type I and type II see 2.4. Gauge Capability).

---

[1] DIN ISO 21747

## 2.2. Machine Capability

Machine capability, also known as short term capability, is the capability of a piece of equipment to produce a product within certain tolerances, i.e. within the upper and lower specification limit. Machine capability studies are typically performed
- prior to purchasing a new piece of equipment
- after major maintenance work
- prior to starting high volume manufacturing (volume production release)

Boundary Conditions

The purpose of the machine capability study is to find out the natural variation of the process result which is only caused by the machine itself. In order to achieve sufficient accuracy typically 50 parts (minimum 25 parts) are produced in a row under stable boundary conditions:

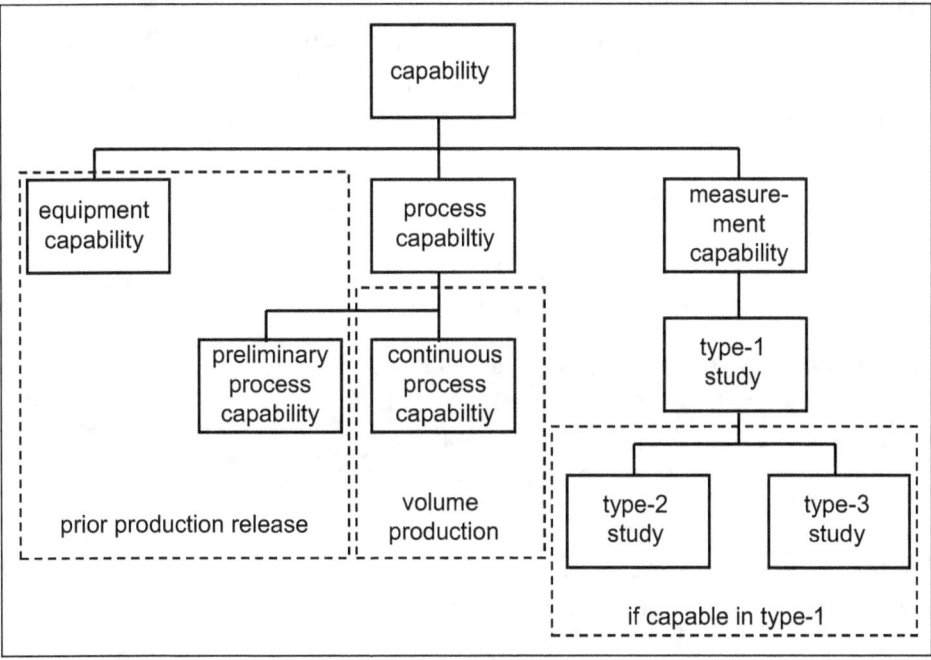

Fig. 2.1 Overview of various kinds of capabilities

(1) The machine has to be in a typical production state, i.e. warmed up and running
(2) The operator has to be an experienced operator who knows how to operate the machine

(3) The raw material, if possible, has to be from one supplier and from one production lot in order to minimize material variations and it must not be scrap material

(4) The production process has to be the standard process, i.e. same rate, temperature, etc, without any atypical interruptions or changes of tools during the study

(5) The environment also has to be standard production environment, i.e. temperature, humidity etc. has to be as specified for the actual production process

If those boundary conditions are violated, the study could result in wrong or misleading results.

Calculation[1]

From the sample of 50 pieces the average values $\bar{x}$ as well as the spread s (standard deviation or variability) are determined. For normal distributions with upper and lower specification limits the following equations apply for the calculation of the relevant machine capability index and the critical machine capability index.

Mean value:

$$\bar{x} = \frac{1}{n} \cdot \sum_{i=1}^{n} x_i$$

with
n  : number of samples
$x_i$ : value of piece i

Spread (standard deviation) of the distribution:

$$s = +\sqrt{s^2} = +\sqrt{\frac{1}{n-1} \cdot \sum_{i=1}^{n}(x_i - \bar{x})^2}$$

The resulting machine capability index $c_m$ is the ratio of the tolerance of the process characteristic T and the production spread 6s:

$$c_m = \frac{T}{6 \cdot s} = \frac{USL - LSL}{6 \cdot s} \geq 1{,}67$$

with
T    : tolerance
USL : upper specification limit
LSL : lower specification limit

Typically, the following definitions apply:

---

[1] Bosch Heft Bd. 9 (2004)

$$c_m < 1{,}33 \quad \text{not capable}$$
$$1{,}33 \leq c_m < 1{,}67 \quad \text{conditionally capable}$$
$$c_m \geq 1{,}67 \quad \text{capable}$$

In some cases, customers might ask for $c_m$ at or even higher than 2.

The machine capability index, however, only considers the spread of the distribution compared with the tolerance of the characteristic. It does not consider a shift of the distribution away from the target value (fig. 2.2).
Therefore, in addition to the capability index the critical capability index $c_{mk}$ is required in order to be able to judge the machine capability. For the calculation of $c_{mk}$ first of all auxiliary values $c_{mu}$ and $c_{ml}$ have to be calculated:

$$c_{mu} = \frac{USL - \bar{x}}{3 \cdot s} \qquad c_{ml} = \frac{\bar{x} - LSL}{3 \cdot s}$$

Due to the fact, that a somewhat shifted distribution is likely to hit the closer specification limit the smaller value of those is called critical machine capability index $c_{mk}$:

$$c_{mk} = \min(c_{mu}; c_{ml}) \geq 1{,}67$$

As for the capability index the following definitions apply:

$$c_{mk} < 1{,}33 \quad \text{not capable}$$
$$1{,}33 \geq c_{mk} < 1{,}67 \quad \text{conditionally capable}$$
$$c_{mk} \geq 1{,}67 \quad \text{capable}$$

If the values are not normally distributed or if there is only a one-sided distribution an alternative approach has to be chosen to calculate the capability indices. The alternative approach is described in the process capability section and the reader is referred to that section.
In some cases, a sample size or 50 pieces might not be possible. In this case the required $c_m$ and $c_{mk}$ values should be significantly higher than 1,67 to signal a satisfactory machine capability.

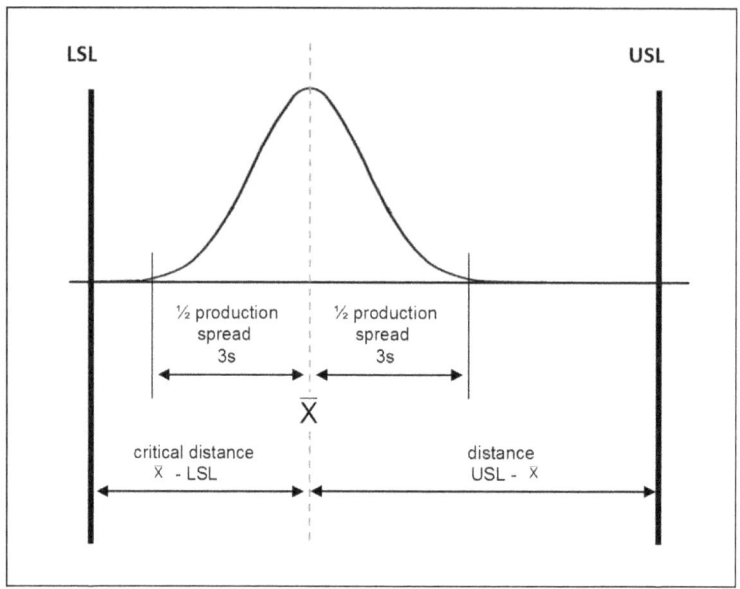

Fig. 2.2 Distribution well within specification limits but shifted towards LSL

## 2.3. Process Capability[1]

In contrast to the machine capability study the process capability study takes into account all parameters, which realistically have an influence on the process variability. Hence it makes sense to consider a sufficiently long period of production for the process capability study in order to test for typical variations in the production environment. Depending on the qualification status of the product, i.e. prior production release and during volume production, the number of pieces available for a capability study is very different. This is accounted for by distinguishing preliminary and continuous process capability.

Influencing parameters
Different from the stable boundary conditions in the machine capability study, now a realistic fluctuation needs to be included, e.g.

(1) more than one piece of equipment to be used
(2) more than one employee to operate the equipment
(3) raw material to be used from various suppliers and from various production lots
(4) material produced in different shifts
(5) etc.

Preliminary process capability

---
[1] Linß, G. (2005)

This study is done prior to volume production release. For the preliminary process capability study 20 to 25 samples with each 3 to 5 pieces have to be taken from production in regular time intervals (target at least 125 parts). It strengthens the validity of the study if the samples are taken in a one to two weeks timeframe. However, since volume production release is not granted at that point in time, it might be difficult to collect the above mentioned number of pieces for the study. In this case, again, a smaller number of pieces can be used if the index limits are increased.

The preliminary process capability indices are $p_p$ and $p_{pk}$, sometimes also $c_p$ and $c_{pk}$ and are calculated the same way as the final process capability indices (see next section).

Continuous process capability

After the volume production release the process capability study is a process over spread over a certain time. The observation time has to be adequate for the particular product and production process, but should generally be at least 20 working days and include at least 25 samples with 5 pieces each[1]. This period of time ensures that all typical influencing parameters are included, e.g. material changes, shift changes, week days, weekend and so forth.

Hence this study provides realistic information about the capability of the production process. Again, mean value and standard deviation of the distribution have to be calculated. For many processes the measurement values are not available as individual values but as averages of several individual samples. For instance if the wafer to wafer variation of a layer thickness is the object of the investigation there are mean values of several wafers with each having at least five individual measurements, which is why in the following section both aspects of the type of distribution are presented.

## 2.3.1. Normally Distributed Data with Upper and Lower Specification Limit

Two sided specification limits[2]

Assuming a normal distribution with upper and lower specification limits the following equations apply.

Mean value of individual values:

$$\mu = \frac{1}{n} \cdot \sum_{i=1}^{n} x_i$$

with

n : number of pieces

---

[1] Dietrich, E., Schulze, A. (2014), p. 230
[2] Bosch Heft Bd. 9 (2004)

$x_i$ : value of piece i

Mean value of samples with sample size > 1:

$$\mu = \frac{1}{m} \cdot \sum_{j=1}^{m} \bar{x}_j$$

with

m : number of samples

$\bar{x}_j$ : average value of sample j

Standard deviation based on individual values:

$$s = +\sqrt{s^2} = +\sqrt{\frac{1}{n-1} \cdot \sum_{i=1}^{n} (x_i - \bar{x})^2}$$

Overall standard deviation for sample sizes > 1:

$$\bar{s}^2 = \frac{1}{m} \cdot \sum_{j=1}^{m} s_j^2$$

$$\sigma = \sqrt{\bar{s}^2}$$

with

$s_j$ : standard deviation of sample j

The process is capable, if the ratio of the tolerance to the process spread is larger than 4/3. Just like the machine capability the process capability index $c_p$ is calculated as follows

$$c_p = \frac{T}{6 \cdot \sigma} = \frac{USL - LSL}{6 \cdot \sigma} \geq 1{,}33$$

with:

T : tolerance
σ : standard deviation
USL : upper specification limit
LSL : lower specification limit

Due to the fact that process capability includes all relevant influencing parameters, the limits are often somewhat more relaxed than for machine capability as compared to machine capability[1].

$c_p < 1{,}33$     not capable
$c_p \geq 1{,}33$     capable

---

[1] Dietrich, E., Schulze, A. (2014), p. 386

In the semiconductor industry typically $c_p$ values of at least 1.67 are expected by the customers. The process capability index $c_p$, however, is not sensitive to a shift of the entire distribution, i.e. the spread of the distribution could be very small compared to the process spread resulting in a large $c_p$ value, but the distribution could be significantly off target. Hence a second index, the critical process capability index has to be considered, which includes the distance of the mean value from the specification limits $C_{pk}$:

$$c_{pk} = \min\left[\frac{USL - \mu}{3\cdot\sigma}; \frac{\mu - LSL}{3\cdot\sigma}\right]$$

The process mean value shall be located in between the lower and upper specification limits and shall have a distance of 4 σ to the closest specification limit.

$c_{pk} < 1{,}33$      not capable
$c_{pk} \geq 1{,}33$      capable

Since in the preliminary process capability study the influencing parameters are less than in the continuous case, the limits are typically tighter:

$p_{pk} < 1{,}67$      not capable
$p_{pk} \geq 1{,}67$      capable

Again, in the semiconductor industry a minimum value of 1.67 for the $c_{pk}$ value is required by the customers.

<u>σ-Quality interpretation</u>
Assuming the specification limits to be multiples of the standard deviation the respective percentage of good and bad parts as well as the respective $c_p$ value is given in table 2.1.

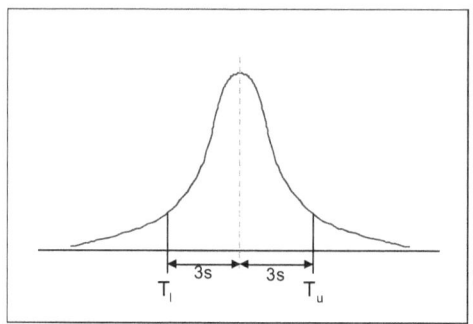

Fig. 2.3 Normal distribution with 3sigma-Quality illustration

| Quality class | Good parts (%) | Scrap parts (%) | Scrap parts (ppm) | $C_p$ |
|---|---|---|---|---|
| 1σ-Quality | 68,27 | 31,73 | 317.300 | 1/3 |
| 2σ-Quality | 95,45 | 4,55 | 45.500 | 2/3 |
| 3σ-Quality | 99,7300 | 0,2700 | 2.700 | 1 |
| 4σ-Quality | 99,9937 | 0,0063 | 63,3 | 4/3 |
| 5σ-Quality | 99,99994 | 0,00006 | 0,6 | 5/3 |
| 6σ-Quality | 99,9999998 | 0,0000002 | 0,02 | 2 |

Table 2.1 Quality classes based on specification limits (symmetric distribution)[1]

## 2.3.2. Non-Normally Distributed Data and One-Sided Specification Limits

First of all before following an alternative approach to calculate the process capability indices a conclusion of the central limit theorem[2] needs to be considered: "the mean values of sufficiently large samples are normal distributed" (see 1.3.1).

This is true independent of the distribution of the original values and of sample sizes of at least four measurements.

Hence process capabilities of mean values are calculated using the equations for the normal distribution.

Example

A simulated experiment was performed with 50 samples with 10 data points per sample. Taken individually they follow a log-normal distribution (fig. 2.4).

---

[1] Bosch Heft Bd. 9 (2004)
[2] Hedderich, J., Sachs, L., 2012, p. 248

The corresponding frequency diagram (fig. 2.5) is shown as reference for the sample distributions. Then for each sample the mean value $\bar{x}_i$ is calculated as well as the overall mean $\bar{\bar{x}}$. These values are then summarized in another frequency diagram showing an almost normal distribution (fig. 2.6).

Now for the mean values, the regular equations for normal distributions can be applied. However, if this is not possible two methods to estimate the process capability shall be shown in this section.

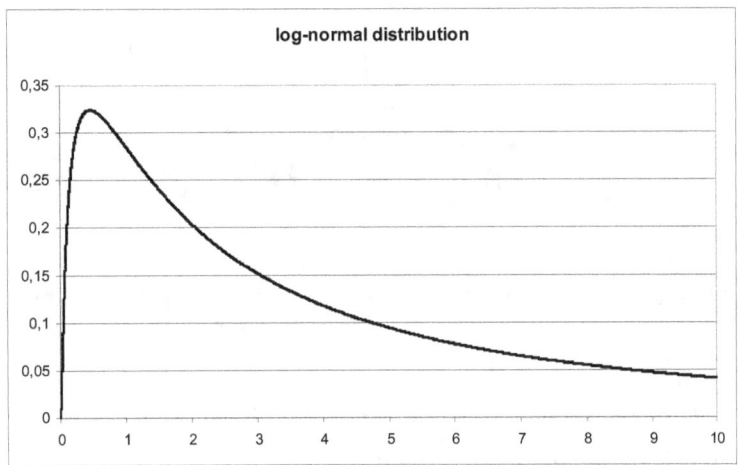

Fig. 2.4 log-normal distribution (created using office software)

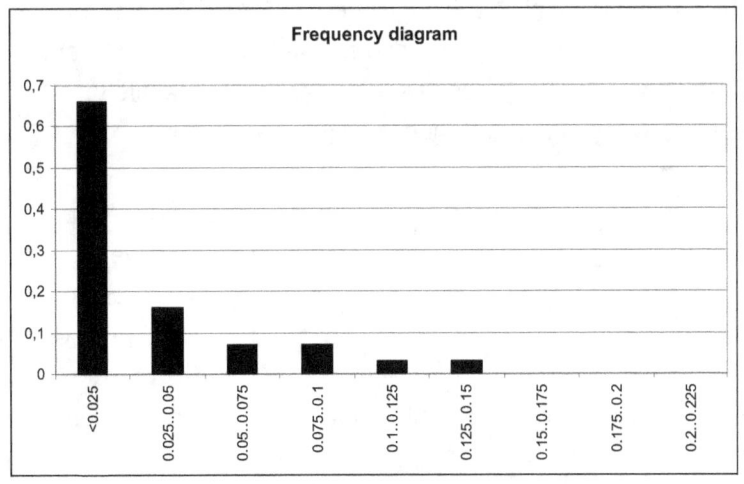

Fig. 2.5 frequency diagram of a log-normal distribution

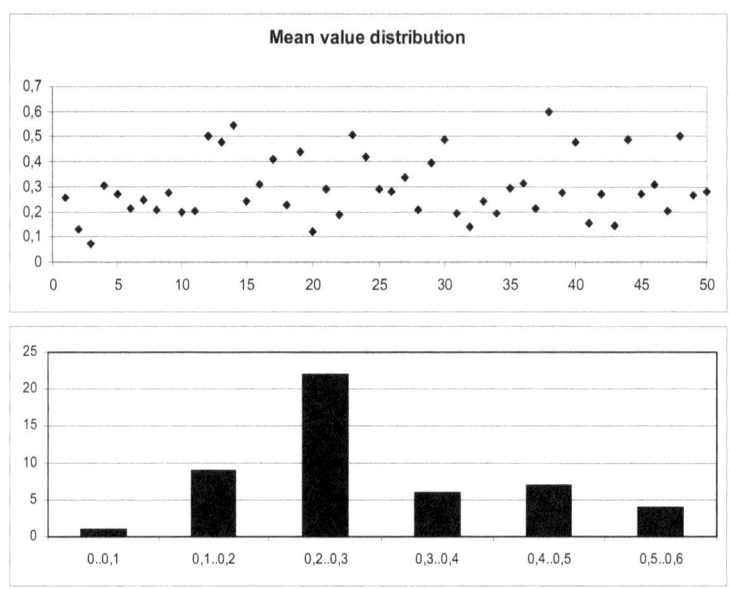

Fig. 2.6 Distribution of mean values of samples with normal distribution

Percentile method[1,2]

In case the kind of distribution is not known or if there is no valid distribution the calculation of capability indices has to be based on percentiles.

The basic idea is to follow the normal distribution approach, hence to divide the tolerance by the process spread 6σ. Since σ is not known, the 6σ range has to be replaced by distribution limits which include 99,73% of the values of the distribution. Therefore the lower limit (-3σ) is replaced by the $L_{p3} = x_{0,135\%}$ percentile and the upper limit (+3σ) is replaced by the respective $U_{p3} = x_{99,865\%}$ percentile (fig. 2.7).

The process capability index $c_p$ is then calculated as follows:

$$c_p = \frac{USL - LSL}{U_{p3} - L_{p3}}$$

with

USL : upper specification limit
LSL : lower specification limit
$U_{p3}$ : $x_{99,865\%}$ percentile
$L_{p3}$ : $x_{0,135\%}$ percentile

The respective critical process capability index $c_{pk}$ is calculated like:

$$c_{pk} = \min\left(\frac{USL - \mu}{U_{p3} - \mu}; \frac{\mu - LSL}{\mu - L_{p3}}\right)$$

---

[1] Korkusuz, D. (2011)
[2] DIN 21474

In reality neither the real process mean µ nor the process limits $U_{p3}$ and $L_{p3}$ are known. Thus these values need to be estimated from the process data. So the mean value µ is calculated from the arithmetic mean of all m*n sample values.

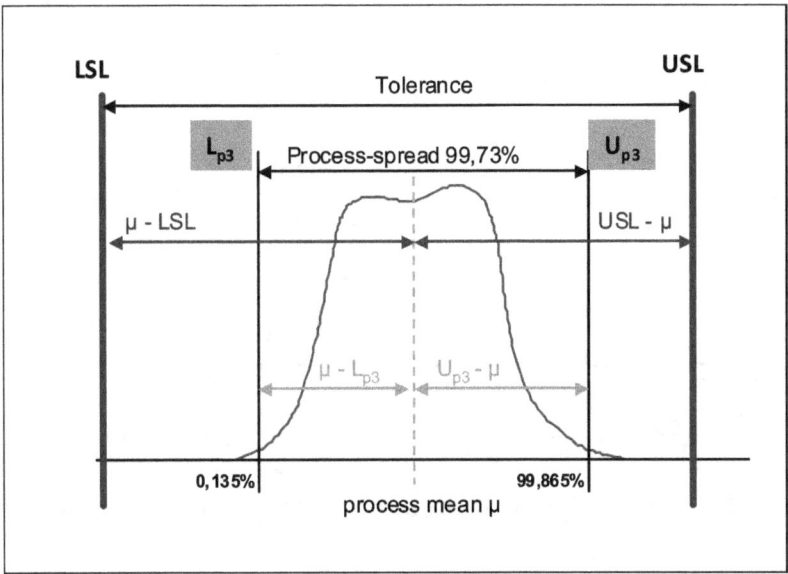

Fig. 2.7 Illustration of the percentile method

Percentile method for one sided specification limits[1]
In many real cases the distribution of measured values is limited by a specification limit on one side and by a physical limit on the other side, e.g. number of particles on a wafer, surface roughness, wafer flatness.
In these cases only the $c_{pk}$ value is calculated using

$$c_{pk} = \frac{USL - \bar{x}}{x_{99,865\%} - \bar{x}}$$

The same approach applies if there is only one specification limit on one side with no lower physical limit. For one-sided specifications only the $c_{pk}$ value exists.

Log-normal distribution
Many distributions in nature or in technology are positively skewed with a steep slope on the left side and an almost flat tail on the right side. Based on the fact that

---
[1] DIN 21474

logarithmizing log-normal data transforms this data into normally distributed values[1] the determination of process capability indices can be done according the following four step process[2]:

(1) Prepare raw data and calculate respective statistical indexes

| sample | sample size | raw data |
|---|---|---|
| 1 | $n_1$ | $X_{11} \ldots X_{1n1}$ |
| 2 | $n_2$ | $X_{21} \ldots X_{2n2}$ |
| 3 | $n_3$ | $X_{31} \ldots X_{3n3}$ |
| ... | | |
| k | $n_k$ | $X_{k1} \ldots X_{knk}$ |

Table 2.2 Raw data

For each set of raw data the mean value, the spread and the range are calculated, e.g. $X_{11}\ldots X_{1n_1}$ → ($\bar{X}_1$, $S_1$, $R_1$)

(2) Transform the raw data:

$X'_{ij} = \ln(X_{ij})$    $USL' = \ln(USL)$    $LSL' = \ln(LSL)$

(3) Apply process statistics

$$N = \sum_{i=1}^{k} n_i \quad \overline{\overline{X'}} = \frac{1}{N}\sum_{i=1}^{k}\sum_{j=1}^{n_i} X'_{ij}$$

with

N : number of raw values
$n_i$ : raw values of sample i

$$S'_{total} = \sqrt{\frac{1}{N-1}\sum_{i=1}^{k}\sum_{j=1}^{n_i}(X'_{ij} - \overline{\overline{X'}})^2}$$

(4) Calculate cp & cpk

---

[1] Hedderich, J., Sachs, L. (2012)
[2] Wagner (2007)

$$c_{pl} = \frac{\bar{X}' - LSL'}{3 \cdot S'_{total}}$$

$$c_{pu} = \frac{USL' - \bar{\bar{X}}'}{3 \cdot S'_{total}}$$

$$c_{pl} = \frac{\bar{\bar{X}}' - LSL'}{3 \cdot S'_{total}}$$

$$c_{pk} = \min(c_{pu}, c_{pl})$$

For one-sided specifications only the $c_{pk}$ value is used.

### 2.3.3. Manual Estimation of the Process Stability and Distribution[1]

A conclusion of the central limit theorems is the fact that mean values of sufficiently large samples (at least four) are normally distributed. The stability of the mean values can be estimated by defining control limits, e.g.:

$$UCL_x = \bar{\bar{x}} + 1{,}3 \cdot \bar{s}$$

$$LCL_x = \bar{\bar{x}} - 1{,}3 \cdot \bar{s}$$

$$UCL_s = 2{,}1 \cdot \bar{s}$$

A violation of the upper or lower control limit is an indication for a trend, e.g. due to wear out of tools.

The determination of the statistical distribution can be done by means of a graphical illustration of the data (e.g. a histogram or probability net). Quantitative tests for distribution can be performed, too.

## 2.4. Gauge Capability[2]

---

[1] Bosch Heft Bd. 9 (2009)
[2] MSA Reference Manual, 4th Edition (2010)

## 2.4.1. Introduction

Measurement data are the basis for process stability characterization and hence for decision making in production. Therefore, the quality of the measurement gauge, which results in reliable measurement data regarding e.g. accuracy, repeatability, reproducibility, linearity etc. is crucial.

Therefore, various methods are employed in the industry to ensure, that independent of the particular operator, of the environment, of the measurement object the gauge used delivers reliable and stable results.

In this context the following expressions are explained with the help of the following quotations from the MSA 4$^{th}$ edition[1]:

➢ Bias

"Difference between the observed average of measurements and the reference value. A systematic error component of the measurement system"

The bias of a measurement gauge is determined by calibration, i.e. the estimation of the difference between the measured value and a known reference value traceable to a national standard.

➢ Stability

"The change in bias over time. A stable measurement process is in statistical control with respect to location."

➢ Linearity

"The change in bias over the normal operating range. The correlation of multiple and independent bias errors over the operating range. A systematic error component of the measurement system"

➢ Precision

"'Closeness' of repeated readings to each other. A random error component of the measurement system

➢ Repeatability

"Variation in measurements obtained with one measuring instrument when used several times by an appraiser while measuring the identical characteristic on the same part. The variation in successive (short-term) trials under fixed and defined conditions of measurement"

Often the repeatability is also called machine variation or gauge capability.

---
[1] MSA Reference Manual, 4th Edition (2010)

> Reproducibility

"Variation in the average of the measurements made by different appraisers using the same gauge when measuring a characteristic on one part. For product and process qualification, the error may be appraiser, environment (time), or method"

## 2.4.2. Gauge Capability[1,2] (type I analysis)

Like for process capabilities there are two capability indices for gauge capabilities:

$c_g$    : describes the repeatability of the measurement results

$c_{gk}$  : considers the bias, i.e. the systematic difference between the measured value and a reference value

Here it is assumed, that there is no influence by the operator. For the gauge capability study typically a standard is measured 50 times (minimum 25 times) under constant conditions. This means after the execution of a measurement the standard is removed from the gauge. Then the entire measurement process is repeated from the beginning. The standard must not be changed during the study. Eventually the mean value, the respective standard deviation $s_w$ as well as the gauge capability indices are calculated according to the following equations.

$$\bar{x} = \frac{1}{n}\sum_{i=1}^{n} x_i \qquad s_w = \sqrt{\frac{1}{n-1}\sum_{i=1}^{n}(x_i - \bar{x})^2}$$

with

$$c_g = \frac{T \cdot k_g}{6 \cdot s_w} \geq 1{,}33$$

$$c_{gk} = \frac{k_{gk} \cdot T - |\bar{x} - x_r|}{3 \cdot s_w} \geq 1.33$$

T     : tolerance
$k_g$   : company specific constant
$k_{gk}$  : company specific constant
$x_r$   : reference value of the standard
$k_{gk}$  : company specific constant

$k_g$ and $k_{gk}$ are constants which are defined by the companies. The following table 2.3 shows examples of constants and capability targets.

---

[1] Hofmann, D. (2009)
[2] Linß, G. (2005)

| Company norm | $k_g$ | $k_{gk}$ | $c_g$ | $c_{gk}$ |
|---|---|---|---|---|
| Ford (EU 1880 1997) | 0,15 | 0,075 | > 1,0 | > 1,0 |
| BMW | 0,2 | 0,1 | > 1,33 | > 1,33 |
| Q-DAS Audi Bosch Opel / GM Daimler | 0,2 | 0,1 | > 1,33 | > 1,33 |

Table 2.3 overview of capability constants and targets[1]

### 2.4.3. Gr&R Study[2,3] (type II analysis)

Once the gauge capability is > 1.33 the measurement system analysis is performed. This is also called gauge r and R (Gr&R) study.

Repeatability r covers variations and changes in the environmental conditions, operator handling, measurement etc., reproducibility R includes operator to operator variation, too. Depending on whether there is an operator influence or not, a Gr&R or Gr study needs to be performed. Typically the Gr&R study is executed with 3 operators, 3 trial runs and 10 parts or 3 operators, 2 trial runs and 10 parts, respectively.

Preparation

(1) Select ten parts from production. These parts shall be representative for the typical production spread and shall be picked at equal intervals. The parts shall not be produced in a consecutive row. The parts have to be numbered randomly.

(2) Define the operators (A, B, C). They should be the persons which later on use the gauge in production.

(3) The resolution of the gauge has to be an order of magnitude better than the process spread.

Execution

(1) Operators A, B and C measure 10 parts in random orders in order to avoid any potential drift.

(2) In order to avoid any unwanted influences the operators shall not be able to see the part numbers. A fourth person is to know which numbered part is being measured and documents the results accordingly.

---

[1] Linß, G. (2005), p. 311
[2] Wagner, T., MSA (2008)
[3] MSA Reference Manual, 4th Edition (2010)

(3) The second (or third) trial run starts again in varying different sequences for A, B and C.
(4) Insert all data collected in the spreadsheet (fig. 2.8)

Calculation
(1) Calculate mean values of trial runs and operators, e.g. for operator A

$$\overline{\overline{X}}_a = \frac{\overline{X}_{a1} + \overline{X}_{a2} + \overline{X}_{a3}}{3}$$

with
$\overline{X}_{a1}$ : mean of first trial run
$\overline{X}_{a2}$ : mean of second trial run
$\overline{X}_{a3}$ : mean of third trial run

(2) Calculate average ranges for all operators, e.g. for A

$$\overline{R}_a = \frac{R_{a1} + R_{a2} + ... + R_{a10}}{10}$$

with
$R_{a1}$ range of part one
$R_{a2}$ range of part two
etc.

(3) Calculate overall average range $\overline{\overline{R}}$

$$\overline{\overline{R}} = \frac{\overline{R}_a + \overline{R}_b + \overline{R}_c}{3}$$

(4) Calculate operator to operator variation $\overline{X}_{Diff}$

$$\overline{X}_{Diff} = \max(\overline{X}_a, \overline{X}_b, \overline{X}_c) - \min(\overline{X}_a, \overline{X}_b, \overline{X}_c)$$

(5) Calculate repeatability r (machine variation EV)

$$r = EV = \overline{\overline{R}} \cdot K_1$$

with
$K_1$ = 0,8862 (2 trial runs)
$K_1$ = 0,5908 (3 trial runs)

(6) Calculate reproducibility R (operator variation or appraiser variation AV)

$$R = \sqrt{(\overline{X}_{Diff} \cdot K_2)^2 - \left(\frac{r^2}{n \cdot t}\right)}$$

with
$K_2$ = 0,7071 (2 operators)
$K_2$ = 0,5231 (3 operators)
r = EV
n : number of parts
t : number of trial runs

|  |  | Part |  |  |  |  |  |  |  |  |  | Average |
|---|---|---|---|---|---|---|---|---|---|---|---|---|
| Operator | Trial run | 1 | 2 | 3 | 4 | 5 | 6 | 7 | 8 | 9 | 10 | |
| A | 1 | | | | | | | | | | | |
| | 2 | | | | | | | | | | | |
| | 3 | | | | | | | | | | | |
| Average | | | | | | | | | | | | $\bar{X}_a =$ |
| Range | | | | | | | | | | | | $\bar{R}_a =$ |
| | | Part | | | | | | | | | | Average |
| Operator | Trial run | 1 | 2 | 3 | 4 | 5 | 6 | 7 | 8 | 9 | 10 | |
| B | 1 | | | | | | | | | | | |
| | 2 | | | | | | | | | | | |
| | 3 | | | | | | | | | | | |
| Average | | | | | | | | | | | | $\bar{X}_b =$ |
| Range | | | | | | | | | | | | $\bar{R}_b =$ |
| | | Part | | | | | | | | | | Average |
| Operator | Trial run | 1 | 2 | 3 | 4 | 5 | 6 | 7 | 8 | 9 | 10 | |
| C | 1 | | | | | | | | | | | |
| | 2 | | | | | | | | | | | |
| | 3 | | | | | | | | | | | |
| Average | | | | | | | | | | | | $\bar{X}_c =$ |
| Range | | | | | | | | | | | | $\bar{R}_c =$ |

Fig. 2.8 MSA data sheet (Gr%R)

(7) Calculate Repeatability & Reproducibility GRR

$$GRR = r \& R = \sqrt{r^2 + R^2}$$

(8) Calculate %GRR

$$\%GRR = Gr \& R\% = 6\frac{r \& R}{T} \cdot 100\%$$

with    T    : tolerance

The results are interpreted as follows:

Gr&R < 10%          : the measurement system is acceptable

10 ≤ Gr&R ≤ 30% : the measurement system might be acceptable depending on the application; the use of the system should be approved by the customer

Gr&R > 30% : the measurement system is not acceptable and has to be improved

### 2.4.4. Gr Study (type III analysis)[1]

In case the operator does not have any influence on the gauge, a repeatability study may be sufficient to qualify the gauge. The type III analysis is a subgroup of the type II analysis.

For this study
- one operator
- measures 25 positions / parts
- two times

Then the data is inserted in a table (table 2.4) and range and spread are calculated for each part. In order to have reliable measurement over the entire spread of the process the data should cover the entire spread of the process.

| count | repeated | | calculated | |
|---|---|---|---|---|
| j | 1 | 2 | $R_j$ | $\overline{x_j}$ |
| 1 | $X_{1,1}$ | $X_{1,2}$ | $R_1$ | $\overline{x_1}$ |
| 2 | $X_{2,1}$ | $X_{2,2}$ | $R_2$ | $\overline{x_2}$ |
| 3 | .. | .. | .. | .. |
| 4 | .. | .. | .. | .. |
| .. | | | | |
| .. | | | | |
| 25 | $X_{25,1}$ | $X_{25,2}$ | $R_{25}$ | $\overline{x_{25}}$ |

Table 2.4 MSA data sheet (Gr study)

Calculation

(1) Calculate the range $R_P$ of the mean values

$$R_P = \max(\overline{x_j}) - \min(\overline{x_j})$$

(2) Calculate the mean range

$$\overline{R} = \frac{1}{25}\sum_{j=1}^{25} R_j$$

---
[1] Linß, G. (2005), 316

(3) Calculate the machine variation EV

$$EV = \overline{R} \cdot K_1$$

with K₁ = 0.8862

(4) Calculate the process variation PV

$$PV = R_P \cdot K_3$$

with K₃ = 0.25

(5) Repeatability & reproducibility (GRR) equals EV. With that the capability index Gr is then

$$\%Gr = 100\% \cdot \frac{EV}{T} = 100\% \cdot \frac{GRR}{T}$$

For the interpretation of the results see 2.4.3!

## 2.5. Business processes for capability studies

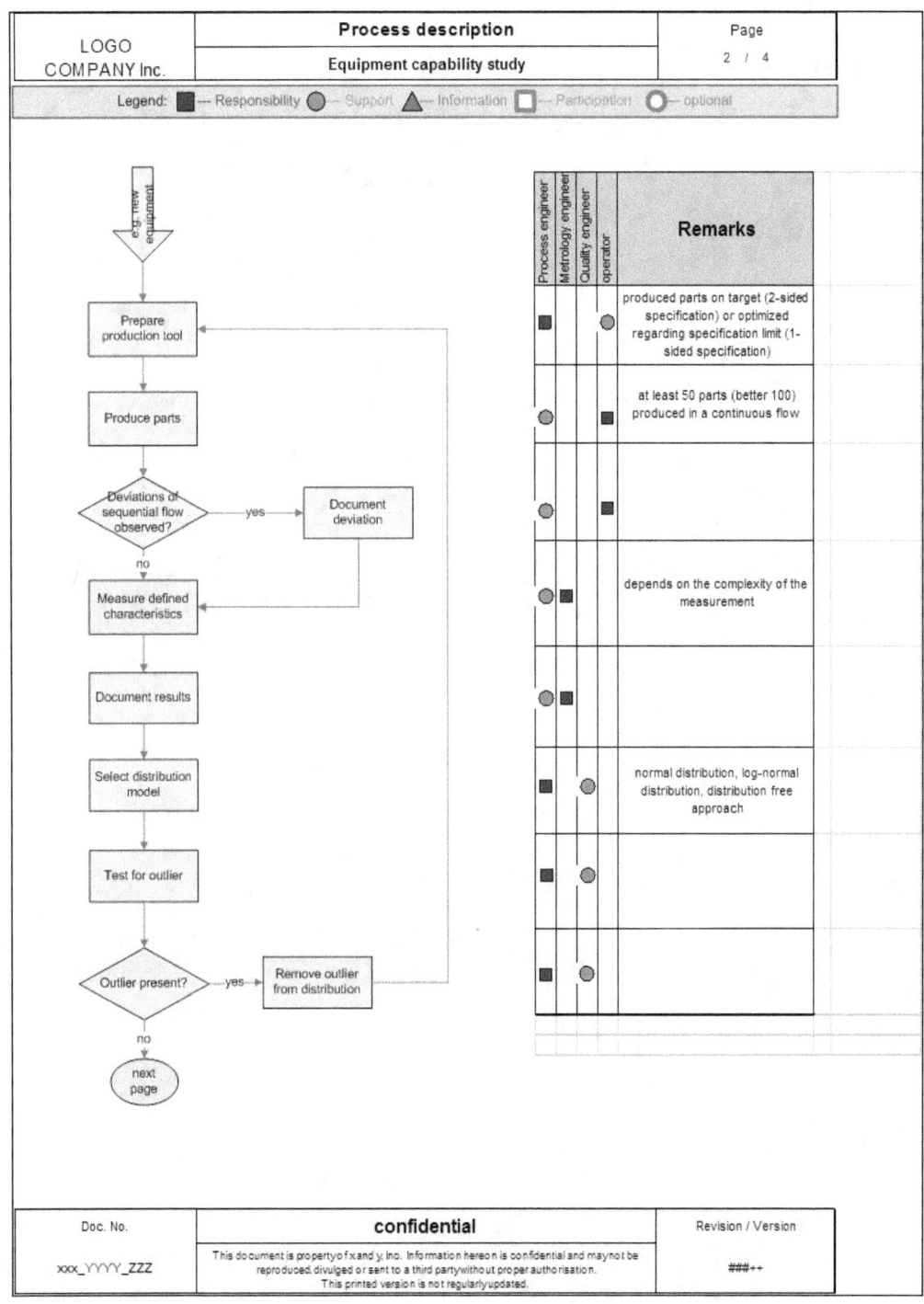

Fig. 2.9 Generic process "machine capability study" – part 1

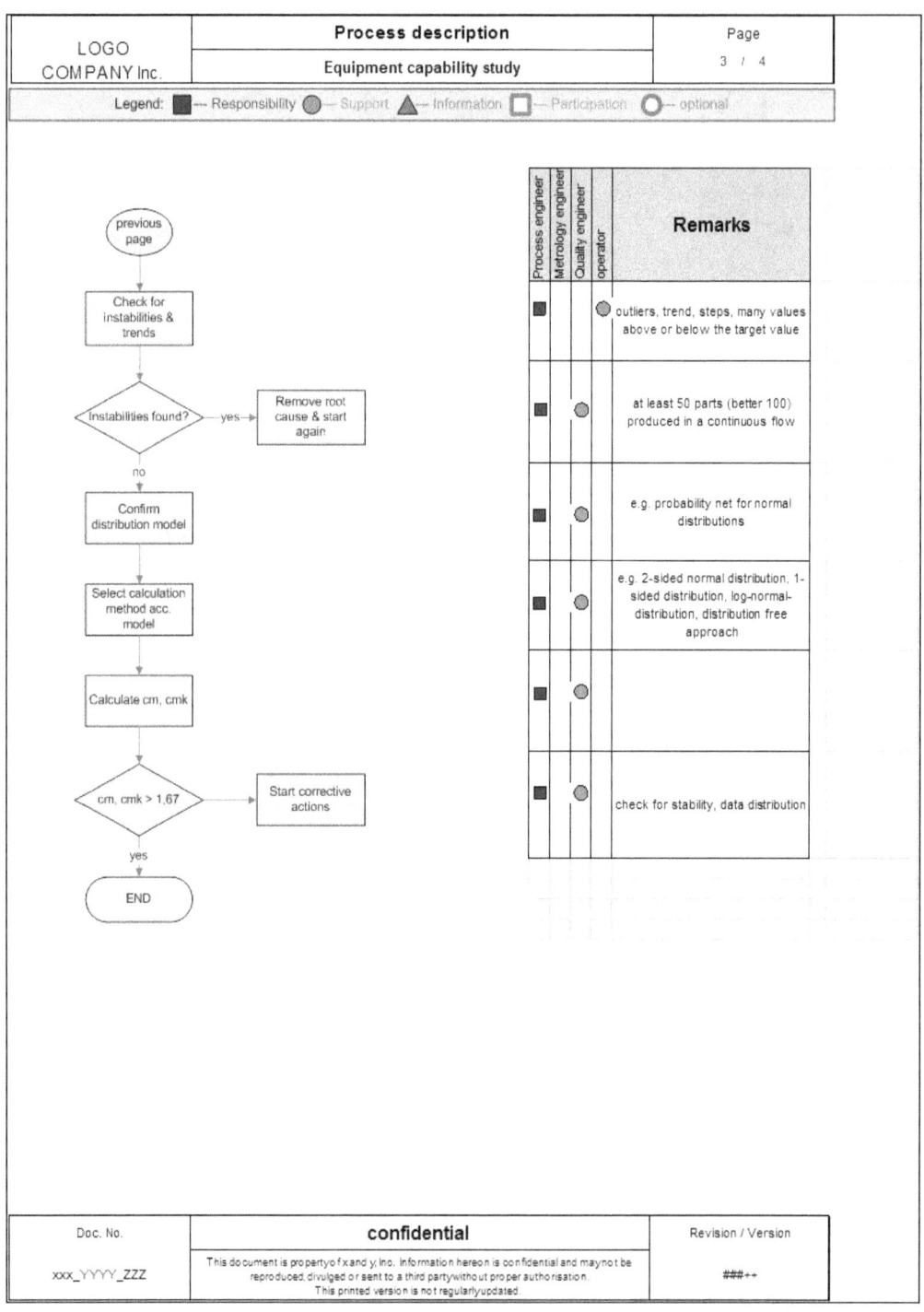

Fig. 2.10 Generic process "machine capability study" – part 2

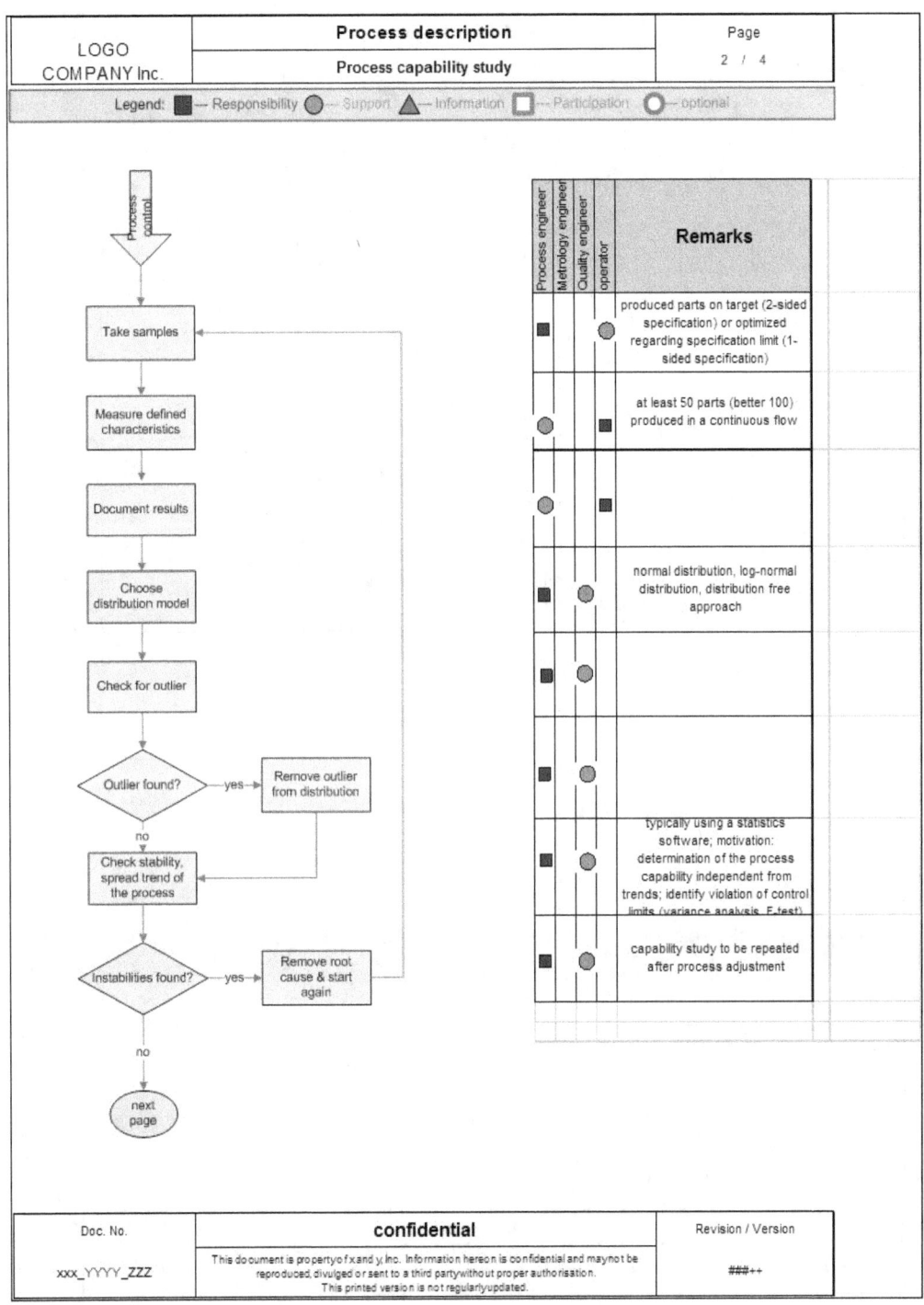

Fig. 2.11 Generic process "process capability study" – part 1

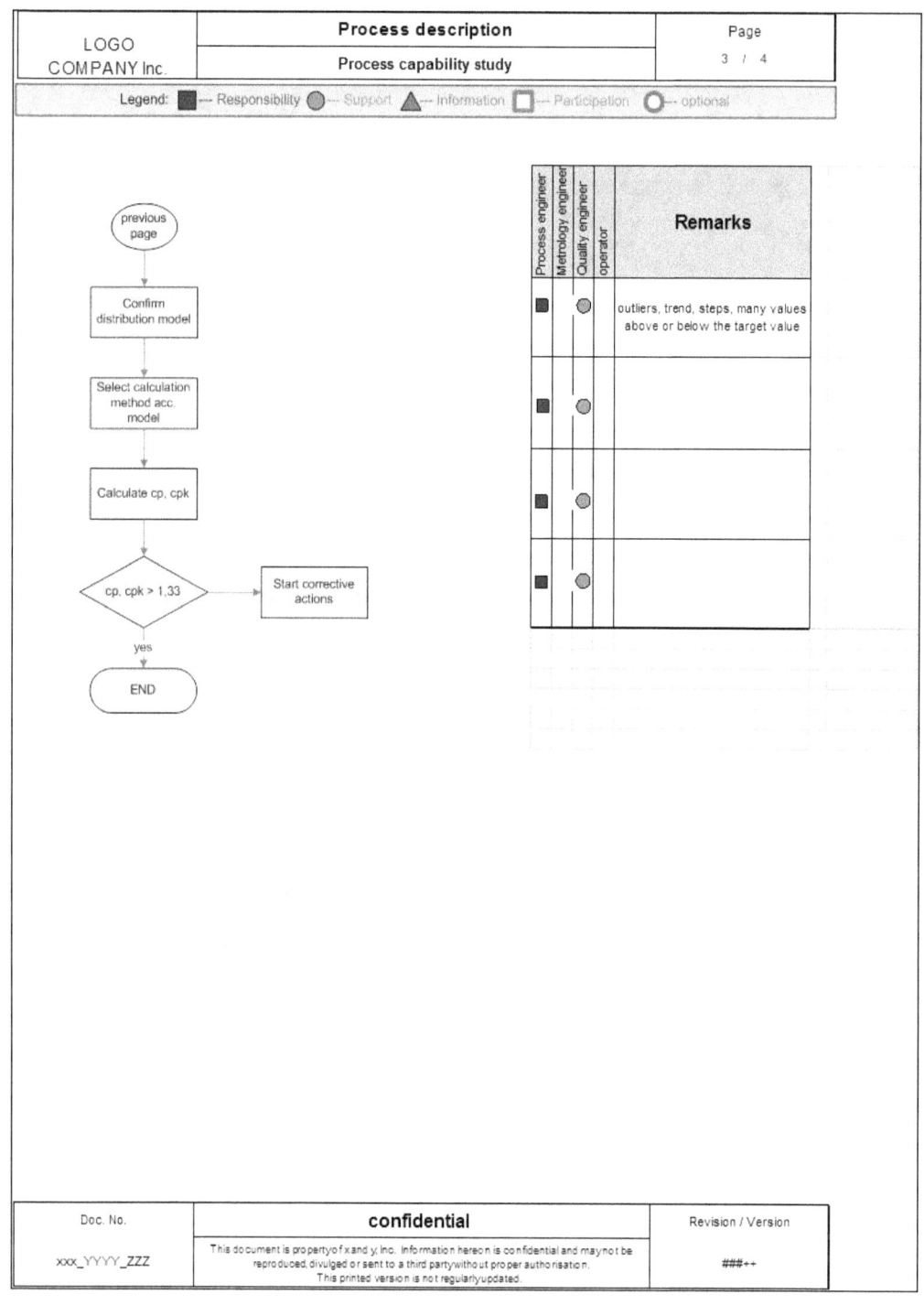

Fig. 2.12 Generic process "process capability study" – part 2

# 3. Statistical Process Control

## 3.1. Motivation

Manufacturing processes do not yield identical process results each time they are carried out. Due to variations of equipment, material, environment etc. they typically result in randomly varying process results or, e.g. in case of tool wear out, even in an additional trend (fig. 3.1).

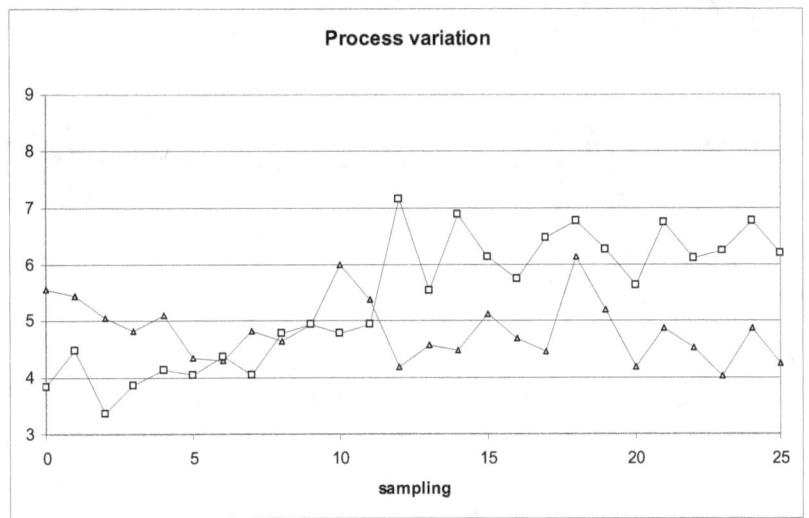

Fig. 3.1 Process with typical process variation and trend

So, unfortunately, variations and trends within the tolerances cannot be avoided and have to be accepted to some extent. However, in order to avoid instable processes, they have to be monitored and corrective actions have to be taken in case of deviations from the target.

One method to keep production processes under control is statistical process control (SPC). In addition to addressing process variations and trends SPC also supports the identification of extreme values and de-centering of the process using statistical methods.

In this context a product is understood as the result of various production processes. Having the particular processes under control is assumed to result in a product within specifications.

The basic idea of SPC, developed by Walter Shewart in the nineteen-twenties[1], is to measure the process results and decide whether or not the process is stable. In case

---

[1] Masing, W. (2014), p. 688

there are indications for an instable process the root cause is investigated and corrective actions are implemented in order to stabilize the process (fig. 3.2)

The driver for those activities is the fact that stable and robust processes lead to predictable process results and therefore to high performing, high yielding and reliable products.

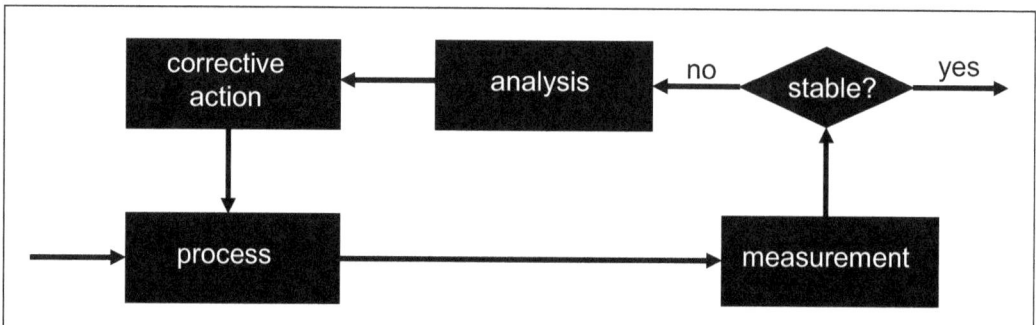

Fig. 3.2 Illustration of the SPC loop

The SPC approach is set up in a way
- that upcoming deviations of critical process parameters are detected very early
- hence, that deviations do not lead to real problems
- that distinguishing between systematic and random problems is supported which helps for the root cause analysis significantly
- that statistical actions limits, i.e. multiples of the standard deviation, lead to a continuous improvement of the process: elimination of deviations leads to smaller standard deviation, smaller standard deviation leads to lower action limits.

In other words, SPC helps to distinguish "common" causes of process variation (which are always present) and "special" causes of process variation which are not normally present and which once identified have to be eliminated. May be more importantly, SPC also prescribes when it is forbidden to "tamper" with the process since this would increase the process variation.

> Statistical Process Control is the basis for the Zero Defect Strategy

Finally, the SPC performance indicators also provide the possibility to measure and compare the production process performance of one or more production facilities, of different suppliers, different pieces of equipment and so forth.

Application

Nowadays statistical process control is applied in many different areas. For instance, in semiconductors the fields of application include[1] not only the production processes themselves, but also product performance parameters, material control, environmental control etc. (fig. 3.3).

| manufacturing facility & material | process environment | inline control | product control |
|---|---|---|---|
| DI water (e.g. resistance) | temperature | layer thickness | sheet resistance |
| chemicals (e.g. particles, pH..) | pressure | pull test results | contact whole resistance |
| wafers (e.g. particles, flatness..) | humidity | layer homogenity | capacity of test structures |
|  | particles | line width | wafer level reliability |
|  |  | alignment |  |

*product related monitoring* (increasing)

Fig. 3.3 Various fields of application for SPC in a semiconductor manufacturing environment (examples only!)

Eventually statistical process control is supposed to reach a process which is capable

"Process Capability is broadly defined as the ability of a process to meet customer expectations"[2]

and stable:

"A product or process is stable when its characteristics (mean, variance, randomness) are constant over a long time period."[3]

## 3.2. Quality Control Chart

### 3.2.1. Introduction

The most frequently used tool for statistical process control is the so called quality control chart (QCC). Control charts are a graphical illustration of process results in order

---
[1] Wagner, T. (2007)
[2] Lean Academy (2012)
[3] Wagner, T., QM Tool Description SPC (2008)

to monitor a process over a longer period of time. Statistical values and deviations are recorded on the chart. It is basically an x-y-function with numbers of samples, dates of sampling etc. on the horizontal axis and the measurement values or statistical aggregation of those on the vertical axis.

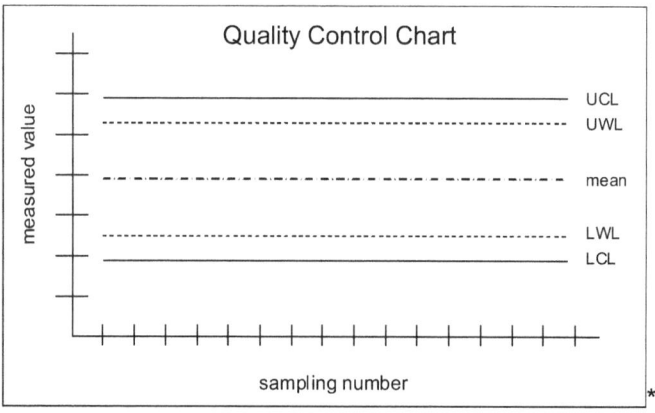

Fig. 3.4 Quality Control Chart

The quality control chart includes statistical or other limits, like upper and lower warning limits (UWL, LWL) or upper and lower control limits (UCL, LCL), or the mean or target values (fig. 3.4).

> **exceeding warning limits leads to increased monitoring, e.g. an additional sample**

The quality control chart reflects the distribution of data points. An accumulation around the mean value and a somewhat reduced occurrence rate towards the control limits results, e.g. in the normal distribution (fig. 3.5).

> **exceeding control limits leads to corrective actions, e.g. root cause analysis for a special cause of variation and corrective action**

If the target value of a distribution is zero, e.g. for surface roughness, particle density or fraction of defective parts, the violation of the statistically determined lower control limit shows an improvement of the process. A root cause analysis is necessary in order to stabilize this behaviour and in order to transfer this knowledge to other processes.

Quality control charts provide the information whether or not the process is under control. It does not provide information about the root cause for being out of control. If a process is completely out of control the quality control chart is not reasonable, in this case a 100% inspection is required instead of sampling which is the basis for SPC.

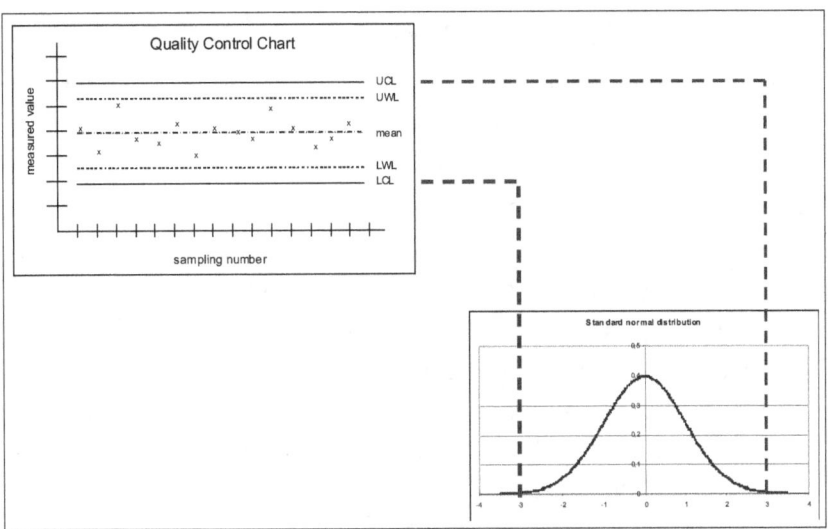

Fig. 3.5 Relationship between control chart and distribution of process results (= probability density function)

## 3.2.2. Sampling

As indicated in fig. 3.5 samples of a defined sample size and in a defined frequency are taken for measurement. The sample size depends on the nature of the quality control chart. For instance, the sample size for individuals control charts is naturally n = 1. For an average control chart typically a sample size consists of 5 to 9 measurement points[1]. The sample size should not be less than 3 and in total at least 125 measurements are desirable[2]. For an attribute control chart, on the other hand, the typical sample size consists of n ≥ 30 parts. Attribute control charts can be used, when the measurement of a particular quality characteristic is not possible and is characterized as "conforming" and "non-conforming".

The decision on how to select the sample is influenced by the process performance. For instance, in the semiconductor world there might be process variations from machine to machine, lot to lot, wafer to wafer and on the wafer (e.g. centre to edge, chip to chip). In case of an only negligible wafer to wafer variation within one lot, measurement on only one wafer is sufficient.

The sampling frequency should be high enough to catch all (or the majority) of problems in the process and low enough to have the benefit of a sampling strategy (for cost reasons). Technically a higher frequency leads to more information and better statistics.

---

[1] Wagner, T., Statistical Process Control (2008)
[2] Schulze, A. (2014), p. 697

In this connection it is also recommended to have a higher number of small sample sizes rather than having one big sample every quarter.

### 3.2.3. Design of Quality Control Charts

Depending on the nature of a characteristic as well as the behaviour of the distribution different kinds of control charts have to be used. A dominant factor for differentiation is if we are dealing with attributive characteristics, e.g. only two process results like good/bad (i.e. conforming / nonconforming) are available, or with variable or measurement characteristics. In the first case only one control chart is used whereas in the latter case typically two control charts are used.[1] Also, it is important to distinguish between a purely statistical approach (e.g. Shewart control charts) and a tolerance based approach (acceptance control charts).

Acceptance control charts are used in case of known process trends, e.g. due to tool wear out, variation of lot mean values or in case of a very high process capability index.

The following steps are required for the setup of a quality control chart:[2]

(1) choose qualitative (attributive) or quantitative (variable) characteristics
(2) collect at least 25 samples[3] with a sample size of n = 5
(3) determine the statistical values (e.g. mean)
(4) determine the appropriate quality control chart (fig. 3.6)
(5) calculate the respective warning and control limits

---

[1] Brunner, F., Wagner, K. (2011), p. 186
[2] Linß, G. (2005), p. 208
[3] DGQ 16-32 (1995)

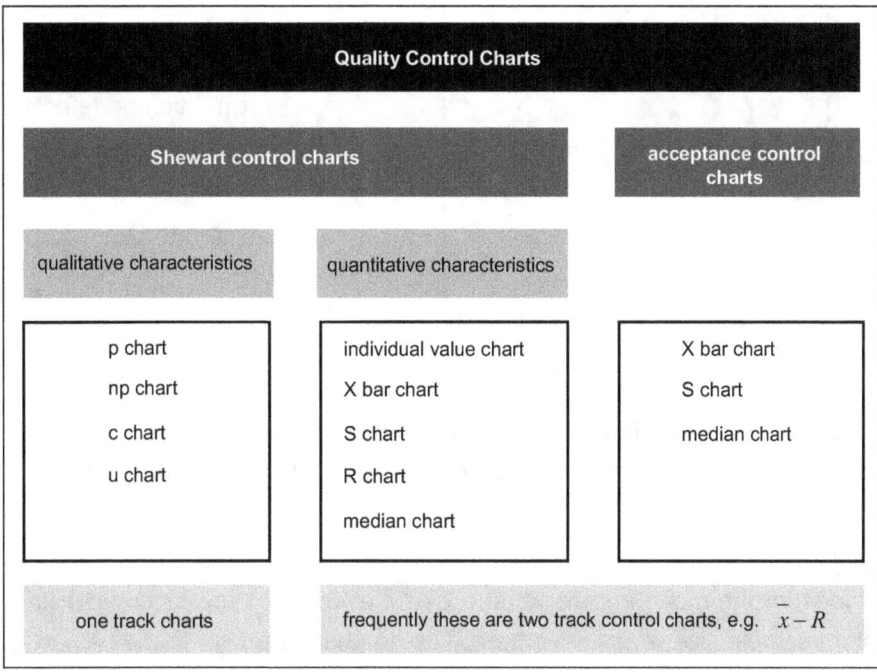

Fig. 3.6 Different kinds of quality control charts[1]

## 3.2.4. Shewart Control Charts for Attributive Characteristics[2]
### 3.2.4.1. p-Chart

P stands for "proportion". "The proportion or fraction nonconforming (defective) in a population is defined as the ratio of the number of nonconforming items in the population to the total number of items in that population"[3]. The sample size is varying. The p-chart deals with parts which are either good or bad, hence the basis is the Binomial distribution.

For instance, that could be number of parts not meeting the specification.

> **p charts are applied to proportions or fractions of defective parts in a population**

With a centre line $\overline{p}$ from a preliminary run the upper and lower control limits can be calculated (normal distribution assumed):

---
[1] Dietrich E., Schulze A. (2014)
[2] Brunner, F., Wagner, K. (2011), p. 186
[3] Engineering Statistics Handbook

$$\overline{p} = \sum_{i=1}^{m} p_i$$

$$UCL = \overline{p} + u_{1-\alpha/2} \cdot \sqrt{\frac{\overline{p} \cdot (1-\overline{p})}{n}}$$

$$LCL = \overline{p} - u_{1-\alpha/2} \cdot \sqrt{\frac{\overline{p} \cdot (1-\overline{p})}{n}}$$

with

m : number of samples
$p_i$ : fraction of defective parts in a sample i
n : samples size of each sub group

In a global business environment typically $u_{1-\alpha/2} = 3$ is used, which correlates to 99.73% acceptance probability, whereas in Europe often a 99% value is chosen.

### 3.2.4.2. np-Chart

np stands for the mean number of successful events np (see binomial distribution). A constant sample size is assumed here. The np-chart is similar to the p-chart, but deals with the number of defective parts instead of proportions.

Assuming an approximately normal distribution $n\overline{p}$ and control limits are calculated as follows:

$$n\overline{p} = np/k$$

$$UCL = n\overline{p} + u_{1-\alpha/2} \cdot \sqrt{n\overline{p} \cdot (1-\overline{p})}$$

$$LCL = n\overline{p} - u_{1-\alpha/2} \cdot \sqrt{n\overline{p} \cdot (1-\overline{p})}$$

with

np  number of defective parts
k   number of sub groups
n   sample size
p   fraction of defective parts

In a global business environment typically $u_{1-\alpha/2} = 3$ is used, which correlates to 99.73% acceptance probability, whereas in Europe often a 99% value is chosen.

### 3.2.4.3. c-Chart

C stands for "count", i.e. the number of defects, for instance the number of scratches on a wafer. The sample size is constant, the product might still be usable. The basis for this chart is the Poisson distribution.

> c charts are applied to the number of defective parts or non conformances in one unit

After calculation of $\bar{c}$ upper and lower control limits are calculated as follows assuming an approximation by the normal distribution:

$$\bar{c} = c/k$$

$$UCL = \bar{c} + u_{1-\alpha/2} \cdot \sqrt{\bar{c}}$$

$$LCL = \bar{c} - u_{1-\alpha/2} \cdot \sqrt{\bar{c}}$$

with
- c : number of defects
- k : number of sub groups

In a global business environment typically $u_{1-\alpha/2} = 3$ is used, which correlates to 99.73% acceptance probability, whereas in Europe often a 99% value is chosen.

### 3.2.4.4. u-Chart

U stands for "unit", i.e. the number of defects per unit. However, the sample size is not constant, for instance the number of defects in a lot with varying lot size. The product might still be usable. The basis for this chart is the Poisson distribution.

After calculation of $\bar{u}$ upper and lower control limits are calculated as follows assuming an approximation by the normal distribution:

$$\bar{u} = c/n$$

$$UCL = \bar{c} + u_{1-\alpha/2} \cdot \sqrt{\frac{\bar{u}}{n}}$$

$$LCL = \bar{c} - u_{1-\alpha/2} \cdot \sqrt{\frac{\bar{u}}{n}}$$

with
- c : number of defects
- n : sample size of each subgroup

In a global business environment typically $u_{1-\alpha/2} = 3$ is used, which correlates to 99.73% acceptance probability, whereas in Europe often a 99% value is chosen.

## 3.2.5. Shewart Control Charts for Variable Characteristics[1]

### 3.2.5.1. Individual Value Chart

In case the individual values are approximately normally distributed an individual value chart can be used[2]. In the individual value chart case at each sampling point a number of n (= sample size) values are plotted in the control chart.

In Europe control and warning limits are defined in a way that the individual values of a randomly taken sample have a 99% or 95% probability to be within those limits. Thus the limits are calculated as follows:

$$UCL = \mu + u_{\frac{1+\sqrt[n]{0.99}}{2}} \cdot \sigma$$

$$UWL = \mu + u_{\frac{1+\sqrt[n]{0.95}}{2}} \cdot \sigma$$

$$LWL = \mu - u_{\frac{1+\sqrt[n]{0.95}}{2}} \cdot \sigma$$

$$LCL = \mu - u_{\frac{1+\sqrt[n]{0.99}}{2}} \cdot \sigma$$

with
- n : sample size
- u : quantile of the standard distribution
- σ : standard deviation
- μ : mean value

The quantiles of the normal distribution can be taken from the respective statistical table, see 13.2.[3] In a global business environment only control limits are used. Using the 99.73% approach for the control limits they are calculated as follows[4]:

$$UCL = \overline{\overline{x}} + E_2 \cdot \overline{R}$$

$$LCL = \overline{\overline{x}} - E_2 \cdot \overline{R}$$

with
- $\overline{R}$ : average range
- $\overline{\overline{x}}$ : overall average

The statistical constants $E_2$[5] can be taken from table 13.4:

---

[1] Schulze, A. (2007), p. 696
[2] DGQ 16-32 (1995)
[3] DGQ 16-32 (1995)
[4] Dietrich, E., Schulze, A. (2014), p. 275
[5] Institute of Quality & Reliability (2008)

### 3.2.5.2. X Bar Chart

As a consequence of the central limit theorem at sample sizes n ≥ 5 the mean values are approximately normally distributed. Hence the Shewart chart can also be used for non normally distributed values. The respective mean values and standard deviation values are calculated as

$$s = \sqrt{\frac{\sum_{i=1}^{n}(x_i - \bar{x})^2}{n-1}}$$

$$\bar{x}_j = \frac{1}{n}\sum_{i=1}^{n} x_i$$

$$\hat{\sigma} = \sqrt{\frac{1}{k}\sum_{i=1}^{k} s_i^2}$$

$$\hat{\mu} = \bar{\bar{x}} = \frac{1}{k}\sum_{j=1}^{k} \bar{x}_j$$

In Europe the warning and control limits are defined in a way that the mean value of a randomly taken sample is found to be within those limits with 95% or 99% probability, respectively. The process mean value is calculated from an initial run or could also be the target value.

$$UCL = \hat{\mu} + u_{1-\alpha/2} \cdot \frac{\hat{\sigma}}{\sqrt{n}} = \hat{\mu} + A_E \hat{\sigma}$$

$$LCL = \hat{\mu} - u_{1-\alpha/2} \cdot \frac{\hat{\sigma}}{\sqrt{n}} = \hat{\mu} - A_E \hat{\sigma}$$

$u_{1-\alpha/2}$ is the quantile of the normal distribution for the confidence level 1-α (see 13.2)

$$UWL = \hat{\mu} + u_{1-\alpha/2} \cdot \frac{\hat{\sigma}}{\sqrt{n}} = \hat{\mu} + A_W \hat{\sigma}$$

$$LWL = \hat{\mu} - u_{1-\alpha/2} \cdot \frac{\hat{\sigma}}{\sqrt{n}} = \hat{\mu} - A_W \hat{\sigma}$$

AE and AW can be taken from table 13.3:

> X bar charts are typically combined with a s-chart to control variance

In a global business environment x bar charts are used either as x bar / R-chart or as x bar / s-charts. Again, only control limits are used.

For the x bar / R chart, based on the 99.73% approach for the control limits they are calculated as follows[1]:

$$UCL_{\bar{x}} = \bar{\bar{x}} + A_2 \cdot \bar{R}$$

$$LCL_{\bar{x}} = \bar{\bar{x}} - A_2 \cdot \bar{R}$$

$$UCL_R = D_4 \cdot \bar{R}$$

$$LCL_R = D_3 \cdot \bar{R}$$

with

$\bar{\bar{x}}$ : overall mean value

$\bar{R}$ : average range

For the x bar / s – chart based on the 99.73% approach for the control limits they are calculated as follows[2]:

$$UCL_{\bar{x}} = \bar{\bar{x}} + A_3 \cdot \bar{s}$$

$$LCL_{\bar{x}} = \bar{\bar{x}} - A_3 \cdot \bar{s}$$

$$UCL_s = B_4 \cdot \bar{s}$$

$$LCL_s = B_3 \cdot \bar{s}$$

$$\bar{s} = \frac{1}{k}\sum_{j=1}^{k} s_j$$

with:

$\bar{s}$ : average standard deviation

The respective constants can be taken from table 13.4 (international approach)[3]

### 3.2.5.3. X Bar Chart with Extended Limits

In case the overall mean value is not stable, e.g. when observing a lot to lot trend, extended limits for the control charts should be used:

---

[1] Dietrich, E., Schulze, A. (2014)
[2] Dietrich, E., Schulze, A. (2014), p. 275
[3] Dietrich, E., Schulze, A. (2014)

$$UCL = \hat{\mu} + (u_{1-\alpha/2} \cdot \frac{\hat{\sigma}}{\sqrt{n}} + u_{1-\alpha'/2} \cdot \hat{\sigma}_A)$$

$$LCL = \hat{\mu} - (u_{1-\alpha/2} \cdot \frac{\hat{\sigma}}{\sqrt{n}} + u_{1-\alpha'/2} \cdot \hat{\sigma}_A)$$

with

$u_{1-\alpha/2}$ : being a quantile of the normal distribution:

for 1-α = 0.99 applies: $u_{1-\alpha/2}$ = 2.578
for 1-α' = 0,8664 applies: $u_{1-\alpha'/2}$ = 1.500
(from experience[1])

$\hat{\sigma}_A$ = standard deviation between the sample mean values

### 3.2.5.4. S Chart

In case the characteristics are approximately normally distributed the one-sided $\chi^2$ distribution is used for the calculation of the s warning and control limits. S stands for standard deviation. This chart provides additional information about the spread of the distribution.

In Europe, again, the warning and control limits are defined in a way that the standard deviation of a randomly taken sample with sample size n is found to be within those limits with 95% or 99% probability, respectively.

The statistical constants $B_{UCL}$, $B_{UWL}$, $B_M$, $B_{LWL}$ and $B_{LCL}$ are found in table 13.5:

$$UCL = \sigma \cdot \sqrt{\frac{\chi^2_{n-1,1-0.01/2}}{n-1}} = \sigma \cdot B_{UCL}$$

$$UWL = \sigma \cdot \sqrt{\frac{\chi^2_{n-1,1-0.05/2}}{n-1}} = \sigma \cdot B_{UWL}$$

$$M = \sigma \cdot B_M$$

$$LWL = \sigma \cdot \sqrt{\frac{\chi^2_{n-1,0.05/2}}{n-1}} = \sigma \cdot B_{LWL}$$

$$LCL = \sigma \cdot \sqrt{\frac{\chi^2_{n-1,0.01/2}}{n-1}} = \sigma \cdot B_{LCL}$$

---

[1] Schulze, A. (2007)

For the international approach see 3.2.5.2.

### 3.2.5.5. R Chart

In case the characteristics are approximately normally distributed the one-sided w-distribution is used for the calculation of the R warning and control limits.

$$UCL = \sigma \cdot w_{n;1-0.01/2} = \sigma \cdot D_{UCL}$$

$$UWL = \sigma \cdot w_{n;1-0.05/2} = \sigma \cdot D_{UWL}$$

$$M = \sigma \cdot D_M = \sigma \cdot d_n$$

$$LWL = \sigma \cdot w_{n;0.05/2} = \sigma \cdot D_{LWL}$$

$$LCL = \sigma \cdot w_{n;0.001/2} = \sigma \cdot D_{LCL}$$

R charts (R stands for Range) is used to show the spread of a distribution and ins mainly used if the size of the sample is equal or smaller than ten.

Again, the warning and control limits are defined in a way that the standard deviation of a randomly taken sample with sample size n is found to be within those limits with 95% or 99% probability, respectively.

The statistical constants $D_{UCL}$, $D_{UWL}$, $D_M$, $D_{LWL}$, $D_{LCL}$ are found in table 13.6.

For the international approach see 3.2.5.2.

### 3.2.5.6. Median Chart[1]

In the case of approximately normally distributed characteristics the location of the distribution can be monitored using the median, too. Due to the way the median is calculated an uneven sample size is recommended.

Typically the variance of the median is smaller than the variance of the respective mean values. Hence the standard deviation of the median $\tilde{x}$ is:

$$\tilde{x} = c_n \cdot \sigma_{\tilde{x}} \hat{\mu}$$

with

n : sample size

The value of $c_n$ depends on the sample size and is found in table 13.7. In Europe, again, the warning and control limits are defined in a way that the standard deviation of a

---
[1] DGQ 16-32 (1995)

randomly taken sample with sample size n is found to be within those limits with 95% or 99% probability, respectively.

Limits and mean value are calculated as follows:

$$UCL = \mu + u_{1-0.01/2} \cdot c_n \cdot \sigma_{\bar{x}}$$

$$UWL = \mu + u_{1-0.05/2} \cdot c_n \cdot \sigma_{\bar{x}}$$

$$M = E(\tilde{x}) = \mu$$

$$LWL = \mu - u_{0.05/2} \cdot c_n \cdot \sigma_{\bar{x}}$$

$$LCL = \mu - u_{0.01/2} \cdot c_n \cdot \sigma_{\bar{x}}$$

> Median charts are typically combined with a R-chart to control variance

In a global business environment median ($\tilde{x}$) charts are typically used as $\tilde{x}$ / R-chart. Again, only control limits are used. Based on the 99.73% approach for the control limits they are calculated as follows[1]:

$$UCL_{\tilde{x}} = \bar{\bar{x}} + \tilde{A}_2 \cdot \bar{R}$$

$$LCL_{\tilde{x}} = \bar{\bar{x}} - \tilde{A}_2 \cdot \bar{R}$$

$$UCL_R = D_4 \cdot \bar{R}$$

$$LCL_R = D_3 \cdot \bar{R}$$

with
$\bar{\bar{x}}$ : average median value
$\bar{R}$ : average range

### 3.2.6. Acceptance Control Charts[2]

In the case that a variation of the process mean value is not critical within a defined area around the target value also acceptance control charts can be used. Only if the mean value is too far away from the target value, which has to be defined prior setting up the SPC chart, corrective actions have to be started.

---

[1] Dietrich, E., Schulze, A. (2014)
[2] DGQ 16-32 (1995)

Acceptance charts do not have warning limits. The control limits are typically calculated based on the specification limits. For the mean value acceptance chart UCL and LCL are calculated according:

$$UCL = UTL - k_A \sigma$$

$$LCL = LTL + k_A \sigma$$

with

$k_A$ to be taken from the Wilrich-Nomogram[1]

Alternatively, $k_A$ can be calculated:

$$k_A = u_{1-p} + \frac{u_{1-P_A}}{\sqrt{n}}$$

with

$u_{1-p}$ : quantile (standard normal distribution)
p : fraction of defective parts
n : sample size
$P_A$ : probability for no action

Example:
The probability for corrective actions shall be 90% at a fraction of bad parts of 1%. Sample size shall be n = 5. Calculate $k_A$.

$$k_A = u_{1-p} + \frac{u_{1-P_A}}{\sqrt{n}} = u_{1-0.01} + \frac{u_{1-0.1}}{\sqrt{5}} = u_{0.99} + \frac{u_{0.9}}{\sqrt{5}}$$

With $u_{0.99}$ = 2.326 and $u_{0.9}$ = 1.282 from the standardized normal distribution (see 12.1) $k_A$ is calculated:

$$k_A = u_{1-p} + \frac{u_{1-P_A}}{\sqrt{n}} = 2.326 + \frac{1.282}{2.2361} = 2.899$$

### 3.2.7 Pearson Control Chart[2]

The Pearson distribution includes an entire family of various distributions and is usable for single, mean, range and standard deviation values. After calculation of kurtosis and skewness of the respective distribution $Q_{0.135\%}$, $Q_{50\%}$ and $Q_{99.865\%}$ are calculated.

$$UCL = Q_{99.865\%}$$

$$CenterLine = Q_{50\%}$$

$$LCL = Q_{0.135\%}$$

---

[1] DGQ 16-32 (1995)
[2] Wagner, T., SPC (2007)

In the next section, guidelines as to which of the various control chart types is most appropriate for a given situation.

## 3.3. Choice of Charts Based on Time Dependent Distribution Models

As discussed previously, real processes do not yield completely stable and absolutely constant process results. Random and systematic influences cause variations, which can be observed within a sample or between sample means in a random or systematic way.

Thus four main models are used to describe the behaviour of a distribution over time:

|  | distribution model[1,2] |
|---|---|
| (1) capable process under control | A |
| (2) not capable, but under control | B |
| (3) capable, but not under control | C |
| (4) not capable and not under control | D |

In addition to those main categories there are several subcategories depending on the form and the behaviour of the distribution and on the root cause for the "not under control" situation.

Prior setting up the final quality control chart an initial run has to be performed in order to determine the particular distribution model[3].

Model A1

In model A1 each sample distribution equals all other sample distributions and the overall distribution. Model A1 is characterized by

- process (sample) average : constant
- variation : constant
- skewness : constant
- kurtosis : constant
- sample distribution : normal distribution
- overall distribution : normal distribution

---

[1] DIN 21747 (2007)
[2] DIN ISO 22514-2 (2013)
[3] Dietrich, E., Schulze, A. (2014)

Distribution A1 represents an ideal process which is hard to find in practice. Hence fig. 3.7 shows a process which is very close to model A1, but does have minor variations in sample average and distribution width (as expected in reality).

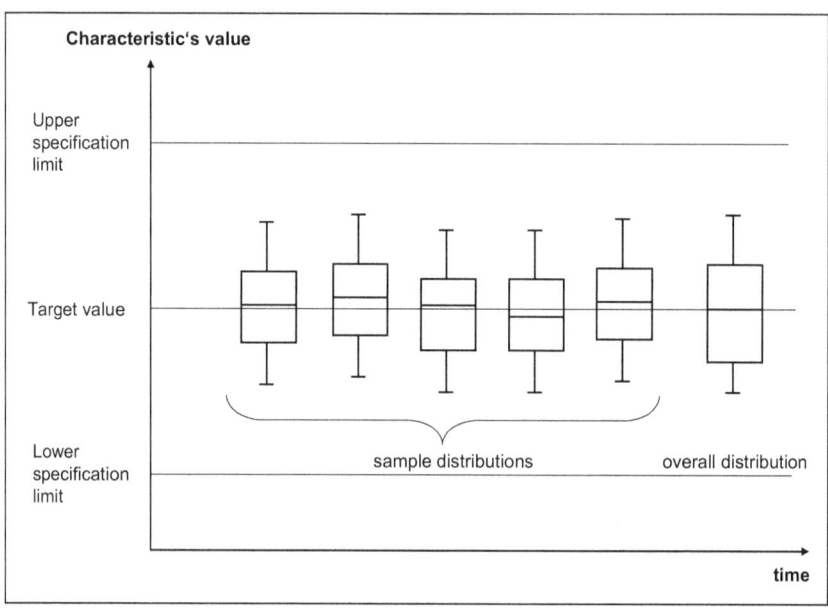

Fig. 3.7 Illustration of the time dependent distribution close to model A1

Model A2

Model A2 is characterized by

- process (sample) average  : constant
- variation  : constant
- skewness  : constant
- kurtosis  : constant
- sample distribution  : skewed, e.g. log-normal or Weibull distribution
- overall distribution  : skewed, e.g. log-normal or Weibull distribution

Typical parameters showing this kind of distribution are parameter with single sided physical limitations like surface roughness, layer thickness or torque. Fig. 3.8 shows a process which is very close to model A2, but does have minor variations in sample average and distribution width (as expected in reality).

## Model B

Model B processes are processes showing variation in the standard deviation with a constant average value. Those distributions are often caused by variations in the preceding raw material.

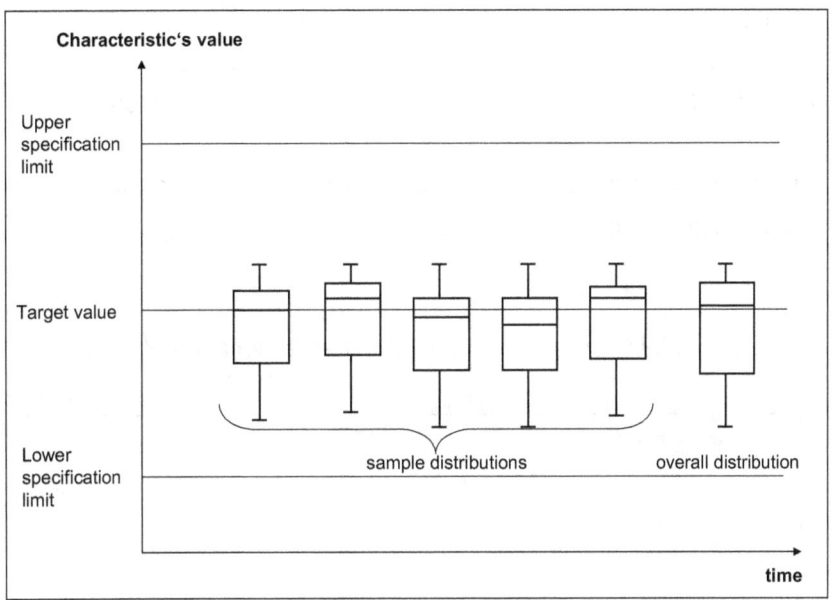

Fig. 3.8 Illustration of the time dependent distribution close to model A2

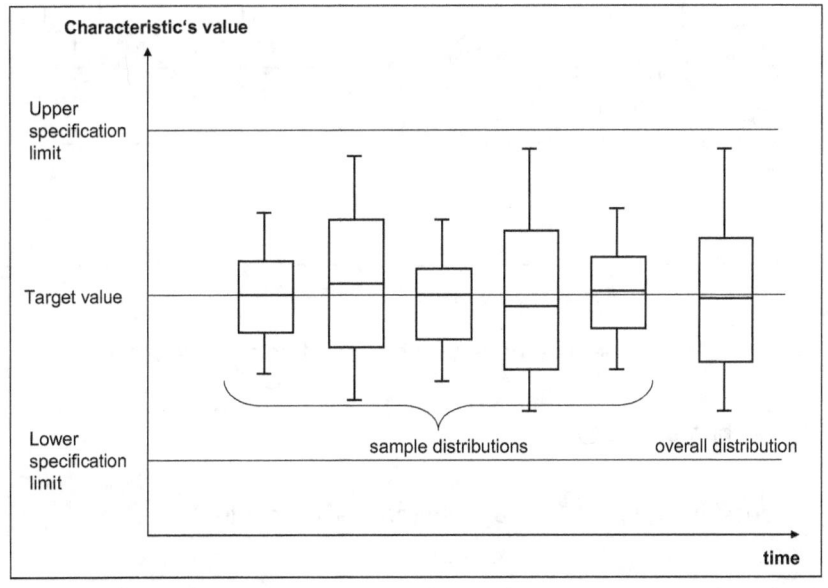

Fig. 3.9 Illustration of the time dependent distribution close to model B

Model B is characterized by

- location : constant
- process (sample) average : not constant – systematic or random
- skewness : constant
- kurtosis : constant
- sample distribution : normal distribution
- overall distribution : no normal distribution

Fig. 3.9 shows a process which is very close to model B, but does have minor variations in sample average and distribution width (as expected in reality).

Model C1

Main feature of model C processes is the fact that the location of samples is varying.

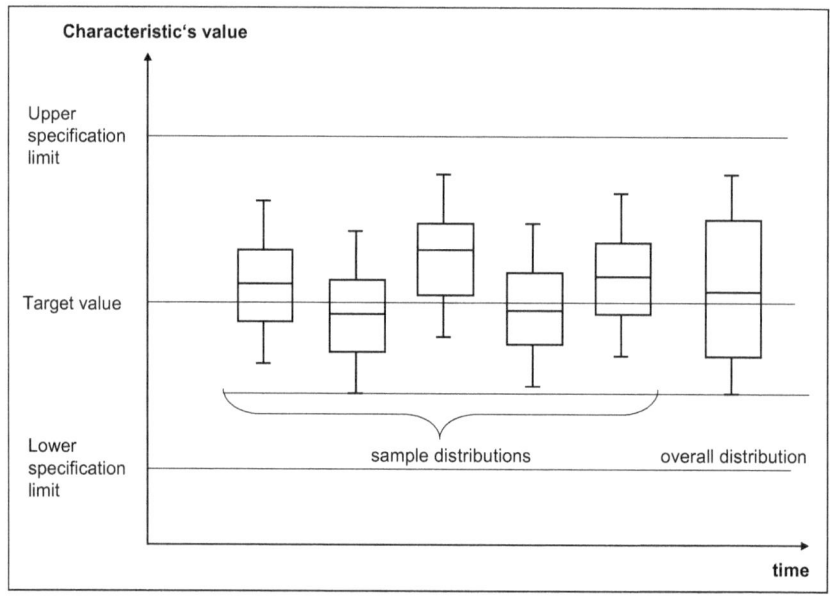

Fig. 3.10 Illustration of the time dependent distribution close to model C1

Model C1 is characterized by

- process average : random (normal distributed)
- variation : constant
- skewness : constant
- kurtosis : constant

- sample distribution : normal distribution
- overall distribution : normal distribution

Potential root cause may be a preceding process influencing the process in question.. Fig. 3.10 shows a process which is very close to model C1, but does have minor variations in sample distribution width (as expected in reality).

## Model C2

Model C2 processes behave similar to model C1 processes except for the fact that the resulting overall distribution is not normally distributed.

Model C2 is characterized by

- process average : random
- variation : constant
- skewness : constant
- kurtosis : constant
- sample distribution : normal distribution
- overall distribution : no normal distribution

Potential root cause may be a preceding process influencing the process in question.

## Model C3

Model C3 is characterized by a systematic shift of the sample mean values.

Model C3 is characterized by

- process average : systematic change
- variation : constant
- skewness : constant
- kurtosis : constant
- sample distribution : normal distribution
- overall distribution : no normal distribution

A potential root cause is tool wear out. Fig. 3.11 shows a process which is very close to model C3, but does have minor variations in sample average shift and distribution width (as expected in reality).

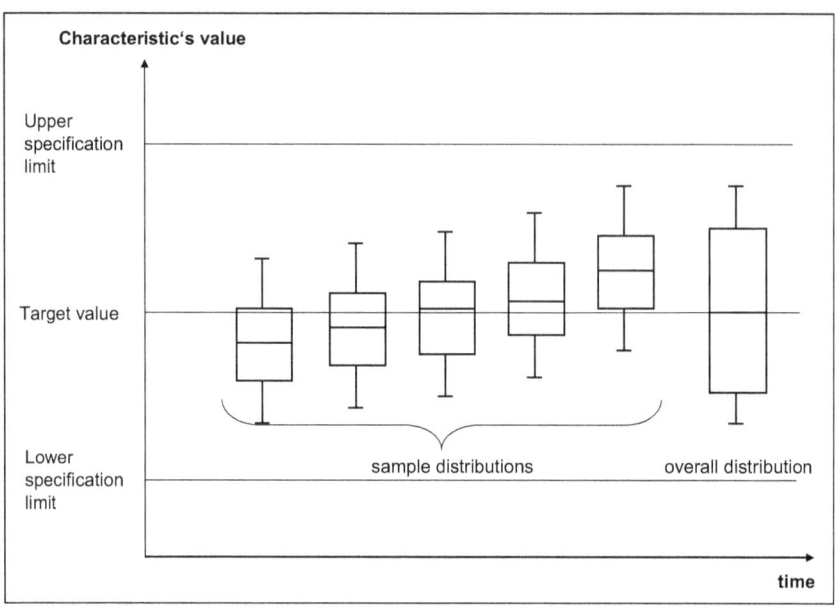

Fig. 3.11 Time dependent distribution close to model C3

Model C4

The main feature of model C4 processes is a systematic or random shift of the sample average, e.g. due to change of lot or tool.

Model C4 is characterized by

- process average : systematic and non-systematic changes
- variation : constant
- skewness : constant
- kurtosis : constant
- sample distribution : normal distribution
- overall distribution : mix of distributions

Fig. 3.12 shows a process which is very close to model C4, but does have minor variations in sample distribution width (as expected in reality).

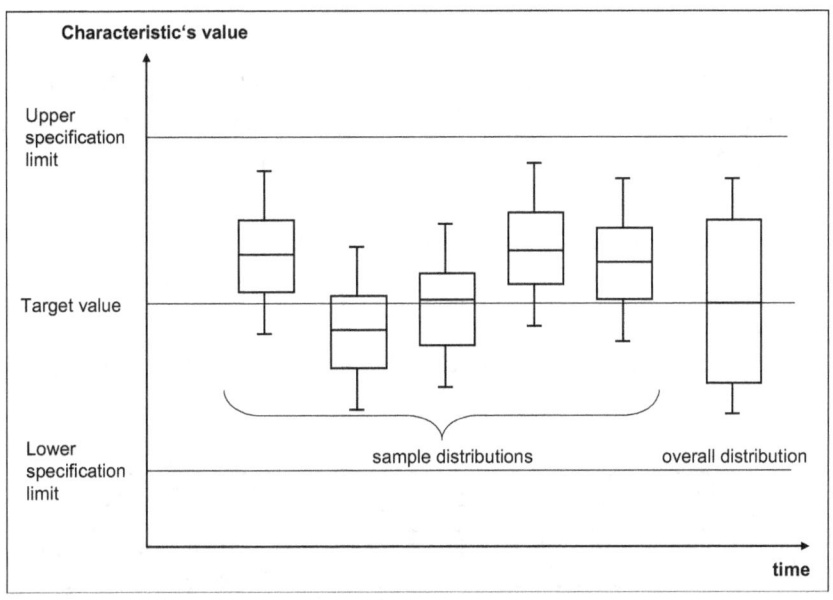

Fig. 3.12 Illustration of the time dependent distribution close to model C4

## Model D

A model D process is the worst case regarding process stability. All parameters may be varying systematically or randomly.

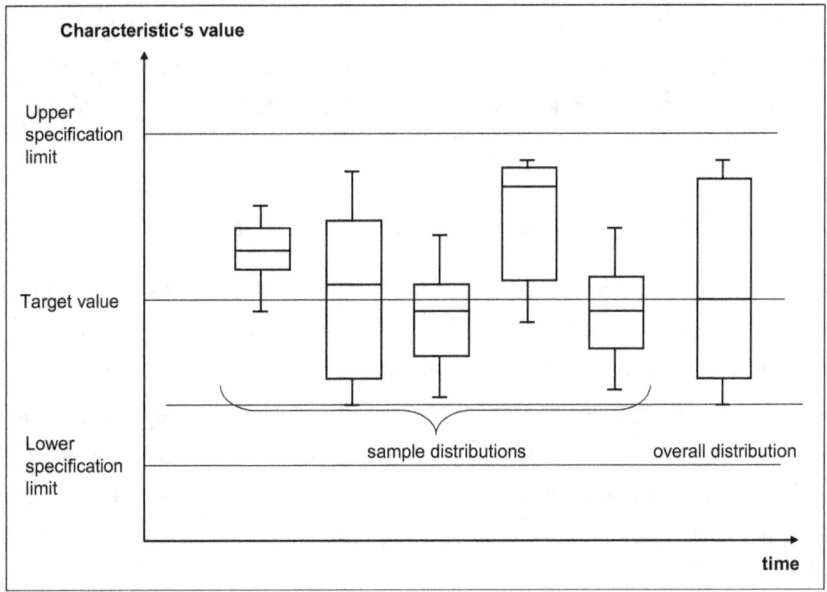

Fig. 3.13 Illustration of the time dependent distribution close to model D

Model D is characterized by

- process average : not constant – systematic and/or random variation
- variation : not constant – systematic and/or random variation
- skewness : not constant – systematic and/or random variation
- sample distribution : arbitrary
- overall distribution : no distribution model

Decision matrix

After the determination of the respective time dependent model the type of control chart can be assigned as follows:

| Time dependend distribution model | Recommended Control Chart |
|---|---|
| A1 | Shewart chart |
| A2 | Shewart chart; alternatively Pearson chart if skewed distribution |
| C1, C2, C4, B, D | Shewart chart with extended limits |
| C3 | Shewart chart with extended limits; alternatively also acceptance charts |

Table 3.1 Assignment of control charts to time dependent distribution model[1]

Table 3.1 is a suggestion only. For instance, it may also be more appropriate to switch to a 100% control for a model D process. After a choice of the control chart type, the next consideration is: Rules how to act upon certain situations. These were historically developed in a company called Western Electric, hence a common name for these is Western-Electric rules or instability rules.

## 3.4. Instability Rules (Western-Electric-Rules)[2]

A process is considered stable as long as the instability rules are not violated. Originally Western Electric defined instability rules for Shewart charts for the 99.73% probability of non interference case, the so called Western-Electric-Rules (WER).

When setting up the quality control chart in addition to the determination of the kind of chart and the calculation of the limits, the instability rules have to be defined. The first one of those rules is mandatory, the remaining rules are implemented as suitable for the particular process:

---

[1] Schulze, A. (2014), p. 704
[2] Wagner, T. (2007)

Rule 1             1 measurement point is outside of the control limit (3 σ)

Rule 2             9 consecutive measurement points are found to be above or below the centre line

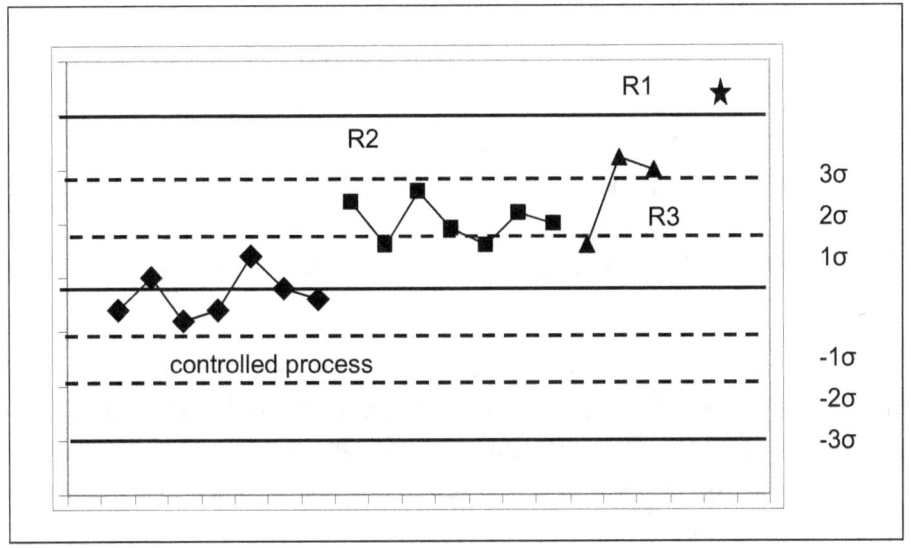

Fig. 3.14 Illustration of a controlled process vs. instable situations with violation of instability rules (1 – 3)

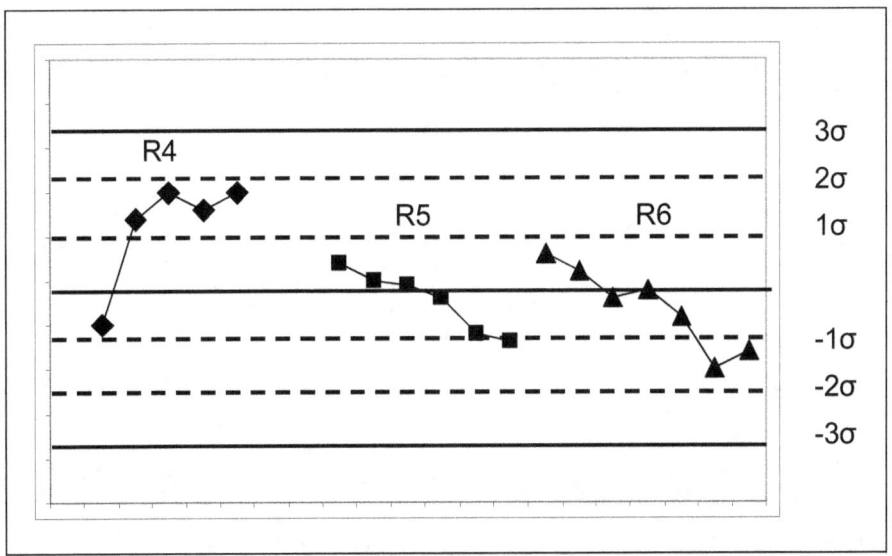

Fig. 3.15 Illustration of instable situations referring to Western Electric rules 4 to 6

| Rule 3 | 2 out of 3 consecutive measurement points are found to be outside the centre line ± 2 σ area |
|---|---|
| Rule 4 | 4 out of 5 consecutive measurement points are found to be outside the center line ± 1 σ area |
| Rule 5 | 6 consecutive measurement values increasing or decreasing (trend) |
| Rule 6 | m consecutive measurement values show a downward or upward trend (regression test) |

## 3.5. Business processes

Besides the above described direct statistical process control, "online SPC", process evaluations using process capability indexes $c_p$ and $c_{pk}$ is executed as "offline" SPC. Situations like outliers, non-centered processes or high process variation are subject of statistical process control characterization by $c_{pk}$. SPC via $c_p$ and $c_{pk}$ reporting does not only focus on the particular values, but also on higher level summaries like number of cp and cpk violations etc.

All those processes regarding the organization of SPC need to be documented in a company. Several examples for SPC related processes are shown in this chapter. As mentioned before, such processes are usually called business processes and can be documented in step by step instructions in a table format and/or as flow charts as shown below.

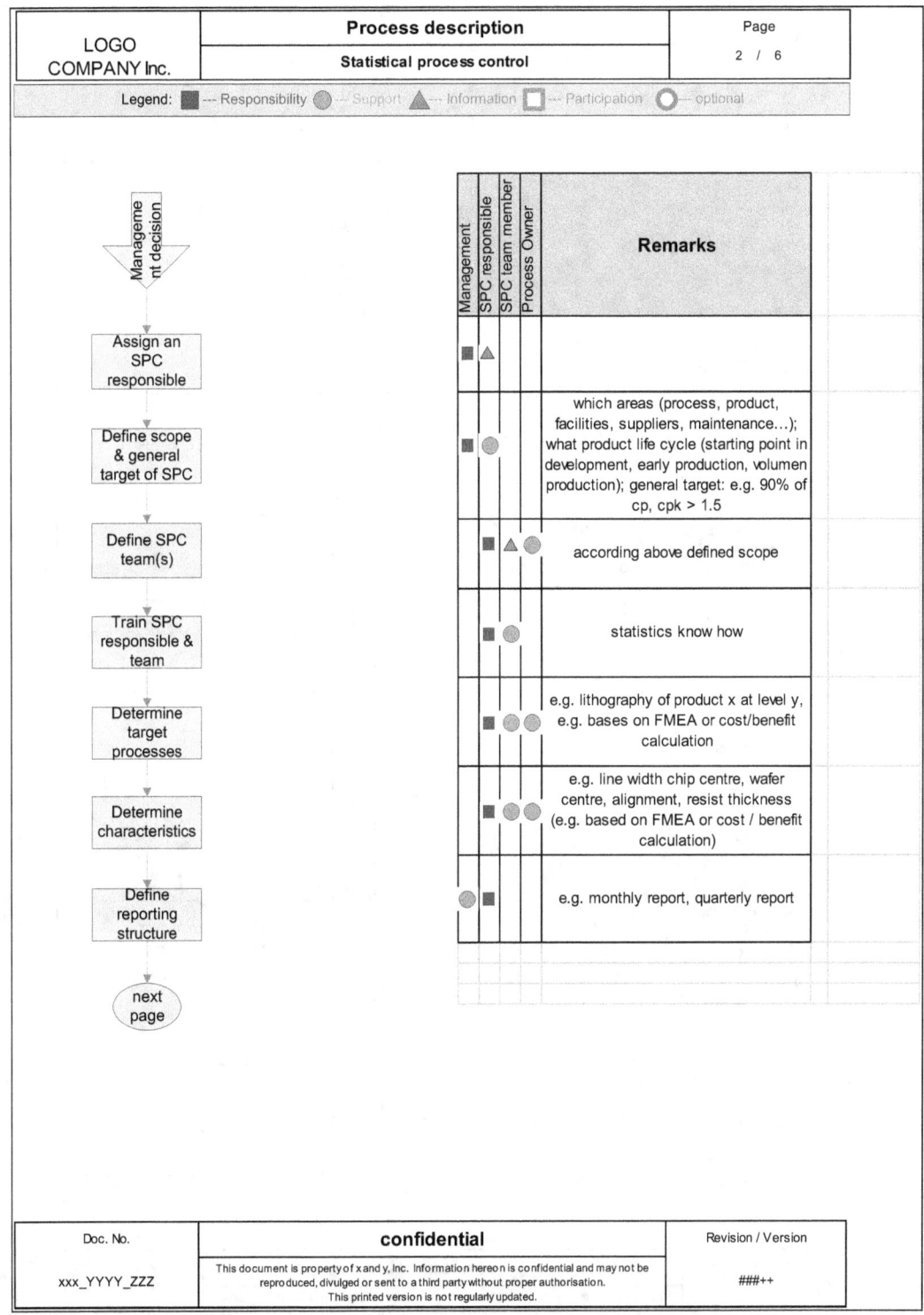

Fig. 3.16 Statistical process control: generic process for the initial phase

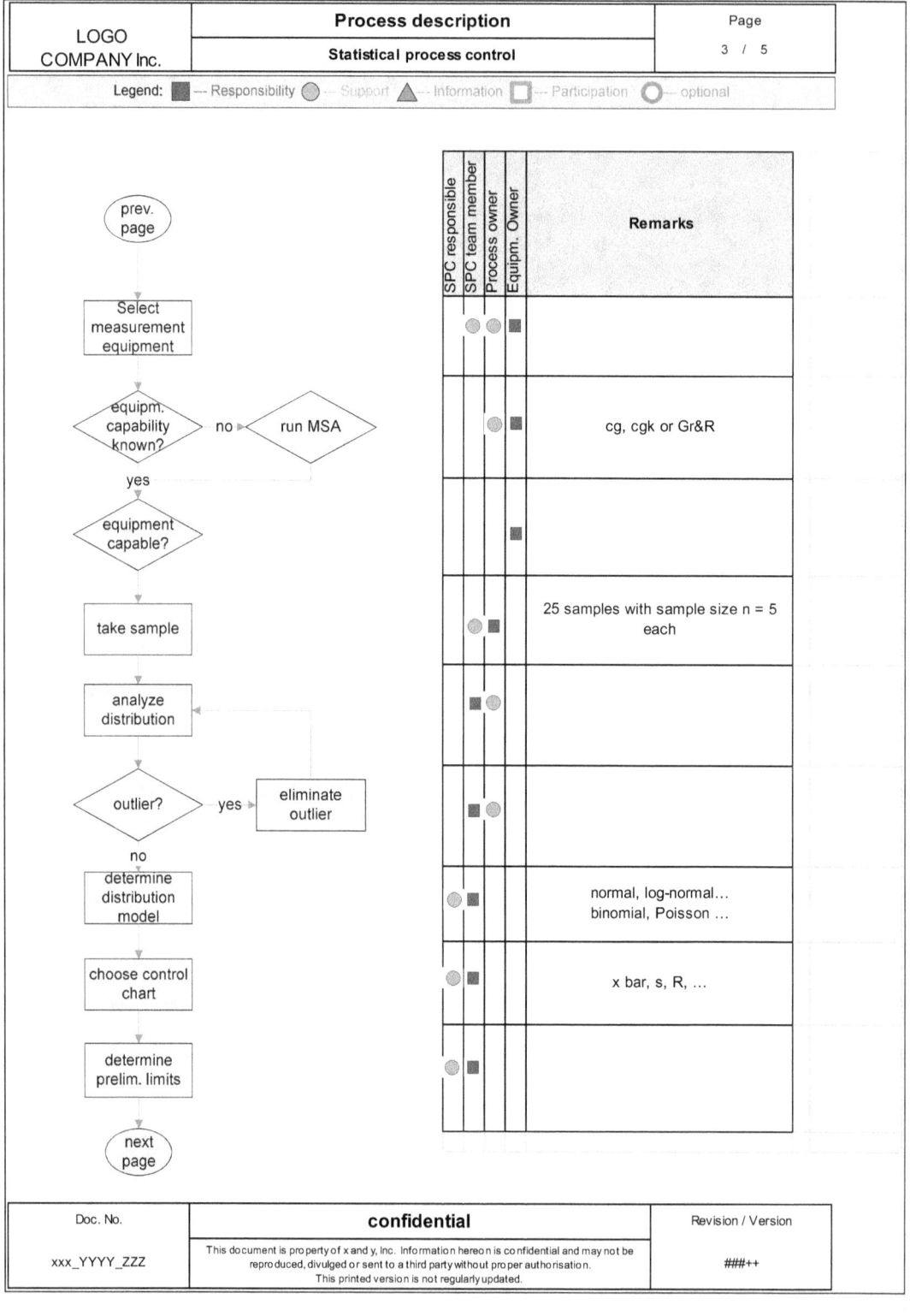

Fig. 3.17 Statistical process control: generic process chart preparation

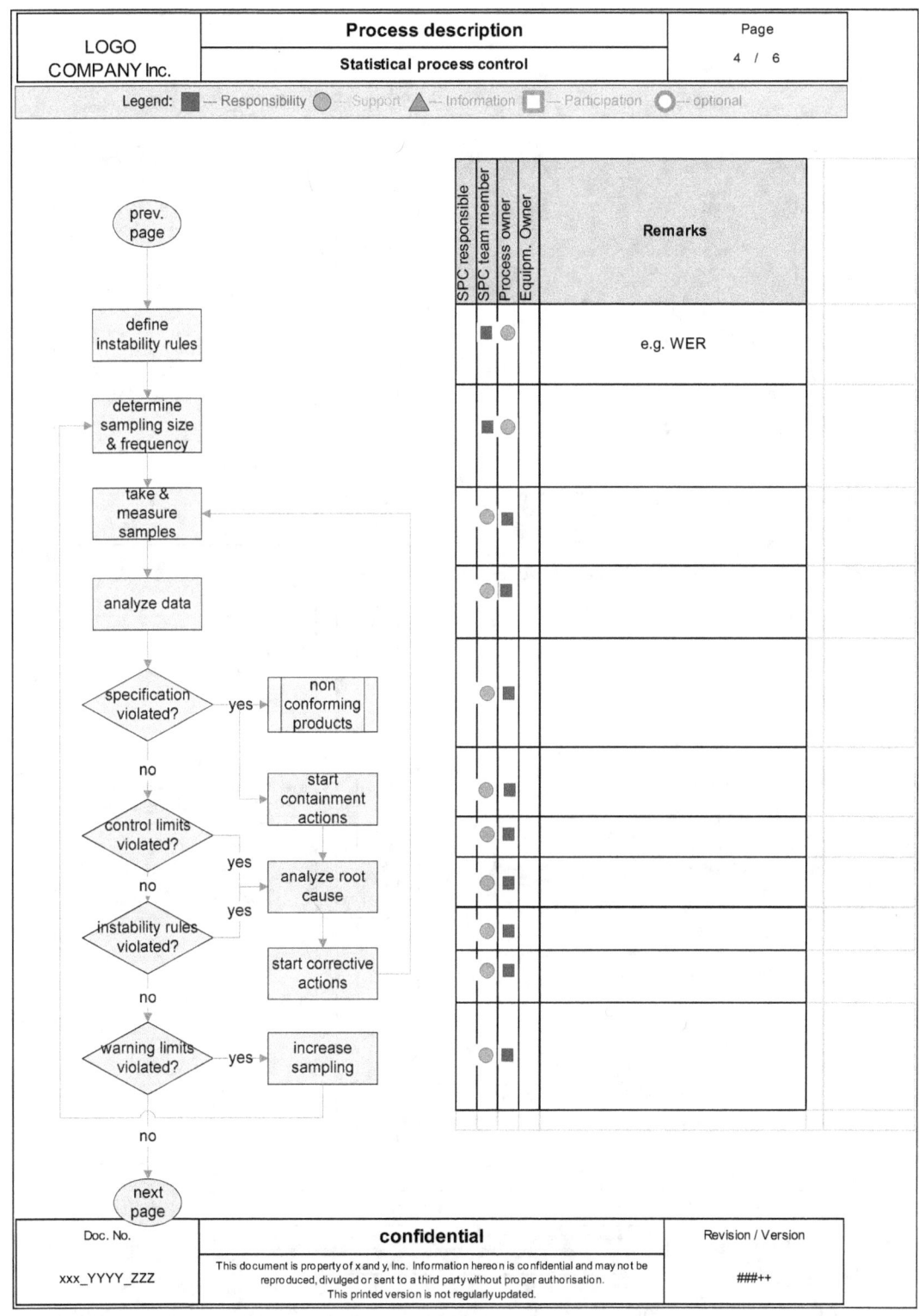

Fig. 3.18 Statistical process control: generic process SPC monitoring

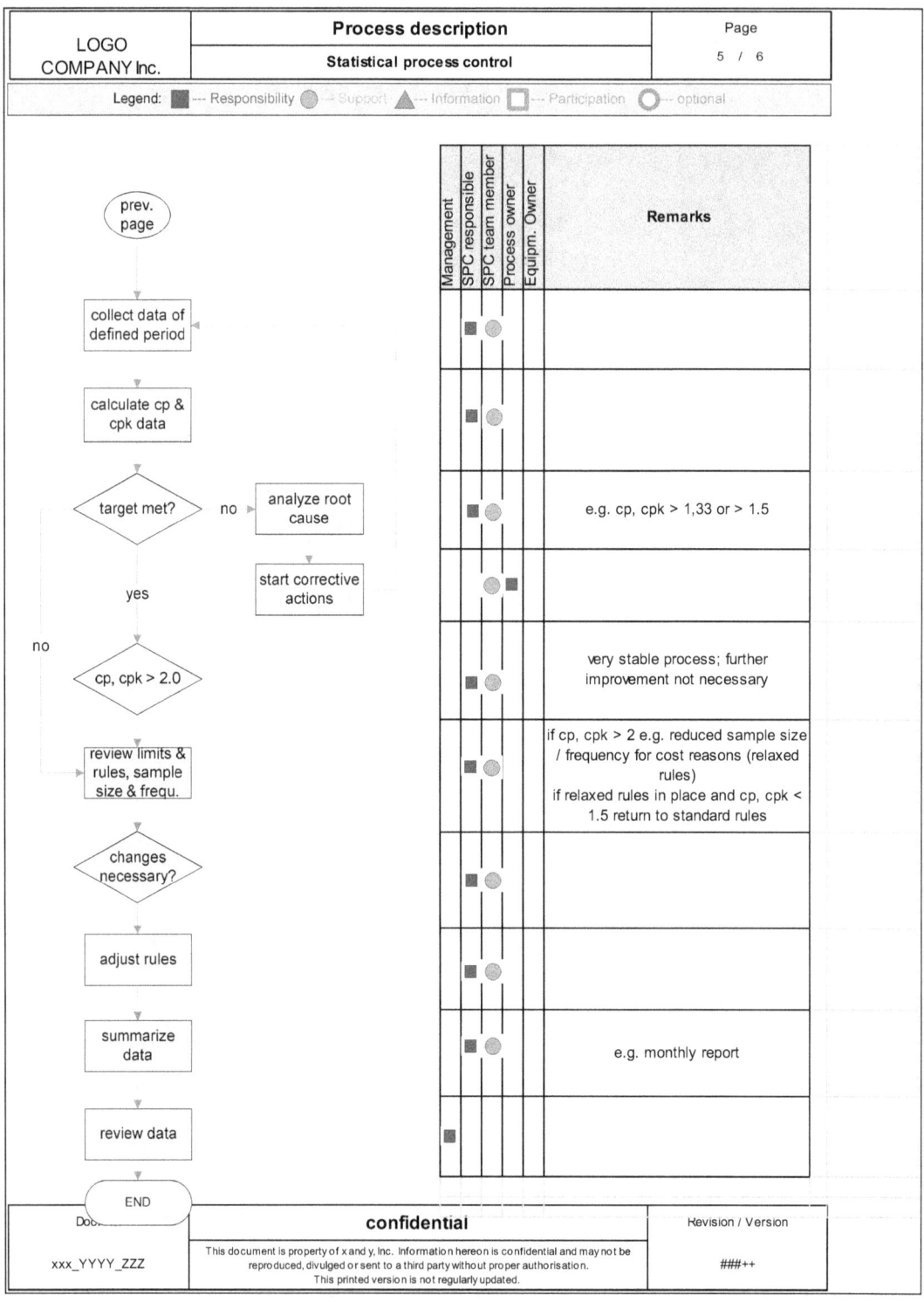

Fig. 3.19 Statistical process control: generic process SPC monitoring & reporting

# 4. Fundamentals of Reliability Engineering

## 4.1. Introduction

> "Reliability is defined as the probability that an item (component, subsystem, or system) performs required functions for an intended period of time under given environmental and operational conditions"

In this context the required function is defined by the product specification, given conditions are application requirements and the given time interval stands for required lifetime.

The respective reliability function R(t) is defined as follows:

> "The reliability function R(t) represents the probability that the system will operate without failures over a time interval [0, t]" [2]

### 4.1.1. Bath Tub Curve

The fail rate of products undergoes three phases over its life:

➤ infant mortality or early fails

This phase is characterized by relatively high fail rates which are decreasing over time. Fails are caused by random events during the manufacturing process which do not manifest themselves in the performance at time = 0, e.g. processing problems and lead to early malfunction. The infant mortality fail rates are expressed in dpm (defects per million, see subsequent sections).

➤ useful life

This phase is characterized by low and constant fail rates as required by the customers. Those fail rates are typically expressed using FIT (failure in time, see subsequent sections)

➤ wear out

The last phase is characterized by a significant increase in failure rate above the accepted level. This increase is caused by material wear out.

---

[1] Song, Y., Wang, B. (2013)
[2] Song, Y., Wang, B. (2013)

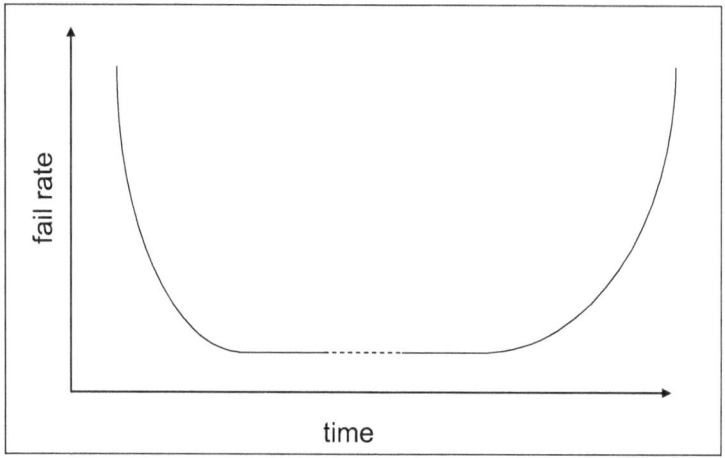

Fig. 4.1 So called bathtub curve illustrating failure rates over time

It is emphasized that

1. each phase is mathematically described by an individual Weibull distribution. Therefore, the bath tub curve is a superposition of three Weibull distributions.
2. the fail rate is not the absolute fail rate f(t), but the conditional fail rate.

$$h(t) = \frac{f(t)}{R(t)}$$

In other words, the rate for failure at t is plotted for those components that have survived until the point t.

### 4.1.2. Failure Levels and Requirements

In the seventies of the last century electronic product failure rates in automobiles in the percentage and sub-percentage range was an accepted level. At that time electronic components started to replace mechanical components with even higher failure rates. As for incoming failures, nowadays the car maker's expectation has moved from a percentage range

$$\boxed{\% \, failure = \frac{number\_of\_failures}{number\_of\_tested\_devices} \times 100\%}$$

> „The underlying failure rate of bimetal flashers were 10% per year and the lifetime of mechanical ignition breakers at 10,000 km"[1]

down to a "defects per million" (dpm) range.

$$dpm\_failure = \frac{number\_of\_failures}{number\_of\_tested\_devices} \times 1.000.000$$

Automotive suppliers are obliged to deliver field failure rates smaller than 10 dpm per year[2]. 10 dpm means that out of one million parts delivered to the customer ten are failing. In order to ensure this level of quality typically a "Zero Defect" approach is chosen.

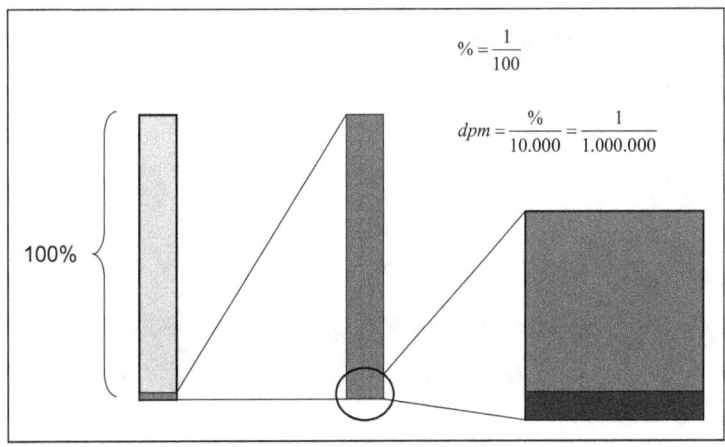

Fig. 4.2 Illustration of Average Outgoing Quality (AOQ)

### 4.1.3. Failure Rate and FIT

The failure rate h(t) or λ(t) can be calculated from the reliability function[3]:

$$\lambda(t) = -\frac{1}{R(t)} \cdot \frac{dR(t)}{dt}$$

---

[1] Winter, R. (2012)
[2] Kallenbach, R. (2015)
[3] Song, Y., Wang, B. (2013)

In contrast to defects per million (dpm) FIT (= failure in time) also take into account failure per time unit:

$$1 \text{ FIT}$$
$$= 1 \text{ fail} / 10^9 \text{ hours}$$
$$= 10^{-9} \text{ fails} / \text{hour}$$

In other words, one system needs to be operated for 1.000.000.000 h. This period of time equals 114155 years, which is neither feasible nor realistic. However, since FIT is a relative figure, this unit applies equally to 1 fail per hour for a population of $10^9$ parts or 1 fail in 1000h for a population of 1 million parts.

FIT rate for a specific lot[1]:

Example
"500 devices were operated over a period of 2000 hours (t) with one failure (f1) after 1000 hours (t1) and one failure (f2) after 1500 hours (t2)".

Calculation of the failure rate λ:

$$\lambda = \frac{2}{(1*1000h)+(1*1500h)+(498*2000h)} = 2*10^{-6} / hour$$

Mathematical relation between FIT and mean time between fail for this example:

$$MTBF = \frac{1}{\lambda} = 500.000h$$

$$FIT\_rate = \frac{10^9}{MTBF} = 2000 FIT$$

*or*

$$FIT\_rate = \lambda * 10^9 = 2000 FIT$$

---
[1] Vishay Semiconductor (2005)

## FIT rates for the population[1]

If the FIT rate for the entire population is to be calculated then the confidence level needs to be considered. The respective confidence factors are taken from a χ²/2 table (see table 4.1). The fail rate λpop is then calculated as follows:

$$\lambda_{pop} = \frac{\chi^2/2}{number\_of\_component\_hours}$$

### Example

Now the previous example with 2000 FIT and 2 failures shall be used to calculate the upper limit for the population's FIT rate with 90% confidence:

$$FIT\_rate = \lambda_{pop} * 10^9 = \frac{\chi^2/2}{no\_of\_comp\_hours} * 10^9 = \frac{5.3}{998.500} * 10^9 = 5308 FIT$$

The fail rates so far have been fail rates determined in the laboratory. In the real life actual field fail rates are not known since customers will not report the time of failure, neither when they started using the product and how intensely the product was used.

| Number of failures | Confidence level | |
|---|---|---|
|  | 60% | 90% |
| 0 | 0.92 | 2.31 |
| 1 | 2.02 | 3.89 |
| 2 | 3.08 | 5.30 |
| 3 | 4.17 | 6.70 |
| 4 | 5.24 | 8.00 |
| 5 | 6.25 | 9.25 |
| 6 | 7.27 | 10.55 |

Table 4.1 χ²/2 confidence factors

The FIT rates have to be calculated using customer return figures, i.e. actual failures in the field. In order to determine field FIT rates the number of fails as well as the accumulated operating hours and date of fail are required:

➢ number of fails

---

[1] Vishay Semiconductor (2005)

It is not very likely that every single fail in the field is being returned by the customer. So, based on experience, an expert judgment needs to be performed on the actual number of fails.

➤ accumulated operating hours

Parts delivered to the customer over time are assumed to have different operating hours

| Daily usage [hours] | 0.75 | 1.25 | 1.75 | 2.25 | 2.75 | 3.5 |
|---|---|---|---|---|---|---|
| Yearly usage [hours] | 273.75 | 456.25 | 638.75 | 821.25 | 1003.7 | 1277.5 |
| Percentage of user | 20 | 40 | 15 | 10 | 10 | 5 |

Table 4.2 Usage of cars estimation

There is also an unknown period of time between delivery and start of use of the product and in addition the intensity of use is not known. Thus, many assumptions have to be made. For example, the following table 4.2 shows an example for an estimated usage of a car.

### FIT Rates from Stress Testing

Typically, semiconductor products have to go through a qualification flow including accelerating stress conditions (see also chapter "Development"). By applying temperature, temperature cycling, humidity and other relevant stress factors the ageing of the parts is accelerated. For different kinds of stress different stress models are available which are used to calculate the acceleration factor.

Most failures in semiconductors are temperature dependent. The following Arrhenius equation is therefore used to calculate expected fail rates, too[1]:

$$AF = \frac{\lambda_{T2}}{\lambda_{T1}} = e^{\left[\frac{E_A}{k}\left(\frac{1}{T1}-\frac{1}{T2}\right)\right]}$$

with

AF : acceleration factor
$\lambda_{T1}$ : failure rate at Temperature 1 (operating temperature)
$\lambda_{T2}$ : failure rate at Temperature 2 (stress temperature)
$E_A$ : activation energy [eV] for activation of the failure mechanism
k : Boltzmann's constant ($8.63 \times 10^{-5}$ eV/K)

---

[1] Vishay Semiconductors (2005)

Activation energies are different for different kinds of failure mechanisms. This has to be considered when calculating the acceleration factors.

Example
Assumed the failure rate in the previous example (FIT rate = 5308 FIT) is the result of a stress test at 150°C. In addition, we assume an operating temperature of 55°C. Then the acceleration factor is calculated to be (at 90% confidence and 55°C operating temperature):

$$AF = \frac{\lambda_{T2}}{\lambda_{T1}} = \frac{\lambda_{(273+55)K}}{\lambda_{(273+150)K}} = 258$$

$$FIT_{328K} = \lambda_{328K} \cdot 10^9 = \frac{\lambda_{423K}}{258} \cdot 10^9 = \frac{5308}{258} = 21 FIT$$

Despite the fact that many semiconductor companies refer to this method in their respective manuals and/or handbooks, this approach is not very useful in areas with very high quality standards in the sub 10 dpm and sub 100 FIT range. The typical qualification volume with 3 lots with 77 parts from each lot does not result in an adequate resolution. In fact, if a fail occurs, the qualification may be considered failed, too. The number of parts required for a reasonable conclusion on the other side would be too high and would not be justified regarding cost, time and resources.

Impact of the Mission Profile[1]
Just like the usage of products in hours also the temperature at use, i.e. the ambient temperature, is subject to a distribution depending on user and environment.
Instead of applying the worst case conditions it is advisable to use the temperature-time-distribution of the use case, i.e. the so called mission profile. An example is given using arbitrary figures in the table 4.3.

The example above results in an overall failure rate of 1.778 FIT in average, whereas the worst case assumption (100% at 85°C) would result in a failure rate of 22.5 FIT. State of the art failure rates of semiconductor devices range from 1 – 10 FIT.

---

[1] Webber, A. (2015)

| Ambient Temperature [°C] | % of time | FIT rate | FIT x % time |
|---|---|---|---|
| -5 | 1 | 0.01 | 0.0001 |
| 5 | 9 | 0.03 | 0.0027 |
| 15 | 12 | 0.09 | 0.0108 |
| 25 | 17 | 0.20 | 0.034 |
| 35 | 22 | 0.53 | 0.1166 |
| 45 | 16 | 1.15 | 0.184 |
| 55 | 12 | 2.56 | 0.3072 |
| 65 | 5 | 4.90 | 0.245 |
| 75 | 4 | 10.70 | 0.428 |
| 85 | 2 | 22.50 | 0.45 |

Table 4.3 arbitrary distribution of temperature dependent FIT rates and percentage of time

### 4.1.4. MTBF & MTTF

In addition to AOQ values in the dpm range and failure rates the mean time between fail for repairable parts and mean time to fail for non repairable parts are often used.

> MTBF
> Mean time between fails
>
> MTTF
> Mean time to fail

For constant fail rates the mean time to fail is the reciprocal of the fail rate $\lambda$. The exponential distribution is the only distribution with constant fail rates. Due to the fact that semiconductor devices are mostly not repairable in this case MTBF equals MTTF. If the mean down time (MDT) of repairable parts is known, then the availability A, i.e. the time the part is operable, can be calculated, too:

$$A = \frac{MBTF}{MDT + MBTF}$$

Finally, the median lifetime is defined to be the period of time to reach the 50% fail level. The mean time to fail can also be expressed as a function of the reliability function R(t)

$$MTTF = \int_0^\infty R(t)dt$$

At constant failure rate λ the equation for MTTF can be simplified:

$$MTTF = \frac{1}{\lambda}$$

## 4.2. Acceleration Models for different stress factors

Nowadays semiconductor devices are subject to a wide range of applications and respective operating times. While operating times for privately used devices are found to be a few thousand hours only, automotive applications are often used over a period of about 12.000 hours[1]. Even significantly higher operating times are observed for industrial grade semiconductors. Texas Instruments industrial grade mission profiles, for instance, assumes an operating time of 131.400 hours over fifteen years at an elevated temperature of 105°C[2] for embedded systems.

Obviously it is not possible to test a significant number of devices for this period of time. Hence accelerated testing becomes necessary. One typical approach is to apply higher temperature to the devices in order to speed up chemical and physical ageing processes. Unfortunately, the failure modes observed at higher temperatures are not necessarily identical to those at operating temperature. Therefore, a sound understanding of the failure mechanisms, i.e. a physical model, is obligatory when designing the stress conditions.

### 4.2.1. Approach

So the basic idea is to apply stress conditions to the device which leads to faster ageing and earlier fails. For this approach an acceleration factor AF is defined:

$$AF = \frac{time\_to\_fail\_under\_use\_conditions}{time\_to\_fail\_under\_stress\_conditions} \qquad AF = \frac{t_{use}}{t_{stress}}$$

The following "recipe" applies:

1. measure time to fail data under stress conditions
2. fit data with an appropriate lifetime distribution (often a Weibull distribution)
3. apply the acceleration model
4. plot the lifetime distribution under operating conditions
5. extrapolate the lifetime distribution to required operating time

---

[1] ZVEI (2013)
[2] Webber, A. (2014)

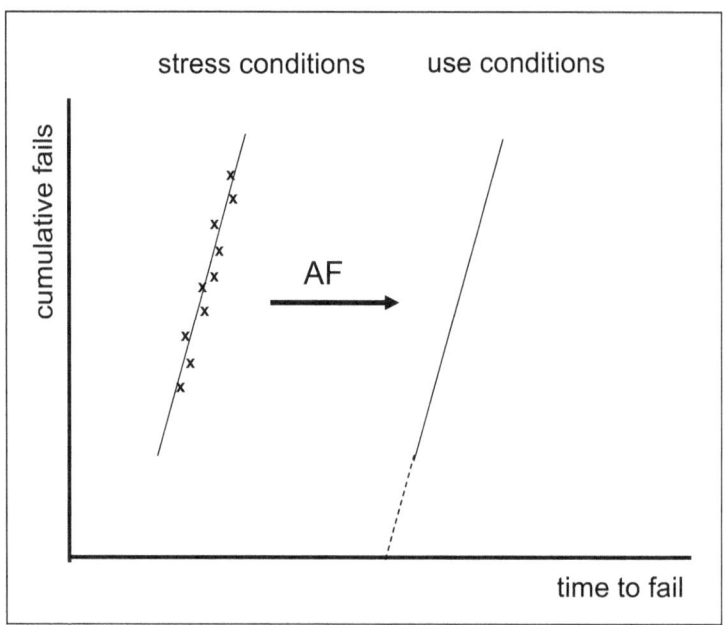

Fig. 4.3 Schematic extrapolation of test data to operating time to fail (AF acceleration factor)

There are different kinds of stress potentially affecting an integrated circuit, such as

- applied voltage (incl. the resulting internal electric fields)
- external current
- mechanical stress
- temperature
- temperature changes
- humidity
- chemicals in contact with the integrated circuit

For different kinds of stress different lifetime models are used (fig. 4.4).

The extrapolation of lifetime typically assumes no mutual interaction of various stress factors and a linear accumulation of damage. Even though this in not true in general for some of the mentioned stresses an acceleration model is useful and will be explained in the following sections.

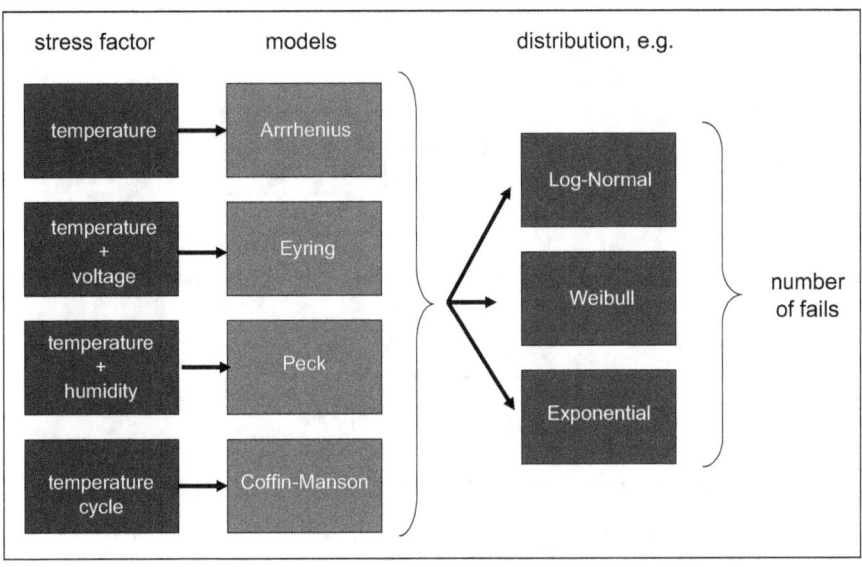

Fig. 4.4 from stress to fail

### 4.2.2. Arrhenius Model[1,2]

First derived by the swedish scientist Svante August Arrhenius (1859 – 1927) the speed of chemical reactions is assumed to be dependent on the temperature applied:

$$R = \frac{dn_0}{dt} = R_0 \cdot e^{-\frac{E_a}{k \cdot T_0}}$$

with
- R : rate of reaction
- $n_0$ : number of parts at $t_0$
- $R_0$ : reaction specific factor
- $E_a$ : activation energy
- k : Boltzmann's constant (~ 8,617*10$^{-5}$ eV/K)
- $T_0$ : absolute temperature

In terms of reliability the speed of the chemical reaction corresponds to the failure rate $\lambda_0$:

$$\lambda_0 = R_0 \cdot e^{-\frac{E_a}{k \cdot T_0}}$$

Assuming another temperature T at a fail rate λ we write the respective ratio:

---

[1] Meyna, A., Pauli, B. (2010), p. 580
[2] JESD122G (2012)

$$\frac{\lambda}{\lambda_0} = \frac{R_0 \cdot e^{-\frac{E_a}{k \cdot T}}}{R_0 \cdot e^{-\frac{E_a}{k \cdot T_0}}} \quad \rightarrow \quad \lambda = e^{\left(\frac{1}{T_0} - \frac{1}{T}\right)\frac{E_a}{k}}$$

The inverse of λ is called mean lifetime. From there the lifetime $t_f$ at temperature T can be estimated:

$$\ln(t_f) = \ln(t_0) - \left(\frac{1}{T_0} - \frac{1}{T}\right) \cdot \frac{E_a}{k}$$

### Example

At a test temperature of 175°C (= 448.15 K) a mean lifetime of t = 2200 hours was measured. The activation energy was determined to be 0.7 eV. The mean lifetime at operating temperature T = 85°C (= 358.15 K) is then calculated as follows:

$$\ln(t) = \ln(2200) - \left(\frac{1}{448.15} - \frac{1}{358.15}\right) \cdot \frac{0.7}{8.63 \cdot 10^{-5}}$$

$$\ln(t) = 7.6962 - (-0.00056073 \cdot 8.111 \cdot 10^{-7}) = 12.244$$

$$t = e^{12.24} = 207814{,}3h = 23.7\ years$$

So at a temperature of T = 85°C a mean lifetime of 23.7 years is estimated.

### Activation Energy

As mentioned above it is well known, that chemical reactions progress much faster at elevated temperatures than in a cold environment. The reactants (atoms and/or molecules) have to gain a minimum amount of energy to start a chemical reaction, the so called activation energy (fig. 4.5).

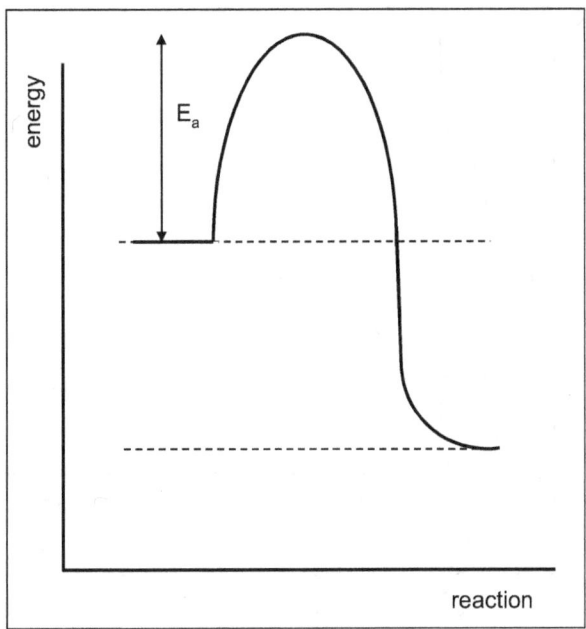

Fig. 4.5 Illustration of required activation energy to start a chemical reaction

Unfortunately, the activation energy is frequently not known and has to be determined experimentally. In order to do so, the Arrhenius equation has to be transformed to a linear equation[1]:

$$R = R_0 \cdot e^{-\frac{E_a}{kT}}$$

$$L = A \cdot e^{\frac{E_a}{kT}}$$

with

R : reaction rate
$R_0$ : reaction rate constant
$E_a$ : Activation Energy [eV]
k : Boltzmann Constant [8.617 x $10^{-5}$ eV/K]
L : time to failure
A : constant

Taking the logarithm on both sides results in:

$$\ln(L) = \ln(A) + \frac{E_a}{kT}$$

With at least three measurements at three different temperatures the plot of the logarithm of lifetime against 1/T results in a straight line (fig. 4.6):

---

[1] Renesas (2013)

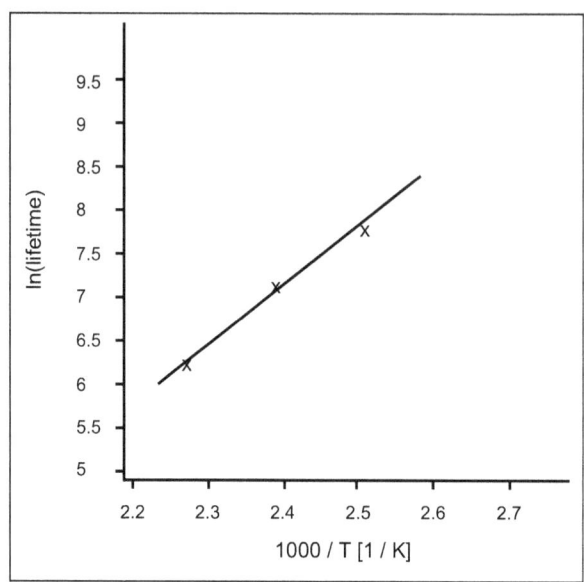

Fig. 4.6 Determination of the Activation Energy $E_a$

Then $E_a$ is equivalent to the slope of the line. On the macroscopic level the Arrhenius equation is typically written using the universal gas constant R:

$$k = A \cdot e^{-\frac{E_a}{RT}}$$

with
k : reaction constant
R : 8,314 J·K⁻¹·mol⁻¹

Then $E_a$ can be calculated, too:

$$y(x = 2.3) = 6.4$$
$$y(x = 2.6) = 8.5$$
$$T_1 = 1000 / 2.3 = 434.78 K$$
$$T_2 = 1000 / 2.6 = 384.62 K$$

$$\ln(k_2) - \ln(k_1) = \left(\ln A - \frac{E_a}{RT_2}\right) - \left(\ln A - \frac{E_a}{RT_1}\right) = \frac{E_a}{k}\left(\frac{1}{T_1} - \frac{1}{T_2}\right)$$

$$E_a = \frac{R \cdot (\ln(L_2) - \ln(L_1))}{\left(\frac{1}{T_1} - \frac{1}{T_2}\right)} = \frac{8.314 J /(mol \cdot K) \cdot (8.5 - 6.4)}{\left(\frac{1}{384.62 K} - \frac{1}{434.78 K}\right)} = 58198 J / mol$$

$$E_a = \frac{58198 \, J/mol}{6.022 \cdot 10^{23} \, /mol} \cdot 6{,}242 \cdot 10^{18} \, eV/J = 0{,}6 eV$$

When looking at failure rates with several different mechanisms with a mechanism specific activation energy for each mechanism the activation energy $E_a$ found here is a mathematical "fudge" value rather than the activation energy of a particular chemical reaction.

| Failure mechanism | Activation Energy [eV] |
|---|---|
| Mechanical wire shorts | 0.3 – 0.4 |
| Diffusion and bulk defects | 0.3 – 0.4 |
| Oxide defects | 0.3 – 0.4 |
| Top-to-bottom metal short | 0.5 |
| Electromigration | 0.4 – 1.2 |
| Electrolytic corrosion | 0.8 – 1.0 |
| Gold-aluminum intermetallics | 0.8 – 2.0 |
| Gold-aluminum bond degradation | 1.0 – 2.2 |
| Polysilicon to metal defect | 0.3 |
| Silicon junction defect | 0.8 |
| Ionic contamination | 1.02 |
| Corrosion | 0.3 – 1.1 |
| Hot Carrier | -1.0 |
| Alloy pitting | 1.77 |

Table 4.4 Activation energies for common failure mechanisms in semiconductor reliability (not exclusively for the Arrhenius model)[1,2]

For a list of activation energies please refer to JESD122.

### 4.2.3. Eyring Model[3]

The Eyring model is an extension of the Arrhenius model[4]. In addition to the temperature stress only described by the Arrhenius model, the Eyring model considers the impact of a non-thermal factor, e.g. humidity, mechanical stress or voltage, in addition. For

---

[1] Vishay Semiconductors (2005), p. 67
[2] Alpha and Omega Semiconductors (2010), p. 8
[3] Meyna, A., Pauli, B. (2010), p. 582
[4] Renesas (2013)

instance, oxidation processes are caused by temperature and humidity. In contrast to the Arrhenius model which is basically an empirical model, the Eyring model is based on principles of quantum mechanics and chemical reaction rate theory. The speed of chemical reactions is assumed to be dependent on the temperature applied for small temperature ranges:

$$R = \frac{dn_0}{dt} = a \cdot T \cdot e^{-\frac{E_a}{k \cdot T_0}}$$

with
- R : rate of reaction
- $n_0$ : number of parts at $t_0$
- $R_0$ : reaction specific factor
- $E_a$ : activation energy
- k : Boltzmann's constant ($\sim 8{,}617 \cdot 10^{-5}$ eV/K)
- T : absolute temperature
- a : temperature related factor

The generalized Eyring model for temperature and one additional stress is expressed in the following form:

$$R = \lambda = \frac{dn}{dt} = a \cdot T \cdot e^{-\frac{b}{k \cdot T} + c \cdot V + \frac{d \cdot V}{k \cdot T}}$$

with
- T : thermal impact factor
- V : non-thermal impact factor
- a, b : temperature related factors
- c, d : non-temperature related factors

a, b, c, d are not known. The acceleration factor AF for two different stress temperatures and two different non-thermal stress conditions is then defined to be:

$$AF = \frac{T}{T_0} \cdot e^{-\frac{b}{k}\left(\frac{1}{T} - \frac{1}{T_0}\right) + c \cdot (V - V_0) + \frac{d}{k}\left(\frac{V}{T} - \frac{V_0}{T_0}\right)}$$

### 4.2.4. Peck Model[1]

In the early days of semiconductor technology hermetic packages were used to encapsulate the actual chip. Being a major cost factor those packages were later on replaced by epoxy moulded encapsulations.

---

[1] JESD122G (2012), p. 45

Unfortunately, those are more sensitive to moisture diffusion through the material or moving along the lead frame. Eventually this leads to corrosion of metal lines in the chip. Peck and Hallberg developed a model in order to deal with this behaviour[1]:

$$\boxed{t_f = A \cdot (\%RH)^{-n} \cdot e^{\frac{E_a}{kT}}}$$ with

| | |
|---|---|
| A | : constant |
| n | : material related constant |
| $E_a$ | : activation energy |

The respective acceleration factor AF for two different temperatures and two different levels of relative humidity (RH) is then:

$$\boxed{AF(RH,T) = \left(\frac{RH_{Stress}}{RH_{use}}\right)^n \cdot e^{\frac{E_a}{k}\left(\frac{1}{T_{use}} - \frac{1}{T_{stress}}\right)}}$$

The Peck model is an empirical model which was developed for the case of aluminium corrosion. Other material systems or electrochemical influence may require another model.

### 4.2.5. Coffin-Manson Model[2,3]

In addition to the constant stress level at high temperature or humidity also temperature cycles, for instance caused by on/off cycles of a device, are resulting in stress in materials, e.g. solder joints. The reason behind is that depending on the material specific thermal expansion coefficients thermal cycles lead to a differing expansion of different materials which results in cycling mechanical stress. In particular for solder joints, which connect different parts within a chip mechanically as well as electrically, it is well known, that they are subject to fatigue due to temperature cycles.

There are various models available to predict failure like "stress-range based", "strain-energy-based" or "fracture-mechanics based". Most common, however, is the "strain-range based" model, which also the basis of the Coffin-Manson-Model. This model was developed by Coffin and Manson independently in the 50s of last century.

The elastic fraction of strain $\Delta\varepsilon_e$ is neglected while the non-elastic plastic fraction $\Delta\varepsilon_p$ depends on the material properties and the number of cycles to failure:

$$\boxed{\Delta\varepsilon_p = A_0 \cdot N_f^{-B}}$$ with

---

[1] Hallberg, Ö., Peck, D.S. (1991)
[2] JESD122G (2012), p. 53
[3] Wen, L, Ross, R. G. (1995)

$$N_f = A_0 \cdot (1/\Delta\varepsilon_p)^B$$

$N_f$ : number of cycles to failure
B : empirically determined exponent
$\Delta\varepsilon_p$ : non-elastic (plastic) strain range
$A_0$ : material specific coefficient

With $\Delta\varepsilon_p \propto \Delta T^k$ the Coffin-Manson-Equation becomes:

$$N_f = C_0 \cdot \Delta T^{-q}$$

with
q : empirically determined Coffin-Manson exponent (specific for each failure mechanism)
$C_0$ : material specific constant

It turned out, that the Coffin-Manson model, originally developed for ductile material, can be applied to brittle material as well. Values for q range from 1 to 3 for ductile metals up to 6 – 9 for brittle material like silicon and dielectrics. From there the Coffin-Manson-Model is used to determine the number of stress cycles based ratio of the stress temperature swing and the use temperature swing[1]:

$$N_{stress} = N_{use} \cdot \left(\frac{\Delta T_{use}}{\Delta T_{Stress}}\right)^k$$

with
$N_{stress}$ : number of cycles during stress
$N_{use}$ : number of cycles in regular use
$\Delta T_{stress}$ : temperature swing during stress
$\Delta T_{use}$ : temperature swing during regular use conditions
k : failure mechanism dependent exponent

The Coffin-Manson model does neglect several physical parameters which might have a significant impact on the validity of the model like:

➢ the level of the average temperature
upper or lower temperature might pass through a phase boundary with significantly changing material properties, e.g. a glass point of a moulding compound

---

[1] Meyna, A., Pauli, B. (2010), p. 587

> the speed of temperature change

stress induced at high temperature cycling frequency may not represent a typical stress behaviour in the use case

> geometry of the device in question

various parts of the device might experience different kind of stress

## 4.2.6. Mission Profile[1]

In real life a product is exposed to varying conditions regarding temperature level, temperature cycle, humidity and so forth.

> "A **mission profile** is a collection of relevant environmental load/stress and operation conditions to which a component will be exposed during its full life cycle." [2]

In order to predict the expected lifetime of a product this has to be considered for the determination of the intended stress test using the so called mission profile.

However, due to the differences of applications and users, it is quite difficult to generate a mission profile. So for application specific integrated circuits (ASICs) there is no other reasonable way than to ask the customer to provide this information. In case a customer information is not possible, e.g. for commodity products like DRAMs, the producer's experts (application engineering, product engineering, marketing etc.) have to make a judgement.

Frequently there are several different mission profiles conceivable. Then the maximum requirement has to be chosen as product requirement. A mission profile should at least contain the following components:

> total expected lifetime

the lifetime could be expressed in years or other appropriate units like km/miles depending on the product or both

> operating time

period of time the product is "on"

> temperature

---

[1] ZVEI, Handbook for Robustness Validation (2013), p. 27
[2] Kanert, W. (2008)

level of temperature at different positions like outside temperature (daily change), internal temperature (e.g. junction temperature) etc. and temperature cycles due to on/off cycles or daily or seasonal temperature shifts

> electrical conditions

voltage and electric current in turning on, power on and/or stand-by mode, possibly also influence of an electrical or magnetical field

> mechanical conditions

vibrations or mechanical shock easily cause breakage of parts

> other conditions

conditions like high humidity level or exposure to chemicals

Eventually the mission profile is supposed to cover stress due to transport, storage, processing and operation in the target application. The more detailed information available the better. The following tables show different mission profiles:

| Service lifetime | 15 years (=131400 h) | |
|---|---|---|
| Mileage | 600000 km | |
| Engine on time | 12000 h | Engine on time is proportional to mileage |
| Engine off time | 119400 h | May be used to determine storage stress test times |
| Engine on/off cycles | 54000 | |

Table 4.5 Vehicle mission profile parameters on vehicle level[1]

After determining the mission profile as exactly as possible the subsequent task is to transfer those requirements into meaningful stress conditions. The following example shows the result for temperature cycles.

---

[1] ZVEI Handbook for Robustness Validation (2013), p. 29

|  | Temperature | Distribution |
|---|---|---|
| Ambient temperature | -40°C | 2% |
|  | 23°C | 18% |
|  | 100°C | 70% |
|  | 130°C | 9% |
|  | 140°C | 1% |
| Humidity | e.g. up to 100% ||

Table 4.6 Partial mission profile (operating time) of "a mechatronic (component) that is the controlling part of an automatic transmission system"[1], for an operating time of 6000 hrs

| Days / year | Ambient temperature |
|---|---|
| 15 | 30°C |
| 25 | 45°C |
| 90 | 55°C |
| 185 | 60°C |
| 35 | 70°C |
| 15 | 80°C |

Table 4.7 Partial mission profile for a solar inverter for a lifetime of 15 years[2]

| ΔT [°C] | Cycles (use) |
|---|---|
| 110 | 9000 |
| 80 | 7500 |
| 70 | 3000 |

Table 4.8 OEM vehicle mission profiles for temperature cycles[3]

Example: stress test conditions for temperature cycles

---

[1] ZVEI Handbook for Robustness Validation (2013), p. 97
[2] Webber, A. (2014), p. 5
[3] ZVEI (2007), p. 15

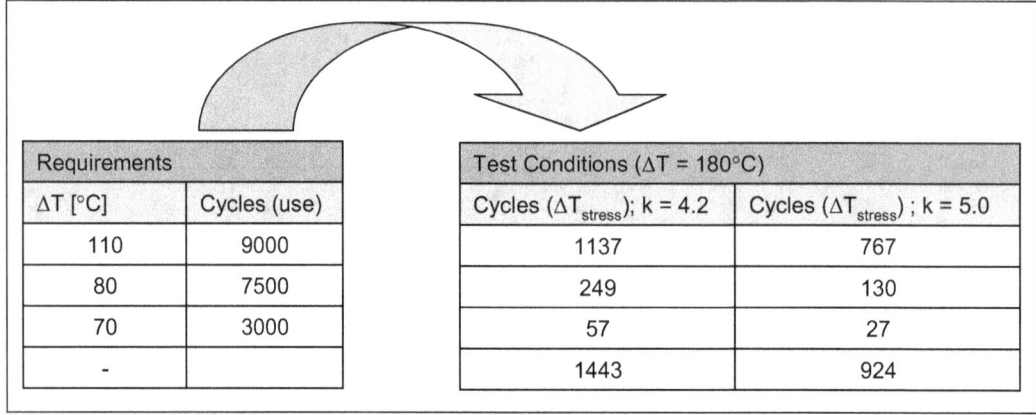

Fig. 4.7 Mission profile transfer into test conditions (temperature cycles)

Each stress condition, i.e. number of cycles at a specific temperature swing, is transferred into a stress condition. Subsequently those are summed up to the number of cycles using the stress temperature swing conditions for different Coffin-Manson exponents.

## 4.3. Reliability Stress Testing

Both standards JESD47I and AEC-Q100 list a major list of stress tests to prove functionality and, in particular, reliability of semiconductor products (devices, packages…). Important reliability related tests are described in this section:

- PC : Preconditioning
- HTSL : High Temperature Storage Life
- TC : Temperature Cycling
- THB : Temperature Humidity Bias
- AC : Autoclave
- (U)HAST : (Unbiased) Highly Accelerated Stress Test
- HTOL : High Temperature Operating Life
- HTRB : High Temperature Reverse Bias
- HTGB : High Temperature Gate Bias

### 4.3.1. Preconditioning (PC)[1]

The purpose of the preconditioning stress is to simulate the solder reflow process at the customer during board assembly.

---

[1] JESD22-A113F (2008)

After electrical test to confirm functionality and visual inspection to ensure no visible damage shipping conditions are simulated using five temperature cycles from -40°C up to +60°C. Whether or not temperature cycling is required depends on the product. Then the component is dried at 125°C for 24 hours and afterwards exposed to a moisture soak environment.

Subsequently three reflow cycles representing first and second pass of a double-sided and double-pass assembly reflow process as well as an additional rework process (one cycle may be used to attach the component on a test board). After cooling down the component is entirely dipped into an activated water soluble flux and cleaned using multiple de-ionized water rinses. Reflow conditions e.g. classification temperatures for various package thicknesses, volumes and materials, are defined in the standard IPC/JEDEC J-STD-020D.1[1].

The failure mechanism addressed by this test is moisture which was absorbed by the package material during transport vaporizes quickly as soon as heat is applied, e.g. during reflow, which might lead to expansion and package cracks. The preconditioning process might also result in a delamination of layers within the integrated circuit and/or the package.

### 4.3.2. High Temperature Storage Life (HTSL)[2]

"The high temperature storage test is typically used to determine the effects of time and temperature, under storage conditions, for thermally activated failure mechanisms and time-to failure distributions of solid state electronic devices, including nonvolatile memory devices"[3].

A set of different storage conditions are available each with a temperature range of -0° to +10°C:

- Condition A    :    + 125°C
- Condition B    :    + 150°C
- Condition C    :    + 175°C
- Condition D    :    + 200°C
- Condition E    :    + 250°C
- Condition F    :    + 300°C

When choosing the temperature level material properties like the melting point and glass transition temperature of materials used in the circuit have to be kept in mind. Ceramic packaged parts are allowed to be stored at high temperature whereas plastic

---

[1] IPC/JEDEC J-STD-020D.1 (2008)
[2] JESD22-A103D (2010)
[3] JESD22-A103D (2010), p. 1

packaged parts are typically limited to conditions A, B and C. Additional storage conditions like moisture are defined in the standard IPC/JEDEC J-STD-020D.1[1].

HTSL is executed without electrical connections (voltage and current) and might lead to destruction of the parts. The Arrhenius model is used for this test to model diffusion caused mechanisms like mobile ion diffusion and intermetallic compound growth or chemical reactions like oxidation or corrosion.

In addition to obvious damages like cracks, chipping or breakage exceeding parametric limits or non-functionality in the subsequent functional test indicate failure of parts.

### 4.3.3. Temperature Cycling (TC)[2]

"This test is conducted to determine the ability of components and solder interconnects to withstand mechanical stresses induced by alternating high- and low-temperature extremes"[3]. Temperature cycling test conditions are defined in JESD22-A104D:

| Test Condition | $T_{min}$ [°C] | $T_{max}$ [°C] |
|---|---|---|
| A | -55 | +85 |
| B | -55 | +125 |
| C | -65 | +150 |
| G | -40 | +125 |
| H | -55 | +150 |
| I | -40 | +115 |
| J | 0 | +100 |
| K | 0 | +125 |
| L | -55 | +110 |
| M | -40 | +150 |
| N | -40 | +85 |

Table 4.9: Various test conditions for Temperature Cycling; tolerance for $T_{min}$ = [0°C;-10°C], tolerance for $T_{max}$ = [+10°C;0°C) for A and N and [+15°C;0°C] all others

Typical component cycle rates are one to three cycles per hour (cph) leading to material fatigue and delamination of layers whereas solder cycle rates are found to be in a range of one to two cycles per hour.

---

[1] IPC/JEDEC J-STD-020D.1 (2008)
[2] JESD22-A104D (2009)
[3] JESD22-A104D (2009), p. 1

The soak mode needs to be determined according to the targeted failure mechanism. The failure modes addressed are fatigue of material, layer delamination, cracking of dies or package, bond lift off etc.

In addition to obvious mechanical damages or warpage exceeding parametric limits or non-functionality in the subsequent functional test indicate failure of parts.

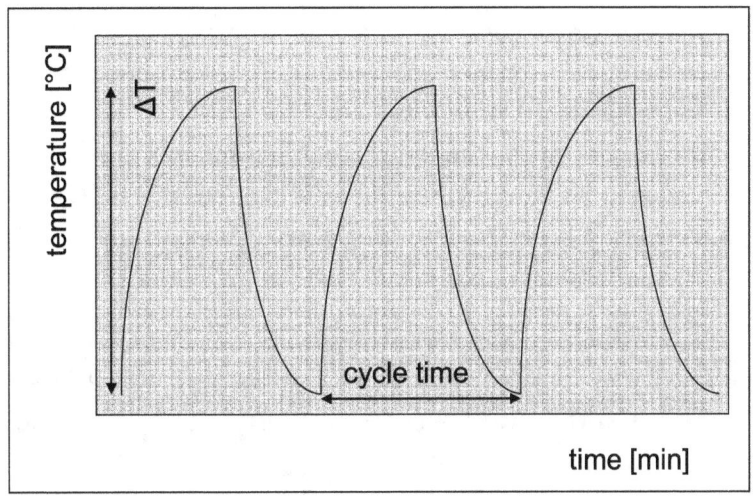

Fig. 4.8 Temperature Profile during TC

### 4.3.4. Temperature Humidity Bias (THB)[1]

This "test is performed to evaluate the reliability of non-hermetic packaged IC devices in humid environments"[2]. With the help of higher temperature and humidity and bias in addition moisture is penetrating through package material or along layer/material interfaces.

Temperature and humidity are set to (85 ± 2)°C and (85 ± 5)%, respectively. Ramp up and ramp down to stress conditions are not to exceed three hours. Continuous and cycled bias are possible. Criteria for choosing continuous or cycled bias are defined in JESD22-A101C.

Relative humidity RH is defined as:

$$RH = \frac{p}{p_{sat}} = \frac{actual\_vapour\_pressure\_of\_the\_air}{saturated\_vapour\_pressure\_of\_the\_air} \cdot 100\%$$

---

[1] JESD22-A101C (2009)
[2] JESD22-A101C (2009), p. 1

THB addresses failure mechanisms like corrosion, electrochemical migration, weakening of adhesion etc., which might lead to leakage currents or layer delamination. The parts are considered fail, if parametric limits are exceeded the part does not pass the subsequent functional test.

### 4.3.5. Autoclave (AC)[1]

"The 'Unbiased Autoclave Test' is performed to evaluate the moisture resistance integrity of non-hermetic packaged solid state devices using moisture condensing or moisture saturated steam environments"[2].

With the help of higher temperature, humidity and pressure moisture is penetrating through package material or along layer/material interfaces. The AC acceleration factor is very high resulting in failure mechanisms not experienced in the application. Hence extrapolation to use conditions might lead to wrong conclusions.

> „The combination of high humidity, high temperature ($>T_g$) and high pressure may produce unrealistic material failures "[3]

Temperature, humidity and vapour pressure are set to (121±2) °C, 100% and 205 kPa, respectively at a typical stress time of 96 hours. Metal corrosion and layer delamination are typical failure mechanisms provoked by this test. The parts are considered fail, if parametric limits are exceeded or the part does not pass the subsequent functional test.

### 4.3.6. (Unbiased) Highly Accelerated Stress Test ((U)HAST)[4]

"The Unbiased HAST is performed for the purpose of evaluating the reliability of non-hermetic packaged solid-state devices in humid environments"[5]. This stress behaves similar to the THB, but at higher temperature, hence higher acceleration and is recommended as an alternative to AC. It is a destructive test with the same failure mechanisms like THB.

Two different test conditions A and B are in place with test condition A being more aggressive at a duration of 96 hours and B at a duration of 264 hours. A conditions are (130±2) °C, (85±5) % and 230 kPa for temperature, relative humidity and vapour

---

[1] JESD22-A102-C (2000)
[2] JESD22-A102-C (2000), p. 1
[3] JESD22-A102-C (2000), p. 1
[4] JESD22-A118A (2011)
[5] JESD22-A118A (2011), p. 1

pressure while condition B applies only (110±2) °C, (85±5) % and 122 kPa. Ramp up and ramp down times shall not exceed three hours.

The parts are considered fail, if parametric limits are exceeded or the part does not pass the subsequent functional test. For the (biased) HAST stress in addition operational bias is applied leading to an additional stress, again similar to THB.

### 4.3.7. High Temperature Operating Life (HTOL)[1]

"This test is used to determine the effects of bias conditions and temperature on solid state devices over time. It simulates the devices' operating condition in an accelerated way, and is primarily for device qualification and reliability monitoring"[2]

HTOL does not address a specific failure mode but intends to determine the intrinsic reliability of the integrated circuits.

Ambient temperature and applied bias shall result in at least 125°C junction temperature, i.e. ambient temperature could be 125°C, 150°C or even 175°C for plastic packaged parts and up to 250°C for ceramic packaged parts. The devices are operated dynamically. The bias conditions have to be chosen in way that the maximum number of possible operating nodes are biased.

The parts are considered fail, if parametric limits are exceeded or the part does not pass the subsequent functional test.

## 4.4. End of Life Testing

Up to now the stress time basically corresponds to the use time with the goal of zero fails within one lot. However, this kind of approach only provides a minimum lifetime and not an overall information about the potential lifetime of the product. If it is necessary to gain more information about the actual expected lifetime the stress test has to be run to fail.

This kind of test is recommended in particular if new materials, new technologies or entirely new applications are subject to a stress test. During the planning phase of end of life testing one should be aware that this approach will increase utilization of the stress equipment significantly.

End of life testing will be discussed later in connection with the area of robust validation and zero defect.

## 4.5. Wear out mechanisms

The term "wear out" describes a failure which is caused by accumulated damage over time under particular stress conditions, for instance electromigration or corrosion. In

---

[1] JESD22-A108D (2010)
[2] JESD22-A108D (2010), p. 1

contrast to wear out the term "overload" is a single overstress situation exceeding the capability of the material, e.g. a fracture or an electrostatic discharge (ESD).

> „We spend too much time in our reliability courses on probability and statistical inference.... that show us how to quantify our ignorance.
> We do not spent enough time on removing that ignorance...the engineering, physics and chemistry of why things fail and why things don't fail"  [1]

A sound understanding of the physical failure mechanism is crucial for the determination of the right acceleration model. For instance the failure mode "open metal line" may be caused by electromigration, stress migration or by corrosion. All of them lead to the same problem, an interrupted metal line, but the physical root cause is completely different. The following sections provide an examplary selection of fail mechanisms for devices, dielectrics and metals.

### 4.5.1. Front End of Line related failure mechanisms[2,3,4,5]

#### 4.5.1.1. Hot Carrier Stress

With decreasing feature sizes in MOS FET devices the electric field close to the drain increases resulting in an acceleration of carriers above kinetic energies of $3/2kT$. Those carriers are known as hot carriers (fig. 4.9).

Carriers exceeding the potential barrier between substrate and gate oxide can be injected into the gate oxide and get stuck there ("trapped").

While moving through the channel impact ionization generates even more electron-hole pairs which lead to interface states, i.e. break up of Si-H bonds[6], and charge trapping in the gate oxide.

> „Defects in the gate oxide are called traps –
> They can trap charges"  [7]

---

[1] Evans, R.A. (1990)
[2] Alpha and Omega Semiconductor (2010)
[3] Renesas (2013)
[4] JEP122G (2022)
[5] Thaduri, A. et al. (2013)
[6] Hu, C. (1985)
[7] Azizi, N, Yiannacouras, P. (2012)

Over time, when more and more carriers accumulate in the gate oxide, the accumulated charge becomes effective by reducing the effective gate length and degrades the MOS FET characteristics, e.g. threshold voltage. In PMOS the injection of electrons into the gate oxide is the main reason for degradation due to hot carriers while in NMOS the generation of interface states is the main reason for degradation.

The use of lightly doped drain (LDD) transistors decreases the potential for hot carriers due to a reduced electric field[1].

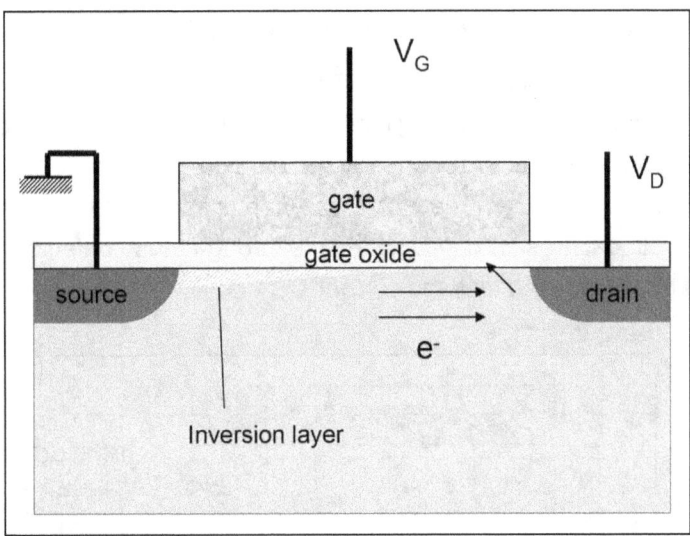

Fig. 4.9 p-channel MOS FET

There are different models in place for n-channel, p-channel (Leff < 250 nm) and p-channel (Leff > 250 nm):

(1) n-channel

$$\tau \propto (I_{sub})^{-N} \cdot e^{(E_a/kT)}$$

with

Isub : maximum substrate current during stress
N : 2 - 4
Ea : activation energy [eV]
k : Boltzmann's constant
T : absolute temperature [K]

---
[1] Alpha and Omega Semiconductor (2010), p. 20

(2) p-channel

$$\tau_{L_{eff}>250nm} \propto (I_G)^{-M} \cdot e^{(E_a/kT)}$$

$$\tau_{L_{eff}<250nm} \propto (I_{sub})^{-N} \cdot e^{(E_a/kT)}$$

IG : maximum gate current during stress
M : 2 - 4
Leff : effective gate length

### 4.5.1.2. Time Dependent Dielectric Breakdown (TDDB)

With decreasing feature sizes in MOS FET devices the gate oxide becomes thinner and thinner which results in an increasing electrical field strength within the oxide layer. High quality oxides withstand up to 10 MV/cm or even more.

Over time however, defects ("electron traps") are generated inside the oxide film by thermal and voltage stress. Tunnelling effects through those defects can be measured as so called stress induced leakage current.

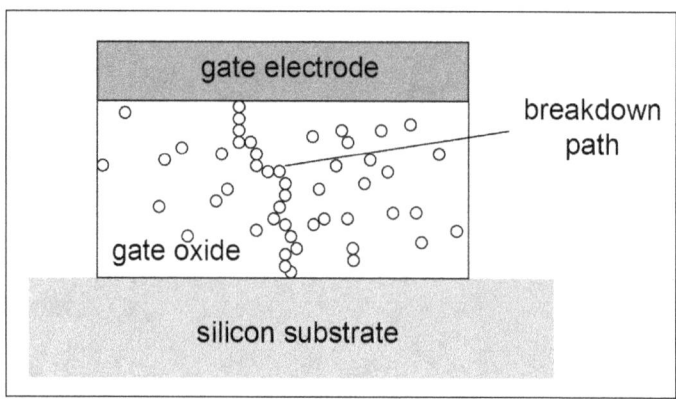

Fig 4.10 Percolation model for the breakdown path in an oxide layer

An increasing number of defects eventually leads to a conducting path through the oxide carrying a large current flow and destroying the oxide. De Graeve et al. proposed a percolation model to describe how the traps form a conducting path considering trap size and oxide thickness[1] (fig. 4.10).

Two models are mainly used to describe the lifetime τ of SiO2 gate dielectric over different oxide thicknesses, the E-model:

---

[1] De Graeve et al. (1998)

$$\tau \propto e^{-\beta \cdot E} \cdot e^{E_a/kT}$$

with
- E : electric field in the oxide
- γ : acceleration parameter
- β : acceleration parameter
- Ea : activation energy [eV]

and the 1/E-model:

$$\tau \propto e^{\gamma/E}$$

The E-model acceleration parameter β itself depends on temperature, too (β ~ 1/T), γ is only slightly dependent on temperature.

For ultrathin gate oxides the temperature dependence is not following the Arrhenius equation:

$$\tau \propto e^{\frac{a(V)}{T} + \frac{b(V)}{T^2}}$$

At higher electrical field strengths (> 10 MV/cm) an additional fail mechanism, the field enhanced thermal bond breakage, appears, too.

### 4.5.1.3. Negative Bias Temperature Instability (NBTI)

The bias temperature instability in CMOS technologies has been known since the 1960s and was explained "over a hydrogen-diffusion controlled mechanism for the creation of interface states"[1] : At the SiO2 / Si interface dangling bonds are saturated by hydrogen atoms forming a Si-H group. In p-MOS FETs temperature and bias stress dissociate these Si-H groups releasing hydrogen which is subsequently diffusing into the oxide layer generating a positively fixed charge. At the same time the dangling bond of the remaining Si turns into an interface state. Over time the number of interface states as well as the number of traps increases and the charge of the traps becomes effective. The combination of both was considered the root cause for parameter degradation, e.g. threshold voltage.

Studies performed in the last years, however, point to a different mechanism for the NBTI failure mode[2], which is more consistent with the experimental data: Depending on its energetic level $E_T$ the charge state of a defect (= the trap) can switch ("switching traps")[3]. $E_T$ stands for thermodynamic energetic level of a trap. Donor like defects become neutral when being charged negatively, i.e. when they are filled by an electron.

---

[1] Grasser, T. et al. (2011)
[2] Grasser, T. et al. (2011)
[3] Grasser, T. et al. (2010)

The defect is positively charged in case it is not filled by an electron (fig. 4.11). Hence the charge state depends on the defect energy level $E_T$ relative to the Fermi energy $E_F$.

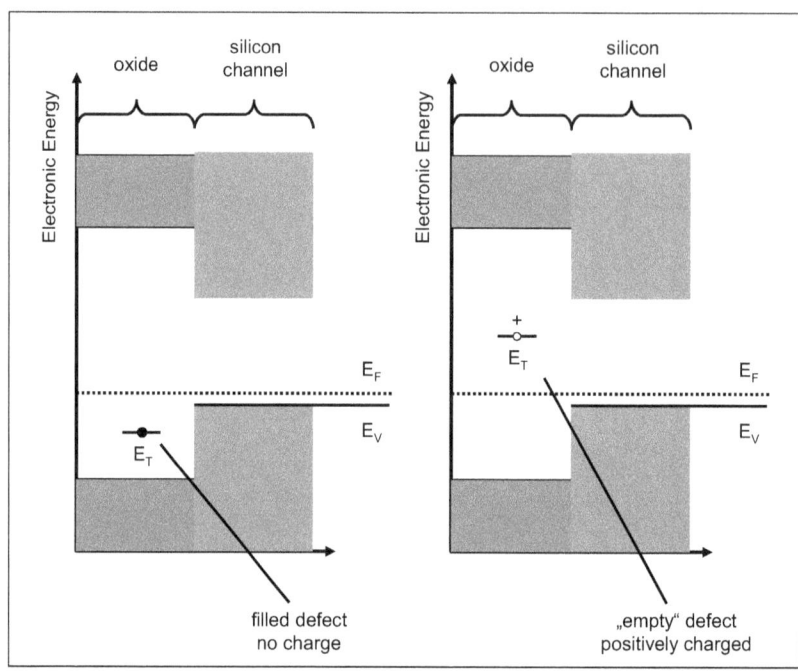

Fig. 4.11 Defect energy level and charging[1]

Independent of the channel length, the lifetime τ is a function of temperature and voltage, i.e. the internal electrical field. Lifetime τ is described in terms of parameter shift Δp by the following equation:

$$\Delta p = A_0 \cdot e^{(\beta V_G)} \cdot e^{(-E_a/kT)} \cdot t^n$$

with
- A0 : factor depending on the gate oxide process and technology
- Ea : activation energy [eV]
- k : Boltzmann's constant
- T : channel temperature [K]

$$\tau = \left[ \frac{\Delta p_t}{A_0} \cdot e^{(E_a/kT_{appl})} \cdot (V_{G.,appl})^\alpha \right]^{1/n}$$

- VG : gate voltage (in inversion)
- α : gate voltage exponent (measured)
- β : gate voltage sensitivity (measured)
- t : stress time

---

[1] Grasser, T. (2013)

n : time exponent (measured)
Tappl : application channel temperature
VG,appl : applied gate voltage
E : electric field

A simplified equation for τ is proposed by the Renesas company:

$$\tau \propto e^{-\beta E} \cdot e^{-E_a/kT}$$

### 4.5.1.4. Electromigration

The fact that intense electric current (> $10^5$ A/cm²) transfers momentum from the electron flow onto diffusing atoms within a metallic phase, e.g. an aluminum metal line, which results in a metal atom movement in the direction of the electron flow is called electromigration (fig. 4.12).

Electromigration results in voids, i.e. opens in a metal line, close to the negative electrode, whereas close to the positive electrode atoms accumulate to so called hillocks. Hillock, i.e. extrusions or whiskers, typically happen at higher temperatures and higher current densities (> $10^5$ A/cm²) and may cause shorting of two adjacent metal lines.

Since electromigration is based on diffusing atoms, the polycrystalline structure of metals is essential for the extent of this phenomenon. It is well known that diffusion along the grain boundaries (Al) or on the surface (Cu) is much faster than within a grain. Diffusion itself increases with increasing temperature.

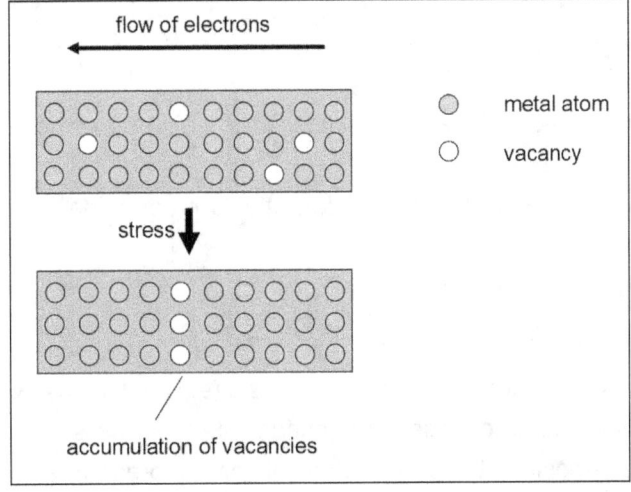

Fig. 4.12 Mechanism of electromigration on atomic level

Black's equation describes the overall the electromigration related lifetime τ :

$$\tau \propto (J - J_{crit})^{-n} \cdot e^{(-E_a/kT)}$$

with
- J : current density [A/cm²]
- Jcrit : minimum current density to activate electromigration
- n : constant (n = 2)
- Ea : activation energy [eV]
- k : Boltzmann constant
- T : absolute temperature [K]

Addition of 2 – 4 % of Cu to an aluminum metal line suppresses the formation of grain boundaries to some extent, hence reduces susceptibility to electromigration down to about 1/50. Typical activation energies are found to be in a range of 0.6 up to 1.0 eV. Temperature does have an influence on the prevailing diffusion mechanisms like grain-boundary diffusion or lattice diffusion. Also the grain structure itself might change with higher temperatures. Hence it is necessary to keep the stress temperature below ~ 250°C.

If the width of the metal line is similar to the average grain size (bamboo structure, fig. 4.13), then electromigration is reduced as compared to wider lines. The reason for this effect is the reduction of grain boundary diffusion along the direction of the metal line.

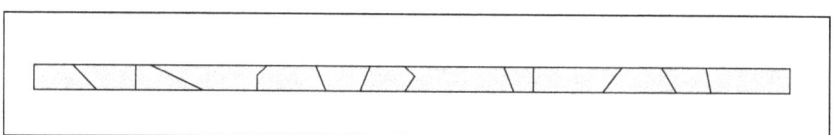

Fig. 4.13 Top down view of metal line with visible grain boundaries

Also differences between electromigration effects in long lines vs. contact holes and vias need to be considered.

### 4.5.1.5. Stress Migration[1]

The interface between metals and other materials (e.g. dielectrics) with different thermal expansion coefficients is exposed to mechanical stress as soon as the temperature differs from the deposition temperature (tensile or compressive stress).

---
[1] Walter, W. (2004)

The mechanical stress leads to material transport, i.e. atomic diffusion, until a sufficient relaxation of the stress has reached. Due to the fact that diffusion is a temperature dependent phenomenon obviously stress migration is temperature dependent, too. Hence the migration rate increases with increasing temperature. Similar to Black's equation the stress voiding lifetime τ can be described as follows:

$$\boxed{\tau \propto (T_0 - T)^{-N} \cdot e^{(-E_a / kT_{stress})}}$$

with

T0 : stress free temperature (~ metal deposition temperature)
T : temperature [K]
N : 2 – 3 for ductile metals
Ea : activation energy (0.5 – 0.6 eV for grain boundary diffusion)

### 4.5.1.6. Metal Corrosion

Corrosion of the bonding pad or within the chip can be caused by moisture and/or contaminants. Aluminum is less corrosive than copper as long as there are no chlorine ions in the water damaging the aluminum native oxide layer. Temperature dependent diffusion of contaminants to the metal and diffusion of metal ions away is required to enable the corrosion process. The corrosion rate in non-liquid environment (e.g. air) is very much dependent on the percent relative humidity. Different models are in use like the reciprocal exponential humidity model:

$$\boxed{\tau \propto e^{(b/RH)} \cdot e^{(-E_a/kT)}}$$

with

b : humidity dependence parameter
RH : relative humidity [%]
Ea : activation energy [eV]
k : Boltzmann's constant
T : temperature [K]

For other models, e.g. power-law humidity model or exponential humidity model the reader is referred to JEP122.

### 4.5.1.7. Alpha Particle Induced Soft Error

Alpha particles, emitted from Uranium and Thorium contained in the chip package and/or wiring material, generate electron hole pairs along their path through silicon. If an electric field happens to be close to this path, e.g. at a pn-junction, the electron hole pairs are split up and electron and holes are moving into opposite directions.

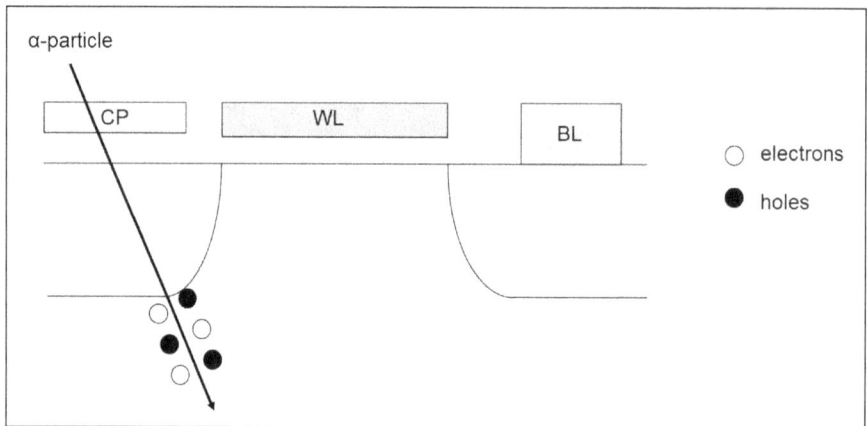

Fig. 4.14 Generation of an electron-hole pair by an α-particle

The accumulation of charges e.g. leads to an inversion of the memory or changes the bit line potential in the specific example of a dynamic random access memory (DRAM), see fig. 4.13. In both cases the memory information (one bit) is lost.

This kind of error is not permanent and is recovered by rewriting the information into the memory cell and subsequent retesting. The determination of soft errors is important for the memory reliability assessment.

In addition to alpha particles also terrestrial cosmic ray induces soft errors in volatile memories like DRAMs have to be taken into account[1].

### 4.5.1.8. Non-Volatile Memory Reliability[2]

A slightly different situation arises in non volatile memories. Non-volatile memories use a floating gate to store electrons, i.e. information.

Programming or writing into the memory cell means injection of electrons into the floating gate by tunnelling or hot electron injection (fig. 4.15 and 4.16). Due to the fact that the floating gate is isolated, the electrons remain there and change the threshold voltage of the transistor which can be sensed later on when the stored information is read.

Several degradation mechanisms for data retention were identified like defects within the oxide, ionic contamination, electrical stress and a high repetition rate of read/write cycles, which are not discussed here in more detail.

---

[1] JESD89-3a (2007)
[2] Willer, J. (2007)

Even though the duration of data retention is roughly ten years or more, it is mainly limited by thermoionic emission: there is always a small probability for an electron to gain enough thermal energy to penetrate the dielectric layer isolating the floating gate.

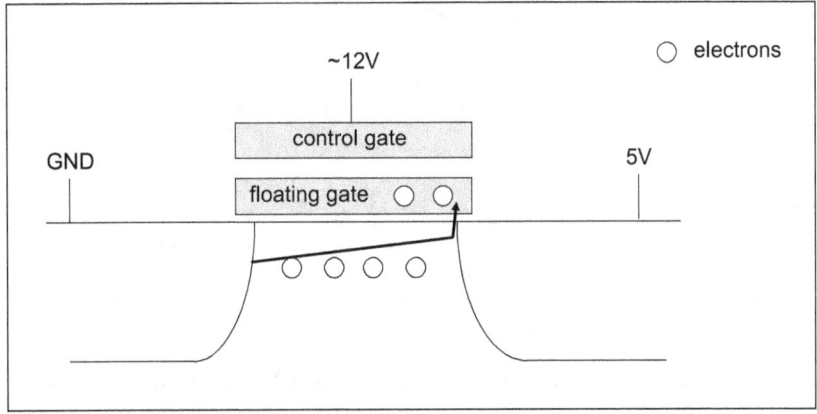

Fig. 4.15 Non-volatile memory programming using hot electrons

Thus this leads to a gradual loss of the stored charge, which constitutes the information stored.

The so called thermionic excitation model describes the loss of electrons[1]:

$$\boxed{\frac{N(t)}{N(0)} = e^{[-v \cdot t \cdot \exp(-E_a / kT)]}}$$

with
- N : number of electrons in the floating gate
- v : relaxation frequency
- $E_a$ : activation energy [eV]
- k : Boltzmann constant
- T : absolute temperature [K]

The ratio of the operating voltage follows the ratio of the number of electrons:

$$\boxed{V_{CC}(t) = V_{CC}(0) \cdot \frac{N(t)}{N(0)}}$$

---
[1] Renesas (2013)

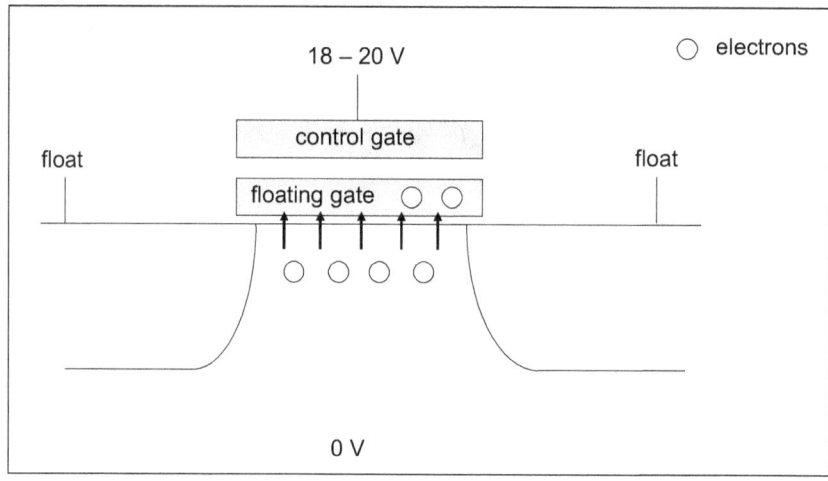

Fig. 4.16 Non-volatile memory programming using Fowler-Nordheim tunnelling[1]

### 4.5.1.9. Surface Inversion

To some extent semiconductor process materials can contain elements like Li, Na or K. The ionized form of those elements moves quickly through silicon dioxide (SiO2) as soon as a weak electric field is applied and temperature is slightly increased (~ 100°C). In case those ions accumulate at the Si/SiO2 interface a surface inversion may be generated which eventually leads to leakage currents and causes device failure.

Being influenced by both temperature and electric field the time to fail follows the relation:

$$\tau \propto \frac{1}{<J_{ion}>} \cdot e^{E_a/kT}$$

with

| | |
|---|---|
| \<Jion\> | : average mobile ion flux |
| Ea | : activation energy [eV] |
| k | : Boltzmann's constant |
| T | : temperature [K] |

## 4.5.2. Assembly Process Related Failure Mechanisms[2,3]

### 4.5.2.1. Wire Bonding Reliability

A typical assembly process is the bonding of a gold wire from the external contact onto an aluminum pad on the surface of the integrated circuit. It is well known that within a Au-Al-joint a number of different forms of alloy with different properties develop, which results in several failure mechanisms[4]:

---

[1] Groeseneken, G. et al. (1998), p. 10
[2] Renesas (2013)
[3] JEP122G (2011)
[4] Renesas (2013), p. 63

- Different thermal expansion coefficients of different alloys lead to a reduction of the strength of the joint.
- Different diffusion coefficients between the joining partners (Au, Al) leads to the formation of voids (Kirkendall effect)
- Presence of bromium initiates Au4Al oxidation resulting in a high resistance layer

Storage at high temperatures provokes those fails. The thickness of the diffusion layer Ldiff depends on temperature and time:

$$L_{diff} = \sqrt{D_0 \cdot e^{-E_a/kT}}$$

with
- $D_0$ : material diffusion coefficient
- $E_a$ : activation energy [eV]
- k : Boltzmann's constant
- T : temperature [K]

In addition to temperature storage also thermal cycling results in thermomechanical stress which, eventually, fatigues up to the point of joint failure. The root cause for this phenomenon is obviously again the mismatch of the thermal expansion coefficients of the original metals as well as various alloys within the joint. The number of cycles to fail Nf is expressed as follows:

$$N_f = C_0 \cdot (\Delta T)^{-q}$$

with
- $C_0$ : material dependent constant
- $\Delta T$ : temperature swing
- q : Coffin-Manson exponent

### 4.5.2.2. Ag Ion Migration

With voltage applied and at high humidity and high temperature, silver ions start to migrate and to accumulate at the cathode side in the form of dendrites or blots. This phenomenon is called ion migration or electrochemical migration. Depending on the size of these extrusions the resistance of the insulator may decrease or metal lines may even be shorted.

Of course electrochemical migration is not limited to Ag ions, but could happen with Cu or Au ions, too. However, most actual problems are observed with silver.

### 4.5.1.3. Moisture Resistance

As described earlier moisture leads to corrosion of metal lines within the integrated circuit and increased leakage[1]. There are two major contributors for moisture to penetrate the inner regions of an integrated circuit:

- Moisture diffusion through the package material
- Moisture penetration along the interface between package material and leadframe/metal wire

These effects depend on the temperature and the percent relative humidity and the quantitative relation is described as follows:

$$\boxed{\tau \propto e^{E_a/kT} \cdot RH^{-n}}$$

with

$E_a$ : activation energy [eV]
$k$ : Boltzmann's constant
$T$ : temperature [K]
$RH$ : relative humidity [%]
$n$ : constant (4 – 6)

## 4.7. System Reliability[2,3]

Single semiconductor devices are typically assembled into a larger system, i.e. an integrated circuit, into a system on a chip or a printed circuit board. These devices may be connected in parallel, in a series or in a more complicated way.
Depending on the way the devices are arranged the reliability of the entire system can be calculated.

### 4.7.1. Serial and Parallel Systems

Frequently systems can be broken down into serial or parallel substructures or combinations of those. In any case the first step is to transfer a real system into a reliability model. A simple example for this activity is illustrated by fig. 4.17. There a circuit diagram for an oscillating circuit is transferred into a block diagram. The oscillating circuit only works if both C and L are working.

---

[1] Van Soestbergen et al. (2010)
[2] Meyna, A., Pauli, B. (2010), p. 169
[3] Renesas (2013)

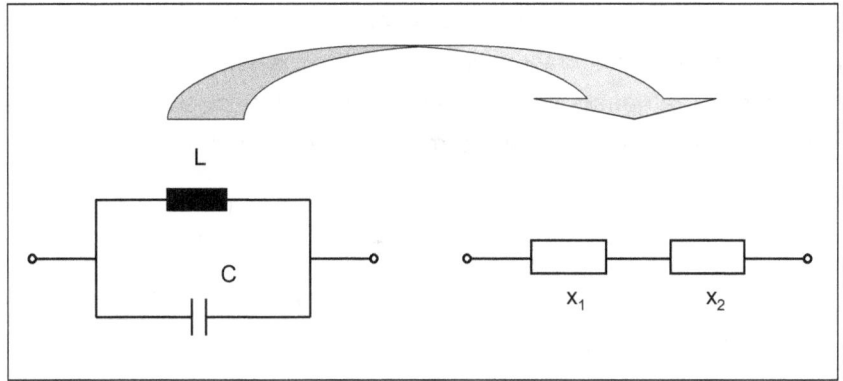

Fig. 4.17 Transfer of a circuit diagram into a serial block diagram

The reliability function R(t) of the serial system is expressed in the following way (under the assumption that failure of the individual system components is entirely independent):

$$R(t) = R_1(t) \cdot R_2(t) \cdot ... R_n(t) = \prod_{i=1}^{n} R_n(t)$$

The respective cumulative failure distribution function F(t) is then:

$$F(t) = 1 - R(t) = 1 - \prod_{i=1}^{n} R_i(t)$$

If the reliability function is described by an exponential function the system fail rate λ is calculated by adding up the component fail rates. The logic relation is an "and" one, i.e. all system components must be functional for the system to be functional. This can be verified by observing, that a product of exponential functions is the exponential of the sum of the prefactors for the time variable.

$$R(t) = \prod_{i=1}^{n} e^{-\lambda_i(t)}$$

$$\lambda(t) = \sum_{i=1}^{n} \lambda_i(t)$$

A parallel system, on the other hand, is characterized by n independent components arranged in a parallel configuration (fig. 4.18). Here it is assumed that the logical relation is an "or" one, i.e. at least one component has to work to enable functionality of the entire system. In other words, those components are redundant.

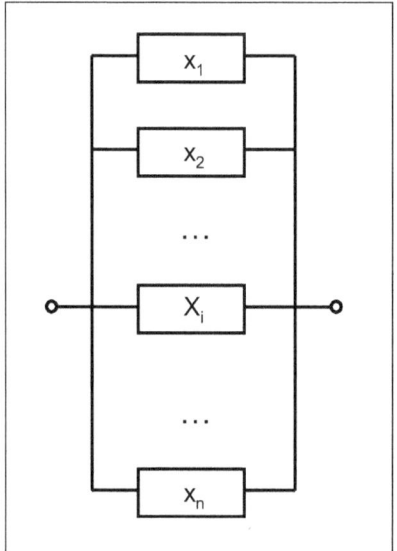

Fig. 4.18 Serial system block diagram

For a two component parallel system, for instance, the reliability function R(t) is expressed as follows for the case that the two events are independent from each other and the events do not exclude each other from happening. In addition it is assumed to be a so called hot redundancy system, i.e. both components are always turned on.

$$R(t) = R_1(t) + R_2(t) - R_1(t) \cdot R_2(t)$$

Assuming $R_1(t) = R_2(t) = e^{-\lambda t}$ the system reliability function becomes:

$$R(t) = 2e^{-\lambda t} - e^{-2\lambda t}$$

Instead of looking at the reliability function it is more convenient to look at the respective cumulative failure distribution function F(t), i.e. the probability of the system to fail:

$$F(t) = F_1(t) \cdot F_2(t) \cdot \ldots \cdot F_n(t) = \prod_{i=1}^{n} F_i(t)$$

The reliability function R(t) then becomes:

$$R(t) = 1 - F(t) = 1 - \prod_{i=1}^{n} F_i(t) = 1 - \left( \prod_{i=1}^{n} (1 - R_i(t)) \right)$$

### 4.7.2. Bridge Structure

Technical systems sometimes cannot be represented using a parallel or serial block diagram. As an example for a more complicated structure the so called bridge structure is explained in this section (fig. 4.19).

Using minimum path set and Boolean operations the system is assumed to survive if at least one of the following paths $P_i$ with the respective reliability functions works:

| | |
|---|---|
| $P_1 = (x_1 \wedge x_3)$ | → $R_{13} = R_1 R_3$ |
| $P_2 = (x_2 \wedge x_4)$ | → $R_{24} = R_2 R_4$ |
| $P_3 = (x_1 \wedge x_5 \wedge x_4)$ | → $R_{154} = R_1 R_5 R_4$ |
| $P_4 = (x_2 \wedge x_5 \wedge x_3)$ | → $R_{253} = R_2 R_5 R_3$ |

The probability of the system to work P is then described using Boolean expression again:

$$P_{System} = P((x_1 \wedge x_3) \vee (x_2 \wedge x_4) \vee (x_1 \wedge x_5 \wedge x_4) \vee (x_2 \wedge x_5 \wedge x_3))$$

Using the failure distribution F(t) to describe the probability of the system to fail R(t) becomes:

$$R(t) = 1 - F(t) = 1 - \prod_{i=1}^{n} (1 - R_i(t))$$
$$R(t) = 1 - ((1 - R_{13}) \cdot (1 - R_{24}) \cdot (1 - R_{154}) \cdot (1 - R_{253}))$$

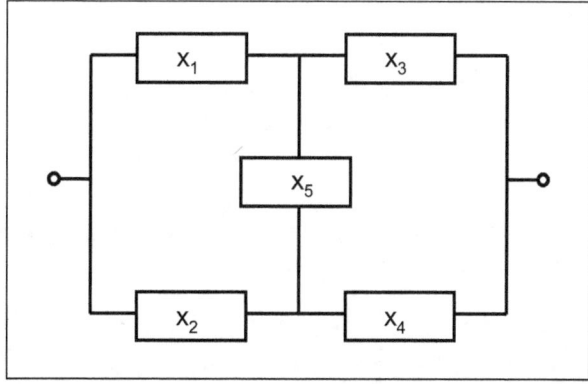

Fig. 4.19 Bridge structure block diagram

### 4.7.3. Different Failure Modes

Often more than one failure mode leads to a failing component. Assuming two failure modes the probability of failure is

$$F = F_1 + F_2$$

with
- F : total probability to fail
- $F_1$ : fail probability, mode 1
- $F_2$ : fail probability, mode 2

The respective probability to survive (reliability function) is then:

$$R = 1 - (F_1 + F_2) = R_1 + R_2 - 1$$

# II. Quality Management: Tools and Operations

# 5. FMEA

In section 5, the failure statistics was described either for one or more than one failure modes / failure mechanisms. To identify a particular failure mechanism, there are two main methods, either physical failure analysis of failed components ("post mortem analysis") or a proactive method once a certain amount of experience has been accumulated. This method is called failure mode and effects analysis, and it constitutes an "ex ante" method to identify failure modes in a proactive way. The purpose of an FMEA analysis is to identify potential failure modes proactively and estimate the risk associated with a particular failure mode. I other words, the philosophy is: prevention is better than cure.

## 5.1. Introduction

FMEA stands for "Failure Mode and Effects Analysis". It is a widespread quality management method for the proactive investigation of potential failures, potential root causes and potential effects of products and processes. The method is based on a team approach targeting for a systematic analysis of problems which could happen in the future.

> A **failure mode** is a category of failure. It describes the way a product or process is failing to perform its correct function.

> An **effect** is the harmful result of a failure affecting the customer. The customer is either an internal or an external customer.

### 5.1.1. History

The history of FMEAs started right after World War II. Military equipment as well as aeronautic / astronautic applications required a high quality level, which is why a systematic approach for the detection of potential failures became necessary. Later on the automotive industry adapted this approach. Today many companies throughout many industry branches apply FMEAs as a proactive tool to assure the required quality levels.

Important steps in the history[2,3,4] of FMEAs are listed as follows:

---

[1] Kmenta, S. (2002)
[2] Catic, D. (2011)
[3] Razak, N. (2011)
[4] Kmenta, S. (2000)

- 1949 : developed by the US military (MIL-P-1629)
- 1950s and 1960s : FMEA method promoted by Boeing
- 1966 : NASA[1] to publish a FMEA procedure to be used in the Apollo-Program
- 1960s : FMEA used in the aeronautics industry
- 1971 : The Ford Motor Company introduces an internal standard regarding the use of FMEAs
- 1980s : FMEAs are implemented in the automotive industry through the Automotive Industry Action Group
- 1985 : The International Electronic Commission (IEC) publishes FMEA standards for system reliability
- 1988 : The Ford Motor Company publishes the "Potential Failure Modes and Effects Analysis in Design and for Manufacturing and Assembly Processes Instruction Manual"
- 1996 : FMEA is described by VDA 5.2 (Verband der Automobilindustrie, Germany)
- Later : up to today new methods are being developed and Additional branches are using the FMEA tool, e.g. off shore wind turbines[2] or in project risk assessment[3]

## 5.1.2. Motivation

The use of FMEAs within a company can be triggered by several independent reasons:
1. Reduction of failure rates in development, production and at the customer

It is well known, that follow up cost caused by failures increases significantly with increasing product life cycle. In the early stages of development a failure can be removed easily without causing a lot of additional cost. If the failure is discovered later, the product is already in production, repairing the misprocessed parts is much more expensive. In the worst case, the product was already sold to many customers, a failure might be observed causing a reliability problem. This would be the most expensive case of follow up cost. This context is also know as "rule of ten", i.e. the cost increases by a factor of 10 through each consecutive stage of the life cycle.

Hence the early detection and avoidance of failures has a large potential to reduce cost.

---

[1] National Aeronautics and Space Administration
[2] Dinmohammadi, F. (2013)
[3] Carbone, T. (2011)

2. Requirements of the respective standards (ISO9001 vs. TS16949)

Even though ISO 9001 does not explicitely require FMEA the FMEA method is one way to predict potential failures and to minimize the effects on the customer (also internal customer).

Hence the use of the FMEA method is one way to meet several requirements of the standard DIN EN ISO9001[1]:

A. Handling of risks

Planning of the Quality Management System must consider risks and opportunities, which includes to avoid or to reduce unwanted effects of potential risks, as well as planning of respective measures. It also requires to assess the effectiveness of those measures.

B. Validation of development

A company needs to ensure that a new product or new process meets the requirements. In addition to various methods of verification the validation can be supported by the use of an FMEA.

C. Validation of production processes

If production processes cannot be verified e.g. by measurement the validation of the process can be supported by the use of a FMEA.

Different from ISO9001 the ISO/TS16949 stipulates the use of FMEAs for design as well as for the production process. Accordingly, the use of FMEAs is mandatory in the automotive industry.

3. Creation of a knowledge base within the company

An FMEA summarizes all potential failures and risks of a product or a process for both internal as well as external customers. Thus it combines the know-how of various disciplines within the company ranging from development and production through a customer near party like sales or project management.

This knowledge can be used for an early risk assessment for new products or for risk management in connection with changes. In particular, if the company feeds back field data into the FMEA, i.e. actual occurrence of a particular failure is recorded and fed back into the FMEA, then eventually this method provides realistic data on risks rather than estimated data. In the standard ISO TS16949 this is mandatory. Failure to collect and analyze field data constitutes a non-conformance which after a certification or second party audit must be corrected within 90 days.

---

[1] DIN EN ISO9001:2015

4. Product liability[1]

If an FMEA is performed in the correct way, then it can be used for exculpation in court proceedings in case of unpredicted failure. FMEAs may be done in various stages of the product design cycle. In addition to the actual execution of the FMEA the correct documentation is mandatory. Otherwise exculpation might not be possible.

5. Customer pressure

Based on product requirements or on the customer's certification, the customer might be requesting FMEAs from a supplier. The customer, however, can also be an internal customer.

### 5.1.3. FMEA as Part of the Company Know-How

The real benefit of FMEAs for a company becomes apparent when they are integrated into the company's processes. This means that at defined points in a particular process the results of a FMEA are required as additional input.

In addition to other areas like project risk assessment or test reduction risk assessment etc., this typically the case in development and production (fig. 5.1).

In the development phase for instance that could mean to implement FMEAs at "several milestones to cope with design-, test- and overall project related failure modes, i.e. risks"[2], for instance at

- Project release
- Chip Size freeze
- Process freeze (ready for qualification)
- Customer sample release
- etc.

This stepwise approach ensures a continuous risk assessment throughout the relevant development phases.

In addition to the development of a new product in many cases major or minor changes of existing products are implemented in order to continuously improve product performance and/or reliability. These product changes require an assessment of the potential risks. If customer requirements like allowed temperature range or expected lifetime change a risk assessment of these particular changes is required, too.

---

[1] See Ricco, R. (2006)
[2] Qimonda, PDHB (2008)

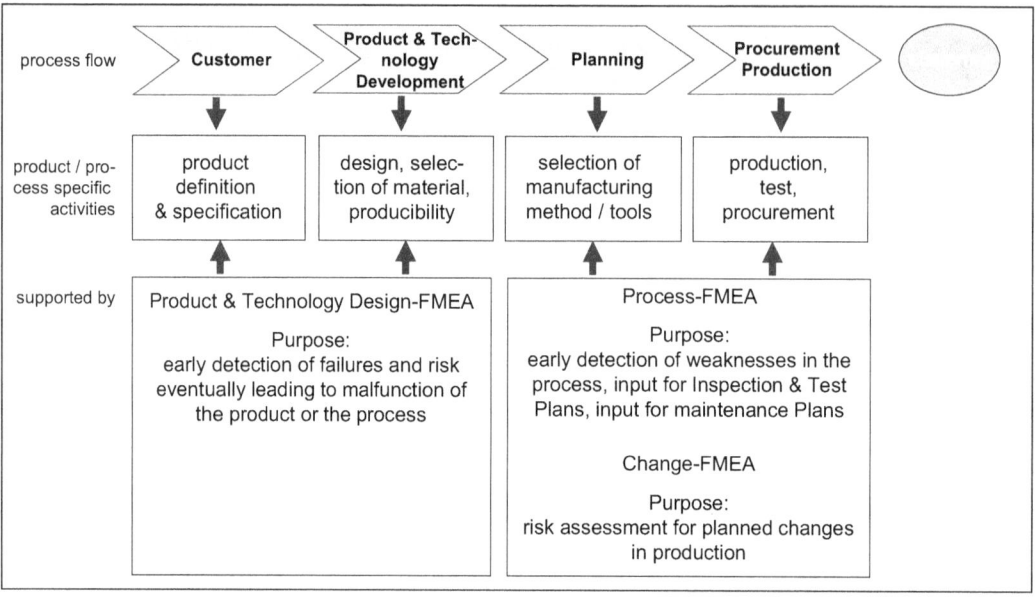

Fig. 5.1 Company Core processes supported by FMEA

Just like in the development process these items need to be implemented in the respective business processes.

On the other hand, a planned deviation from the documented production process ("process of record") also requires the execution of a FMEA. In detail, depending on the sensitivity of the process, the following changes might require a formal risk assessment:

- use of new or different pieces of equipment and tools
- qualification of a new (additional) production site
- transfer of production to another site
- process changes, e.g.
    - layer thickness reduction
    - new material supplier
    - change of material for cost reduction
    - test time reduction
    - etc.

Typically these changes are managed by applying a process change business process.

### 5.1.4. Product- and Process-FMEA

Depending on the focus of the assessment there are two different types of FMEAs. For an assessment of the mode of operation of a product, for instance, a product FMEA is used, whereas for the assessment of a sequence of activities like the relevant product manufacturing process a process FMEA is used.

This approach covers the entire system as well as software, interfaces, engineering and construction, components, the manufacturing process, the assembly process etc.

Product FMEAs
- examine the features of products and systems down to the layout, dimensioning and material of those features and
- determine potential deviations and corrective actions to ensure proper functionality of the product

Process FMEAs
- examines activities and procedures for the production of products and systems down to the requirements for parameters influencing single process steps
- determine potential deviations and corrective actions to ensure the correct flow of the process and eventually the quality of the product

## 5.2. FMEA Execution[1]

### 5.2.1. Preparation

Prior to the execution of the actual risk assessment several activities need to be performed in order to start the FMEA well prepared and to execute the FMEA as efficiently as possible.

#### 5.2.1.1. Responsibilities

First of all responsibilities need to be defined. In addition to the regular team members a sponsor, an FMEA moderator and a person in charge of the FMEA process are required. In many cases the moderator and the person in charge is one person only.

Sponsor
In the first place the sponsor decides if an FMEA needs to be done. He defines responsibilities and supports the collection of information. More important, however, is that the sponsor provides the resources or supports the person in charge in getting sufficient resources from relevant departments.

Person in charge
The person in charge is responsible to collect all the required documents and information. He also organizes the schedule and the respective FMEA meetings, i.e.

[1] VDA (2012)

reserves the room, provides work equipment, selects and invites FMEA members etc. In addition, he defines the topic and scope of the FMEA.

FMEA moderator

The moderator supports the selection of the team members and the schedule. The moderator needs to be experienced in the FMEA method, in moderating a team and in presenting result. In addition, the moderator has to be a team player. The person in charge and the FMEA moderator is typically the same person.

Team members

Primarily the team members possess professional experience of the topic, e.g. the manufacturing process or the product or both. They need to be experts in their respective fields of knowledge. In addition, the team members support the preparation of the FMEA.

### 5.2.1.2. Documents

Depending on the focus of the FMEA the following documents can be useful to be available at the beginning of the FMEA:

- Specifications & Drawings
- System descriptions
- Functional descriptions
- Requirements of authorities
- Customer requirements or customer agreements
- Internal or external quality targets
- FMEAs of different or predecessor products or processes
- etc.

Since all these documents together might be thousands of pages if printed out, in the most cases it makes sense to collect those, but to provide digital access to them during the FMEA. So whenever a question is raised regarding a particular part of the FMEA item, the respective document can be accessed and all members of the team can look at it at the same time and discuss it.

### 5.2.1.3. Steps

The execution of the actual FMEA is based on a stepwise approach prior start of the FMEA including at least:

- clarification of the FMEA type (product or process)

- definition of the person in charge
- definition of team members and their responsibilities
- determination of the schedule
- organization of meetings
- determination of required resources and getting respective management commitment

The steps to be performed during the actual execution of the risk assessment are described in the following chapters.

## 5.2.2. General Approach

The FMEA is performed in five consecutive steps:

Step 1: structural analysis.
The entire system (product or process) is broken down in its modules and components. The result of this analysis is then visualized (fig. 5.2).
The complexity of the system and the available resources define the number of levels of the structural analysis.

Step 2: functional analysis
Each structural element on each level does have at least one functional purpose. Also the respective interface functions are identified in this step.

Step 3: failure analysis
Using brainstorming or other techniques for each function potential failures are identified. The guiding question is: "what can go wrong" at this step of the process or with this part of the product.
Typically, there is more than one potential failure mode for each function. Failure analysis includes identification of potential root causes and also potential effects for each of these failures (fig. 5.3).
Looking at an identified potential failure the FMEA team now has to ask "what happens to the customer, if this failure occurs". The customer could be an internal as well as an external customer. The team also asks "what makes this failure happen" and based on their experience they will come up with root causes.

Since in this manner, a large number of potential failure modes can be identified for a given system, it is obvious that the risk associated with each failure mode has to be quantified in order to be able to prioritize the risk treatment, i.e. to decide on action to mitigate the risk or make an informed decision that the risk is acceptable and no action is required.

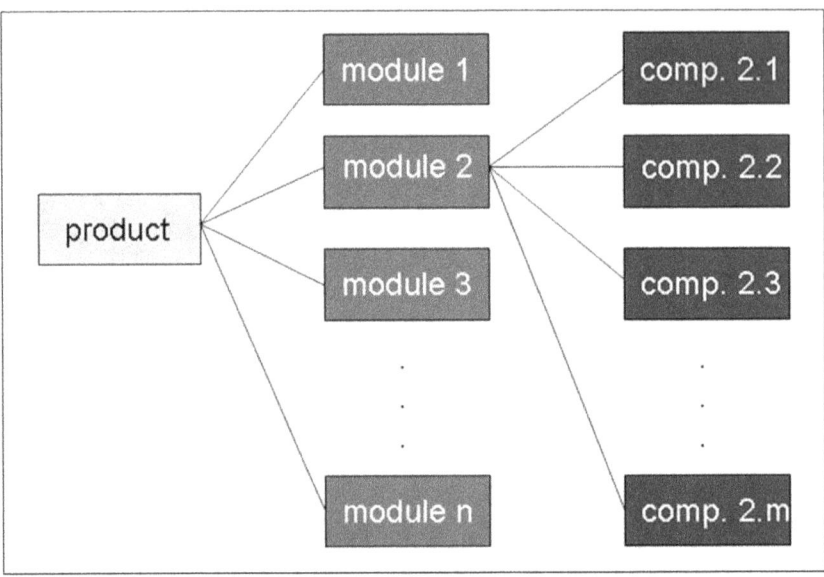

Fig. 5.2 Example: two level structural analysis of a product

In an FMEA, the quantification of the risk is broken down into three quantifiers, the first one is the severity of the risk, the other two are the frequency of occurrence and the detectability.

> **Severity** describes the impact of a failure effect on the customer.

Eventually one component could have k different functions with each n different causes and m different effects resulting quickly in a very large number of cause and effects chains which have to be evaluated regarding occurrence of the failure, severity, i.e. significance of the effect and detection rating.

In order to reduce the number of combinations usually only the most relevant effect is considered. The metric for the assessment relevance of a failure mode and its effect is explained below in detail.

> **Occurrence** is the probability that a cause will happen and lead to a failure within the product lifetime.

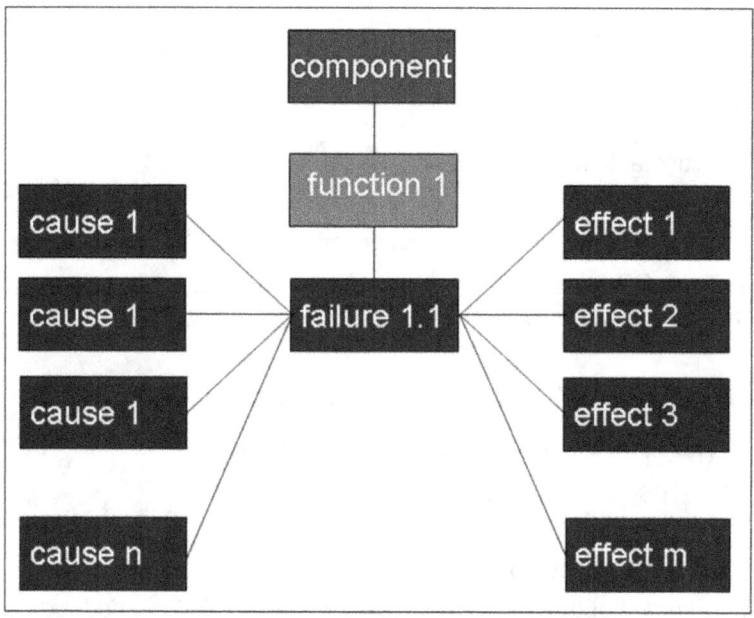

Fig. 5.3 Generic decomposition of failure causes and effects

Step 4: Assessment of risks associated with failure modes
After identification of failure modes, their root causes and their effects, existing "current process checks" are to be documented in the FMEA sheet. This could be checks like visual inspection, functional tests or even reliability testing and so forth.

> **Detection** is the probability that the current check will detect the cause of the failure or the failure itself. [1]

Now occurrence of the failure, severity of the consequences and likelihood of detection needs to be evaluated according a 10-stage scale with "1" being the best case and "10" being the worst case. For the evaluation of the likelihood of detection of a failure the current process check method has to be considered.

From those three figures the so called Risk Priority Number (RPN) is generated by simple multiplication of the three numbers:

---

[1] Kmenta, S. (2002)

> **RPN = Occurrence x Severity x Detection**

Step 5: corrective actions
Based on defined rules, e.g. by RPN limits, the company has the obligation to start corrective actions for those failure modes for which the risk is unacceptably high, as quantified by the RPN and the RPN limits. The corrective action is assigned to a team member and it's purpose is to

- eliminate the root cause, hence to reduce the likelihood of occurrence, or
- increase the detection capability, hence to reduce the likelihood of slippage to the customer

The corrective action does not impact the severity of the failure.
Along with the name of the responsible team member, also the desired completion date needs to be documented on the FMEA sheet.
After completion of the corrective action, another evaluation of occurrence, severity and detection has to be performed in order to confirm that the identified risks are now at an acceptable level. If this is not the case additional corrective actions have to be defined.

### 5.2.3. Example: Product FMEA

As described above the product FMEA is executed in five steps.

Step 1: structural analysis
For the structural analysis the breakdown of the product is done according the hierarchical structure of the product. It starts at the system level through assembly groups down to single components and ends eventually on the characteristics level of those components.
The degree of detail for new products is usually higher than for well known products. It ends when failures are covered by a corrective action on a particular detailing level. Only in case another risk is discovered, further detailing becomes necessary (example fig. 5.4.).
The structural analysis is the basis for the functional analysis of the product and thereby the failure analysis itself.

Step 2: functional analysis
In the next step for each structural unit the respective function is determined. As described above this is not limited to only one function. The main purpose of this step is to identify potential malfunctioning of the system and the subordinated units (fig. 5.5).

It is also useful to combine malfunctioning conditions to a higher structure of malfunctioning on a higher hierarchical level.

Malfunction means e.g.

- deviation from the specified functional target
- limited function
- unintended function
- decentering of the function

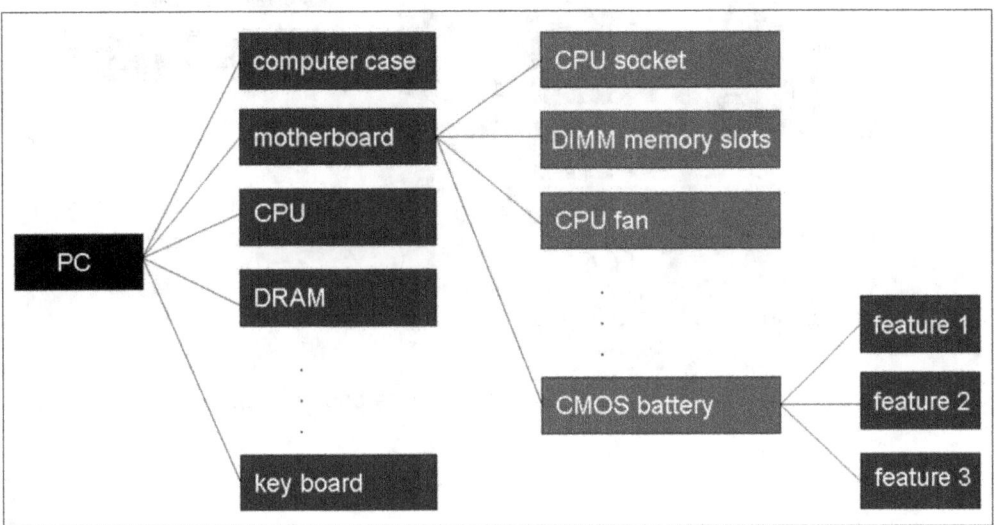

Fig. 5.4 Product structure of a Personal Computer

Step 3: failure analysis
In step 3 potential failures are identified for each of the functions. The description of a failure, i.e. a malfunction, has to be complete and detailed enough.

A malfunction described by the word "defect" cannot be elaborated regarding root cause, effects or e.g. likelihood of occurrence (fig. 5.6), therefore this is not a sufficient description of a failure.

Step 4: assessment
Assuming a correct failure description, root cause and effects as well as occurrence, severity and likelihood of detection are now determined using Risk Priority Numbers (RPN):

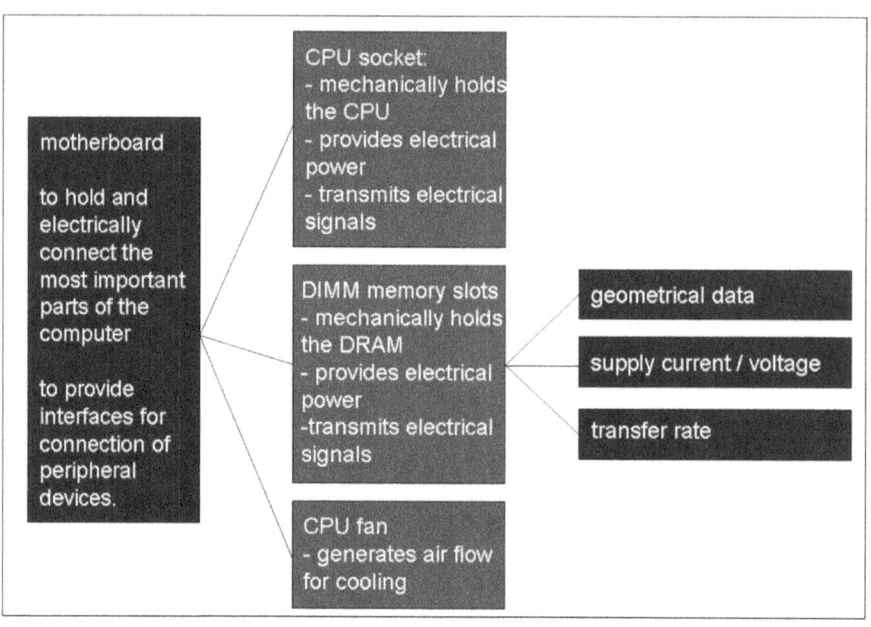

Fig. 5.5 Partial functional analysis of a Personal Computer

Occurrence         O    likelihood of a failure cause to occur

Severity           S    severity of the effect of the failure

Detection          D    likelihood of detection of a failure

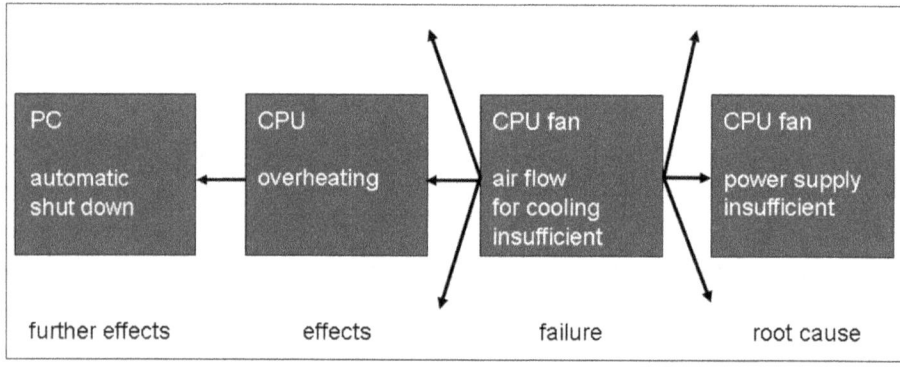

Fig. 5.6 Root cause – failure – effects – chain (product)

O, S and D are evaluated using figures from 1 through 10. 10 always means "highest risk". The meaning of these numbers needs to be defined product specific prior starting the execution of FMEAs. In some cases, the RPNs have to be aligned with the customer

Severity
In "severity" a "9" or a "10" typically means a safety issue or a deviation from legal requirements or a substantial risk for the company. On the other hand, a "severity 1 failure" is a very minor malfunctioning, which is hardly noticeable.
In case the effect of a failure is unknown the severity has to be set 10.

Occurrence
Occurrence describes the likelihood of the failure root cause during the entire product life time considering existing process controls. It can also be described in form of frequency of occurrence.
Here "10" means very high probability of the failure to occur without existing measure to avoid it, whereas "1" stands for the highly unlikely case the failure root cause occurs.

Detection
Detection describes the likelihood of detection of a failure that has occurred considering current process check mechanisms.
Now "10" means that a failure cannot at all or cannot in time be detected, whereas "1" means the failure will be detected with certainty or early enough for timely corrective actions.

Depending on the defined action limit corrective actions are defined with responsible persons and timelines.

Step 5: corrective actions
Corrective actions are either actions to avoid occurrence of a failure, i.e. the design layout has to be adjusted, or they are actions to increase the likelihood of detection of a failure.
In both cases those actions need to be described clearly and comprehensible. The statement "test" or "experiment" is not sufficient.
Also in both cases, responsible persons from the FMEA team as well as timelines for the completion of actions have to be defined.

## 5.2.4. Example: Process FMEA

As described above the product FMEA is executed in five steps.

Step 1: structural analysis
For the structural analysis the breakdown of the process is done according the hierarchical order of the process. It starts at the superior process level through more

and more detailed process levels and ends eventually on the 4Ms – level (Man – Machine – Material – Method).

The degree of detail for new processes is usually higher than for well known processes. It ends when failures are covered by a corrective action on a particular detailing level. Only in case another risk is discovered, further analysis at more detailed level becomes necessary.

The four Ms may e.g. be:

- Man : operator of equipment or project leader or …
- Machine : tools, automatic systems, handlers…
- Material : raw material, chuck material, chemicals…
- Method : process parameters, clean room, particles, temperatures ….

Again the structural analysis is the basis for the functional analysis of the process and thereby the failure analysis itself.

Step 2: functional analysis

In the next step for each process the associated function is determined. As described above this is not limited to only one function. The main purpose of this step is to identify potential malfunctioning of the overall or of the detailed processes (fig. 5.8).

Knowledge of the process as well as of the environment is necessary for the functional analysis. This includes e.g. to know product and process requirements, the cost situation or even safety or environmental issues.

The functional analysis of a process is visualized using the functional structure or a flow chart and shows the respective purpose of the particular process.

Step 3: failure analysis

Again in step 3 potential failures are identified for each of the process steps. The description of a failure, i.e. a malfunction, has to be complete and meaningful. A malfunction described by the word "bad process" cannot be elaborated regarding the root cause, effects or e.g. likelihood of occurrence (fig. 5.9) and is therefore not acceptable.

Potential malfunctions of a process are for instance

- deviations from the specified target of the process, e.g. layer thickness exceeding the maximum value
- incomplete execution of a process, e.g. residues after dry etch due to insufficient overetch
- unnecessary process

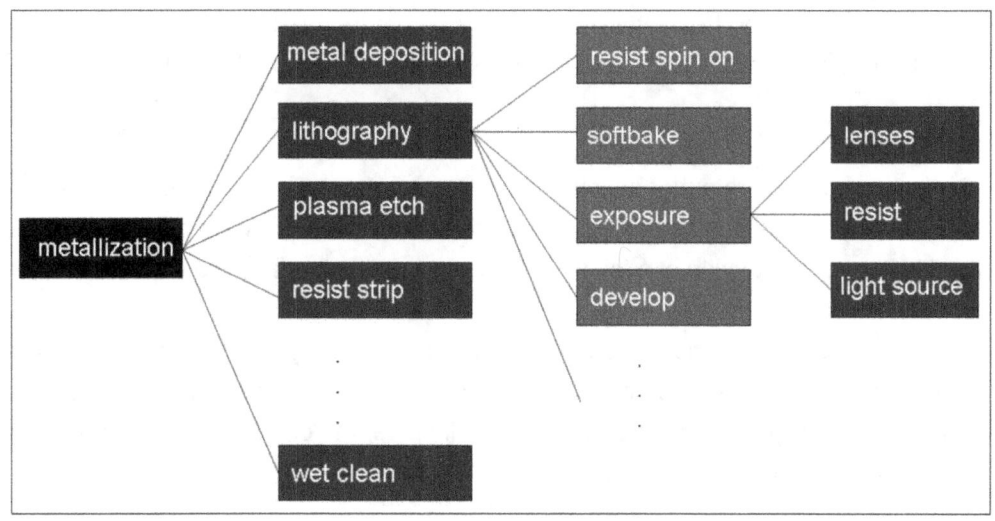

Fig. 5.7 Process structure of the semiconductor process: metal structuring

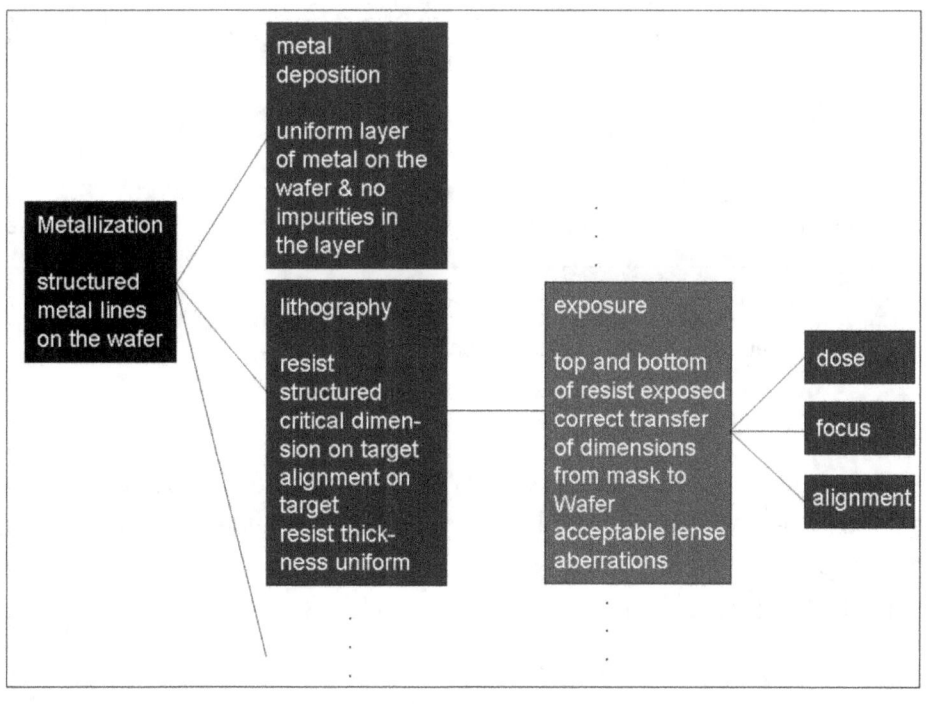

Fig. 5.8 Partial functional analysis of a chip making process: metallization

Step 4: assessment

Assuming a correct failure description, root cause and effects as well as occurrence, severity and likelihood of detection are now determined using the previously defined Risk Priority Numbers (RPN): RPN = O x S x D (see Product FMEA, Step 4).

Step 5: corrective actions

Corrective actions are either actions to avoid occurrence of a failure, i.e. change of the process flow or use of a more stable piece of equipment, or they are actions to increase the likelihood of detection of a failure. Those are e.g. implementation of additional visual control or adding a functional test.

In both cases those actions need to be described clearly and comprehensibly. The statement "test" or "experiment" is not sufficient. Also in both cases, responsible persons from the FMEA team as well as timelines for the completion of actions have to be defined.

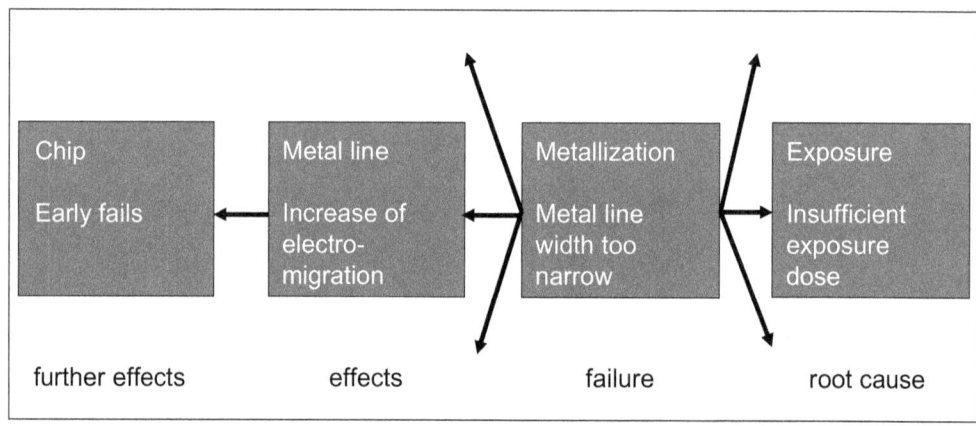

Fig. 5.9 Root cause – failure – effects – chain (process)

## 5.2.5. The FMEA Form

In addition to using professional Quality Management related software for actually performing the FMEA a simple form can be used to fulfil the task as well. Using the previous example (fig. 5.9) a few lines of an FMEA with arbitrary RPNs are filled in for illustration (fig. 5.10).

| # | Function | Potential failure | Effect of failure | S | Potential cause | O | Process control | D | RPN |
|---|---|---|---|---|---|---|---|---|---|
| 1 | | Metal line width too small | Increase of electromigration, eventually early fails | 8 | Insufficient exposure dose | 2 | | 9 | 144 |
| 2 | | Metal line width too small | Increase of electromigration, eventually early fails | 8 | Resist thickness to high, resist residues after exposure | 2 | | 9 | 144 |
| | | | | | | | | | |
| | | | | | | | | | |

Fig. 5.10 Example FMEA without current process control

Both failures (1 and 2) described show high risk priority number, which make corrective actions necessary.

The failure 1 risk priority number for instance can be reduced by avoiding the root cause ("prevention") by the use of a more powerful and stable optical system or it can be reduced by applying additional measurements of the resist line width prior to etching.

In the next version of the FMEA this additional measurement would show up as the current process check and reduce D to maybe 2. The improved optical system would decrease occurrence down to 1. So in both cases the risk priority number would be significantly reduced (fig. 5.11).

| # | Function | Potential failure | Effect of failure | S | Potential cause | O | Process control | D | RPN |
|---|---|---|---|---|---|---|---|---|---|
| 1 | | Metal line width too small | Increase of electromigration, eventually early fails | 8 | Insufficient exposure dose | 2 | measurement of resist line width afer exposure | 2 | 32 |
| 1 | | Metal line width too small | Increase of electromigration, eventually early fails | 8 | Insufficient exposure dose | 1 | | 9 | 72 |
| | | | | | | | | | |
| | | | | | | | | | |

Fig. 5.11 Example FMEA after implementing a preventive action or additional process check

## 5.3. Risk Priority Number

Occurrence, Severity and Detection are integer numbers ranging each from 1 through 10, "1" being the best case, "10" being the worst case regarding potential risks. From those three figures the so called Risk Priority Number (RPN) is generated by simple multiplication:

$$RPN = Occurrence \times Severity \times Detection$$

Even though the FMEA is not an exact mathematical method, it is useful for the company to have defined risk priority numbers. If this is not the case (see table 5.1), then the discussion about interpretation of RPN values will start during the FMEA. This reduces the efficiency of the team. An additional disadvantage is that a different team is more likely to come to a different assessment of the RPN.
The kind of product or process also influences the way RPNs are set up. The following chapters show various examples.[1,2,3,4,5,6]

| RPN | Occurrence | Severity | Detection |
|---|---|---|---|
| 1 | unlikely | no noticeable effect | definitely detectable |
| 2 | rare cases | minor effect | almost definitely detectable |
| 3 | rare cases, but similar cases occurred | minor effect | good chance of detection |
| 4 | occasionally | customer slightly irritated | good chance of detection |
| 5 | occasionally | customer moderately irritated | likely to be detected |
| 6 | cumulative | customer irritation | chance of detection |
| 7 | repeatedly | customer unsatisfied | maybe detectable |
| 8 | repeatedly | customer unsatisfied | unlikely to be detected |
| 9 | constantly repeated | customer very unsatisfied | highly unlikely to be detected |
| 10 | permanent | customer very unsatisfied | not detectable |

Table 5.1 Example for a general definition of risk priority numbers

---

[1] Kenol, J. (2010)
[2] Raytheon Company (2007
[3] Quality-One (2002)
[4] Dovich, R. ( )
[5] Gerber, K. (2009)
[6] Arndt et al. (1993)

RPNs always need to be designed based on company, customer or product specific requirements. The more detailed the number the more knowledge of the process or the product is necessary. Despite having a good knowledge of the FMEA item accurate risk priority numbers, in particular when starting with FMEAs in the company, are dependent on judgements. They should therefore be subject to regular review, e.g. by adjusting occurrence of a particular defect, after learning from field data.

### 5.3.1. Occurrence

Based on actual arbitrary examples the following table 5.2 shows six examples the occurrence risk priority numbers can be defined in different ways.

Example 1
This is a very qualitative description of the probability of a defect occurring. It does not even distinguish between RPN 4, 5 and 6. This kind of description is very general, i.e. during the execution of the FMEA the team members might have a different understanding of expressions like "unlikely" or "almost", leading to lengthy discussions. FMEA results of different teams are hard to compare.

| RPN | example 1 | example 2 | example 3 | example 4 | example 5 | example 6 |
|---|---|---|---|---|---|---|
| 1 | failure is unlikely | 1 failure in 5 years | MTBF > 10000h | < 1 / 15000000 | < 10 ppm | cpk ~ 2,0 |
| 2 | relatively few failures | 1 failure in 3 to 5 years | MTBF 6001 - 10000 h | 1 / 150000 | 100 ppm | cpk ~ 2,0 |
| 3 | relatively few failures | 1 failure in 1 to 3 years | MTBF 3001 - 6000 h | 1 / 15000 | 500 ppm | cpk ~ 1,67 |
| 4 | occasional failures | 1 failure per year | MTBF 2001 - 3000 h | 1 / 2000 | 1000 ppm | cpk ~ 1,33 |
| 5 | occasional failures | 1 failure in 6 months | MTBF 1000 - 2000 h | 1 / 400 | 2000 ppm | cpk ~ 1,17 |
| 6 | occasional failures | 1 failure in three months | MTBF 401 - 1000 h | 1 / 80 | 5000 ppm | cpk ~ 1,0 |
| 7 | repeated failures | 1 failure per month | MTBF 101 - 400 h | 1 / 20 | 10000 ppm (1 %) | cpk ~ 0,83 |
| 8 | repeated failures | 1 failure per week | MTBF 11 - 100 h | 1 / 8 | 2% | cpk ~ 0,67 |
| 9 | failure is almost inevitable | 1 failure in three to four days | MTBF 2 to 10 h | 1 / 3 | 5% | cpk ~ 0,33 |
| 10 | failure is almost inevitable | more than on failure per day | MTBF < 1 h | > 1 / 2 | > 10% | cpk < 0,33 |

Table 5.2 Various examples for "occurrence"

Example 2
In this case occurrence is described using actual probabilities. The scale is based on daily experience. This kind of scale is often used to judge likelihood of a defect in one part, e.g. a manufacturing equipment.

Example 3
Similar to example 2 occurrence is described using actual probabilities. In this case it is based on MTBF (mean time between fails). This kind of scale is often used to judge likelihood of malfunctioning of e.g. a manufacturing equipment.

Examples 4 – 5
This kind of description can be used if a large number of production parts is considered. These numbers are chosen depending on the sensitivity of the product, of production or of the customer.

Example 6
For the $c_{pk}$ based scale the most detailed information is required. A good understanding of the process and its stability is needed. Eventually the cpk value also defines a number of parts which are not within the specification. It makes sense to add the expected defect numbers in order to provide the FMEA team members a comprehensible figure.

## 5.3.2. Severity

The following table 5.3 shows six examples of severity definitions. Again there are very qualitative descriptions as well as number based descriptions. Numbers can be based on yield data or loss of money etc.

Example 1 – 3:
Severity is often described qualitatively since many effects like "customer annoyance" is basically impossible to provide a number for. This makes it necessary to have somebody in the team who knows the potential customer reaction very well. The effects of the failure on the production are neglected in this case.

Example 4
This example describes potential impact of a failure on up- or downtime of e.g. production equipment.

Example 5 – 6:
The number given in example five, for instance, are targeting the effects on production as well as on final yield, whereas in example six only yield numbers are found. The impact on the customer or on the reliability of the product is excluded in both cases.

## 5.3.3. Detection

"Detection" describes the probability a defect is found prior reaching the customer. This section again shows different approaches to define the respective risk priority number.

Examples 1 – 3
Examples 1 to 3 provide a qualitative description of "detection". The actual escape rate is impossible to estimate from this description.

Example 4
This is a qualitative description of detection, too. However, it is related to the way the company performs process checks. It starts at no check and ends at 100% automatic check. If there is no information about the actual escape rate, this kind of description provides at least a somewhat fact based risk assessment.

Example 5 – 6
In these two cases detection is described by the escape rate using concrete numbers. The actual value of these numbers depends e.g. on the product or the customer.

## 5.3.4. Action Limits

The FMEA method asks for corrective actions in case of unacceptable high risk. There are basically two ways to define "unacceptable high risk":

(1) Action limit
I.e. starting at the action limit, e.g. typically RPN = 100, a corrective action needs to be performed in order to reduce occurrence of the defect or in order to increase detection. After the execution of the corrective action a reassessment has to take place. In case the new RPN is still at or above the action limit, another corrective action has to be implemented.

(2) Pareto
In this case starting from the highest level the risks are arranged with decreasing RPN. Then the action limit is defined as a percentage of risks for which a corrective action needs to be done, e.g. the initial 20% high risk topics.
With this method there is the problem that either high risk topics are not considered because there are more than 20% high risk topics, or, in the converse case, corrective actions are executed for low risk topics. In particular the first of these two kinds of risk is not acceptable for the customer.

| RPN | example 1 | example 2 | example 3 | example 4 | example 5 | example 6 |
|---|---|---|---|---|---|---|
| 1 | not noticeable to the customer; would not affect the customer's process or product | no effect | neutral | parameters within upper/lower control limits; no adjustment required | rework of some production parts | yield loss 0% |
| 2 | not be readily apparent to the customer, but would have minor effects on the customer's process or product | defect noticed by discriminating customer | neutral | parameters within upper/lower control limits; adjustment required | rework of several production parts | yield loss 0% |
| 3 | would create a minor nuisance to the customer, but the customer can overcome it in the process or product without performance loss | defect noticed by average customer | moderate customer annoyance | parameters exceed upper or lower control limit; adjustments required | production equipment down | yield loss 0% |
| 4 | can be overcome with modifications to the customer's process or product, but there is minor performance loss | defect noticed by most customers | slight customer annoyance | equipment downtime < 30 min. | rework of some lots | yield loss < 0,5% |
| 5 | creates enough of a performance loss to cause the customer to complain | customer experiences some dissatisfaction | moderate customer annoyance | equipment downtime 30 - 60 min. | production equipment down, bottleneck situation | yield loss < 1% |
| 6 | results in a subsystem or partial malfunction of the product | customer experiences discomfort | customer annoyance | equipment downtime 1 to 3 h | loss of a lot | yield loss < 2% |
| 7 | causes a high degree of customer dissatisfaction | customer dissatisfied | customer dissatisfied | equipment downtime 4 to 7 h | significant yield loss several lots | yield loss < 6% |
| 8 | would render the unit inoperable or unfit for use | loss of primary function | customer dissatisfied | equipment downtime > 8 h | loss of several lots | yield loss < 10% |
| 9 | would create noncompliance with the federal government | failure mode affects safe operation / noncompliance with government regulation with warning | customer highly dissatisfied | safety issue or non-compliance with govenment regulations | significant yield loss | yield loss > 20% |
| 10 | could injure the customer or an employee | failure mode affects safe operation / noncompliance with government regulation without warning | customer highly dissatisfied | safety issue or non-compliance with govenment regulations | massive yield loss | yield loss > 20% |

Table 5.3 Various examples for "severity"

| RPN | example 1 | example 2 | example 3 | example 4 | example 5 | example 6 |
|---|---|---|---|---|---|---|
| 1 | will almost certainly be detected | almost certain detection | Very high: the detection of the existance of a defect is almost certain | The defect is obvious or there is 100% automatic inspection with regular calibration and preventive maintenance | escape rate < 10 ppm | escape rate < 4 ppm |
| 2 | very high chance to detect the failure | almost certain detection | Very high: the detection of the existance of a defect is almost certain | All product is 100% automatically inspected | escape rate 100 ppm | escape rate 4 - 100 ppm |
| 3 | high chance to detect the failure | moderately high chance of detection | High: There are controls to detect defects, but there is a small chance of defects not being detected. | An effective SPC program is in place with process capability ($C_{pk}$) greater than 1.33 | escape rate 500 ppm | escape rate 100 - 500 ppm |
| 4 | moderately high chance to detect the failure | moderately high chance of detection | High: There are controls to detect defects, but there is a small chance of defects not being detected. | High: SPC is used and there is immediate reaction to out-of-control conditions | escape rate 0,1% | escape rate 500 - 2000 ppm |
| 5 | moderate chance to detect the failure | moderate chance of detection | High: There are controls to detect defects, but there is a small chance of defects not being detected. | Some Statistical Process Control (SPC) is used in process, and product is final inspected off-line | escape rate 0,2% | escape rate 0,2 - 0,5 % |
| 6 | low chance to detect the failure | low chance of detection | Low: There is only a small chance of detecting an existing defect. | Product is 100% manually inspected using go-no-go or other mistake-proofing gauges | escape rate 0,5% | escape rate 0,5 - 1 % |
| 7 | very low chance to detect the failure | very low chance of detection | Low: There is only a small chance of detecting an existing defect. | Product is 100% manually inspected in the process | escape rate 1% | escape rate 1 - 2 % |
| 8 | remote chance to detect the failure | remote chance of detection | Low: There is only a small chance of detecting an existing defect. | Product is accepted based on no defectives in a sample | escape rate 2% | escape rate 2 - 4 % |
| 9 | very remote chance to detect the failure | very remote chance of detection | Very low: Controls in place will not gernerally detect the existance of a defect. | Product is sampled, inspected, and released based on Acceptable Quality Level (AQL) sampling | escape rate 5% | escape rate 4 - 6,5 % |
| 10 | will not detect the failure | cannot detect | None: a defect will almost certainly escape detection | The product is not inspected or the defect caused by failure is not detectable | escape rate > 10 % | escape rate > 6,5 % |

Table 5.4 Various examples for "detection"

## 5.4 Weaknesses of the Method[1,2]

Even though the FMEA method is one of the most successful methods of quality assurance in many industrial areas there are some weaknesses which must be considered.

### 5.4.1 Absolute Value of Risk Priority Numbers

The qualitative descriptions of risk priority numbers leave the respective FMEA team a relatively high degree of freedom how to interpret "almost" or "rare cases" etc. Typically, a team will eventually find a way how to handle those RPNs, but the understanding will differ from team to team. This means, that the risk priority numbers of different teams can be hard to compare. Quantitative definitions of S, D and O reduce this shortcoming. On the other hand, it is very difficult to estimate the exact occurrence or detection, too. Estimating an occurrence rate of 200 ppm or 500 ppm is almost impossible, in particular when there is not a lot of experience in the company with the method as well as with the FMEA item.

However, if the FMEA is planned to be used as a standard tool in the company and in a way that the FMEA is seen as a know-how database it makes sense to use the numerical values instead of qualitative descriptions. The numerical values can be subject to regular review and will eventually represent reality-near numbers.

### 5.4.2 Scaling of Risk Priority Numbers

Risk priority numbers do not represent actual and authoritative probabilities of failure occurrence or of failure escape rates to the customer. They do however introduce a meaningful risk ranking scale which helps with decisions about risk management. The same risk priority number may result in different actual overall losses of money or losses of yield. For instance, if we take example 2 of table 5.2 and example 4 of table 5.3 for occurrence and severity, respectively, the same RPN = 24 (= O x S) will result in 1 to more than 8 hours equipment downtime per year (fig. 5.12).

Another example related to scaling of RPNs is the actual overall failure escape rate to the customer O x D, with D expressed as escape rate at the control step (example 5, Table 5.4) based on an assumed occurrence (example 5, Table 5.2).

---

[1] Kmenta, S. (2000)
[2] Kmenta, S. (2002)

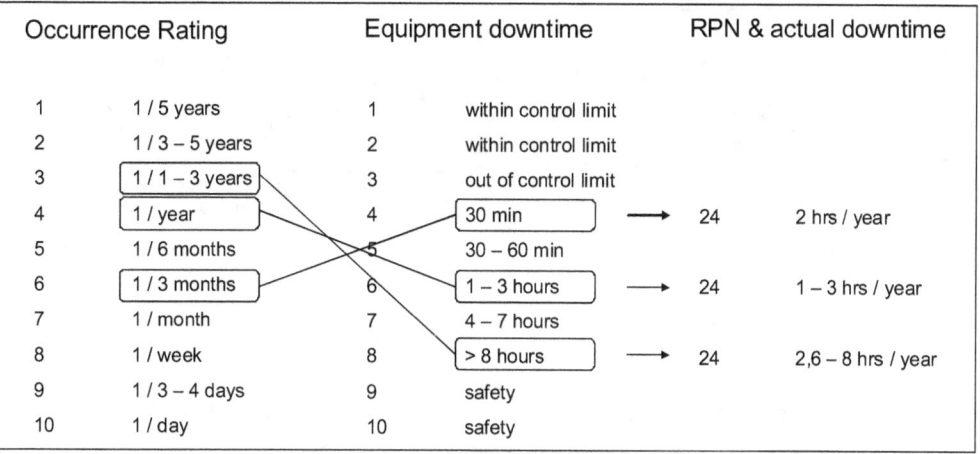

Fig. 5.12 Differing effects for the same RPN

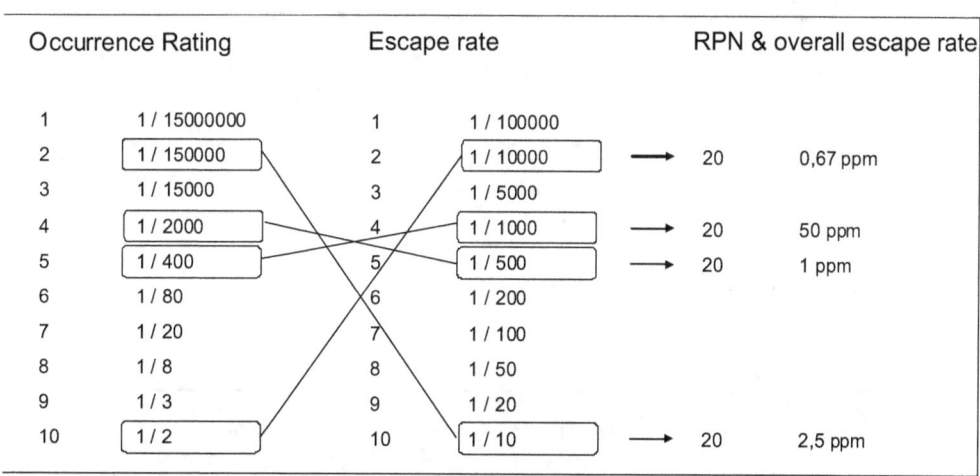

Fig. 5.13 Differing escape rates to the customer for same RPN values

## 5.5. Business Process

Based on various sources this chapter provides a generic process description for the FMEA process.[1, 2, 3]

---

[1] Catic, D. et al. (2011)
[2] Gerber, K. (2009)
[3] Villacourt, M. (1992)

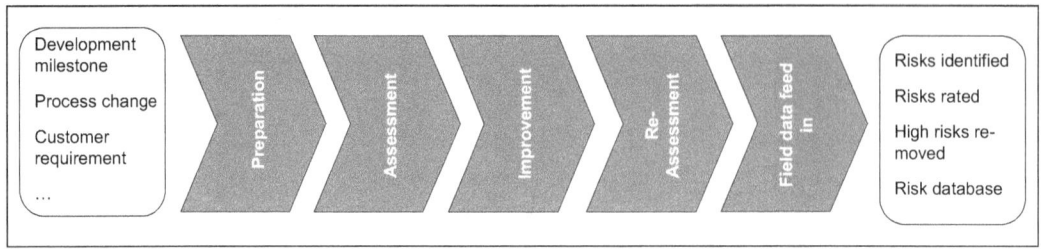

Fig. 5.14 Overall FMEA process

In the following pages for each phase of the FMEA process a generic subprocess is shown as basis for a company specific process.

| Process board: FMEA | |
|---|---|
| Process Owner | Head of Quality Management |
| Resources | FMEA Team, Moderator, Sponsor<br>FMEA template, moderation tools, process description meeting room, tools |
| Frequency of process evaluation | 1 - 2 years |
| Performance Indicators | (1) number of failures escaping to the customer (internal / external) regarding the FMEA item<br>(2) number of remaining risks above action limit<br>(3) risk reduction after improvement phase [%] |

Table 5.5 Process owner and team members and monitoring approach (process board FMEA)

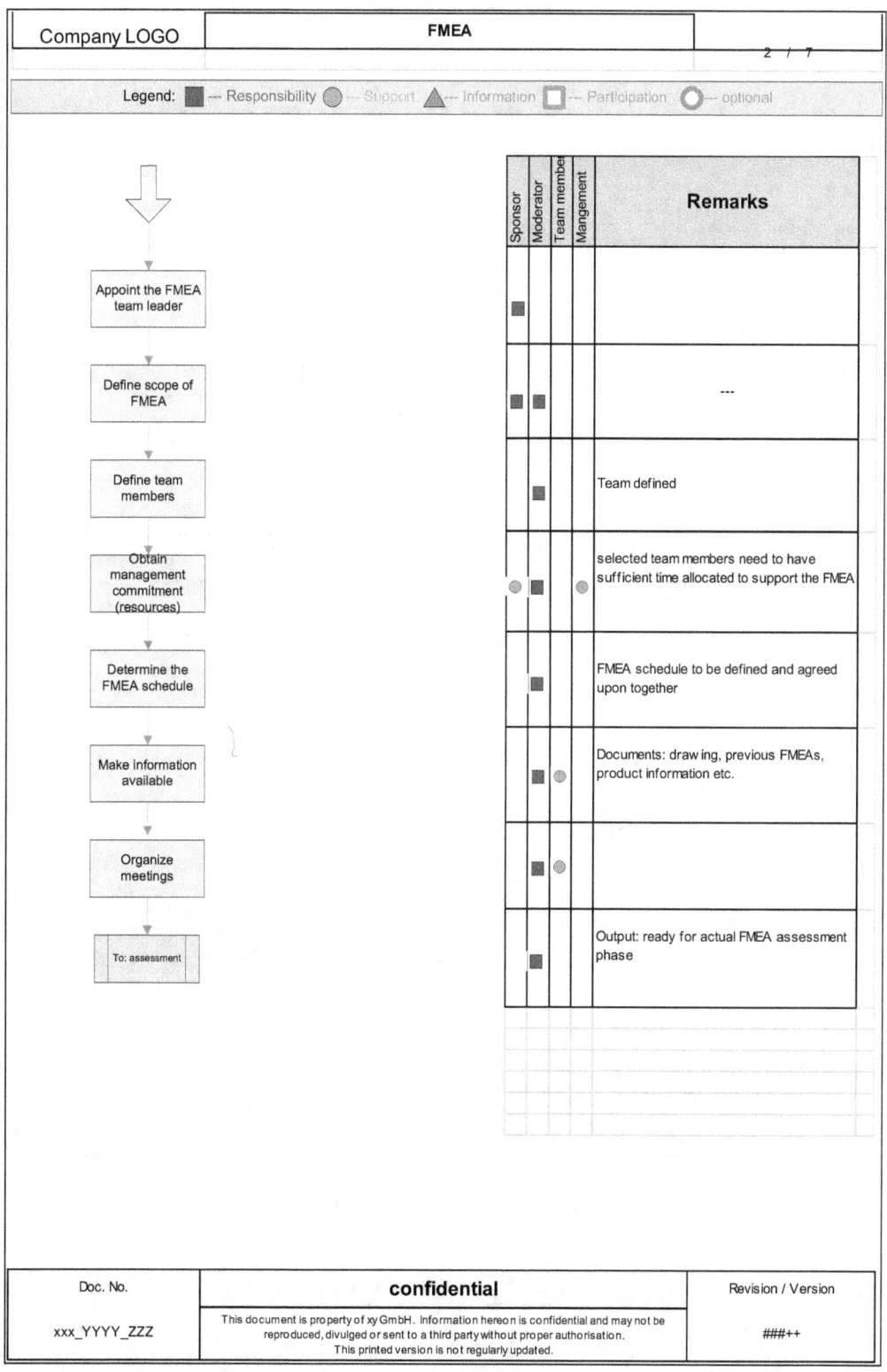

Fig. 5.15 Generic process "FMEA preparation"

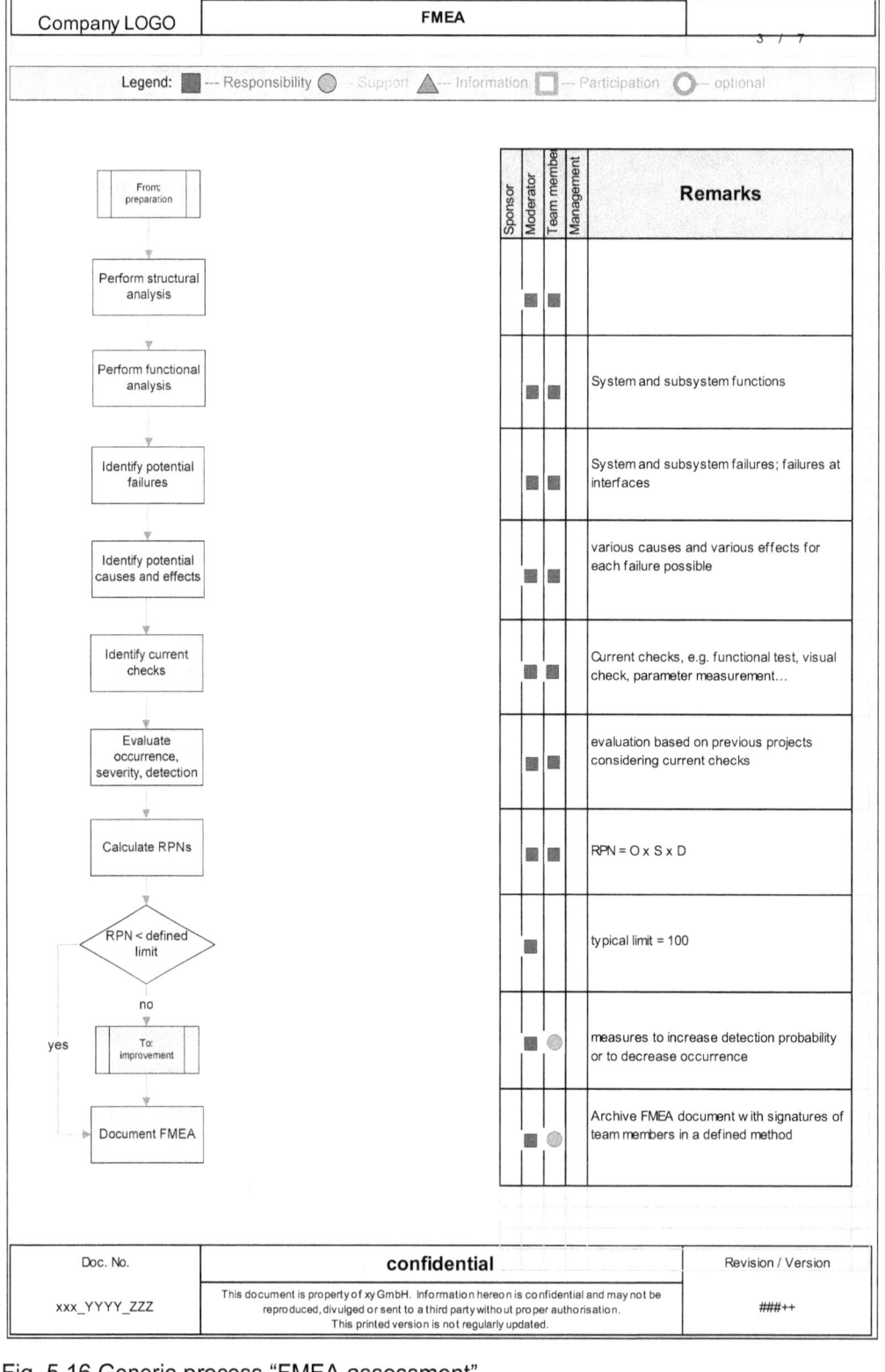

Fig. 5.16 Generic process "FMEA assessment"

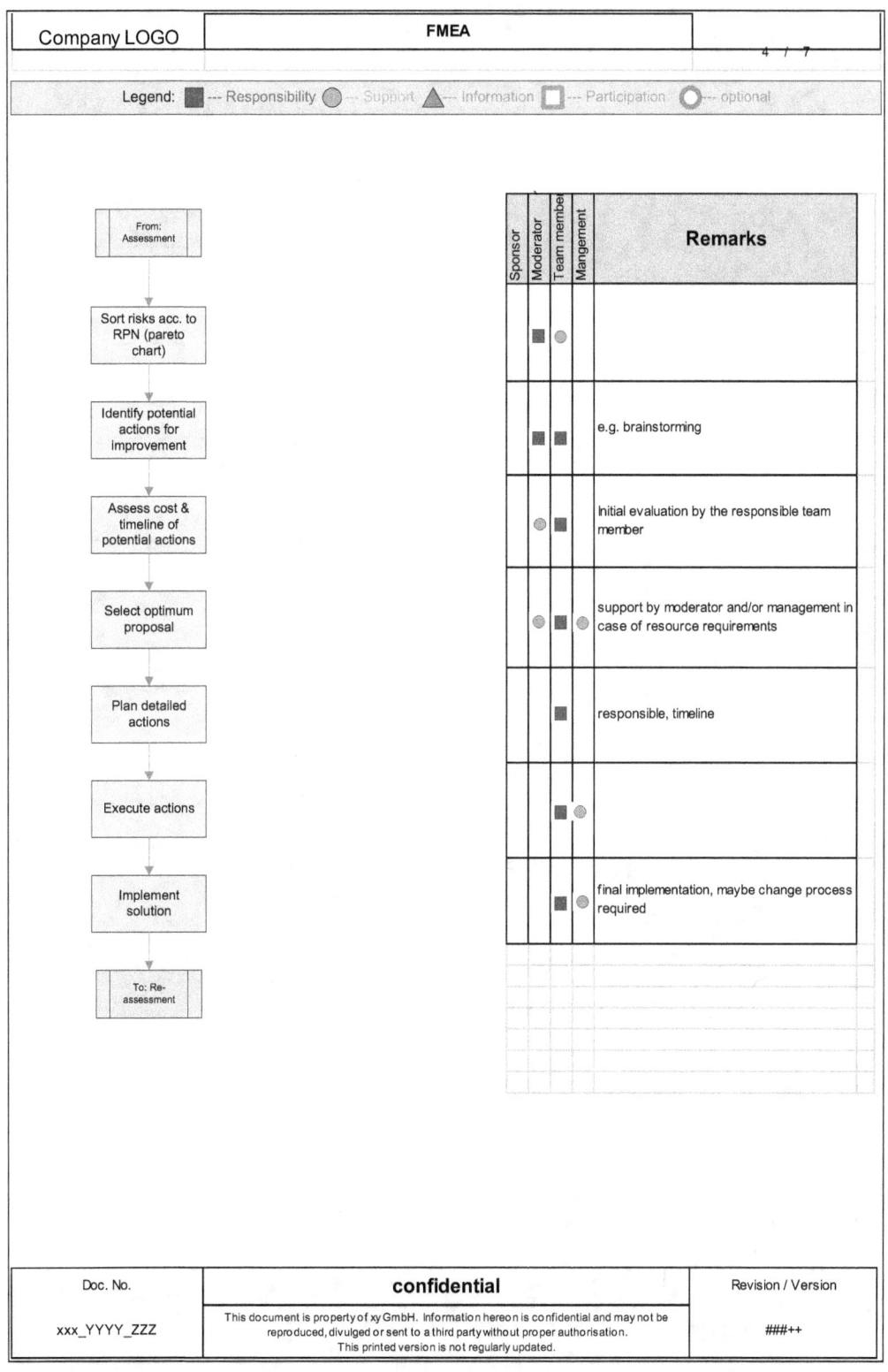

Fig. 5.17 Generic process "FMEA improvement"

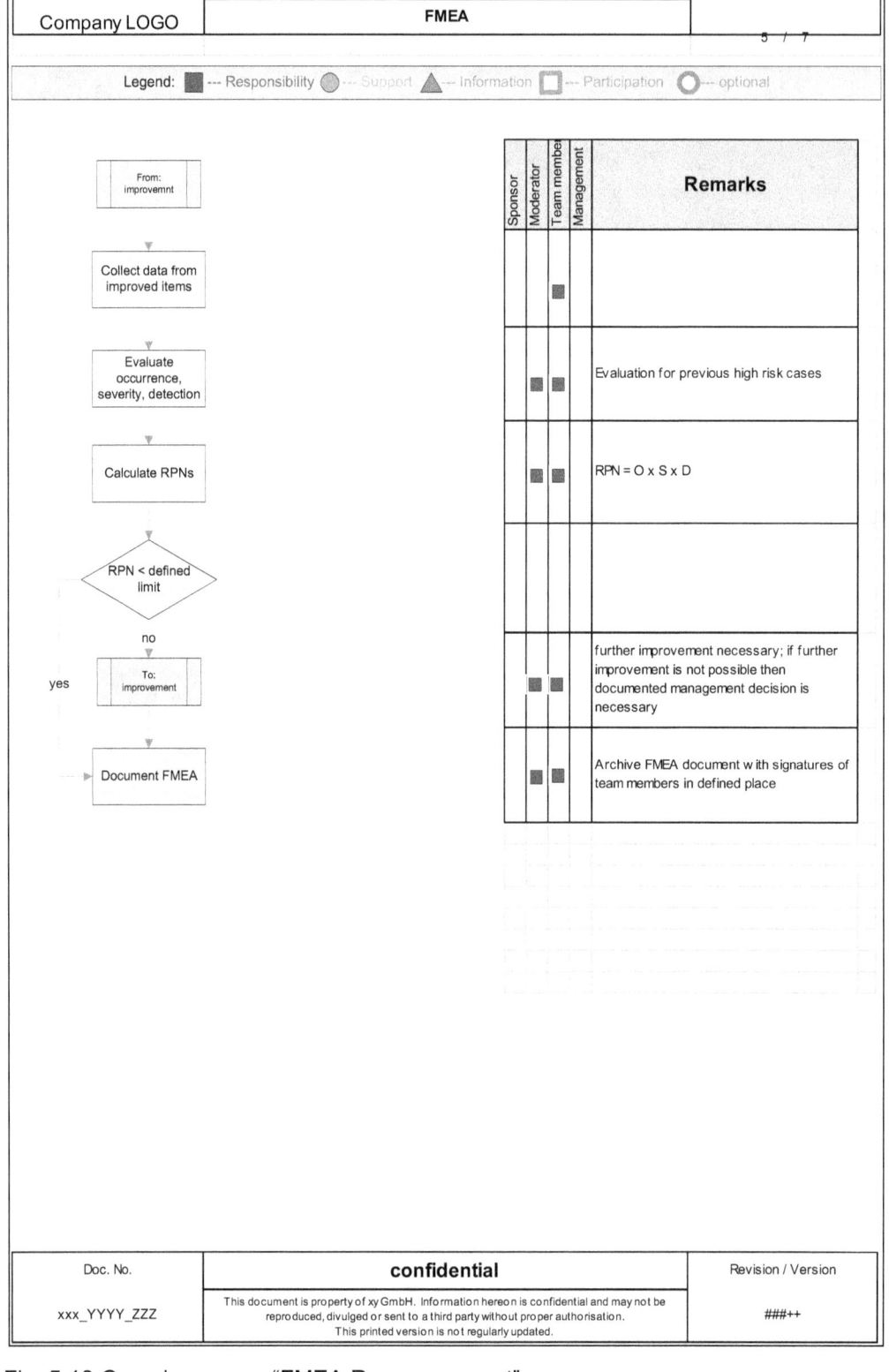

Fig. 5.18 Generic process "FMEA Re-assessment"

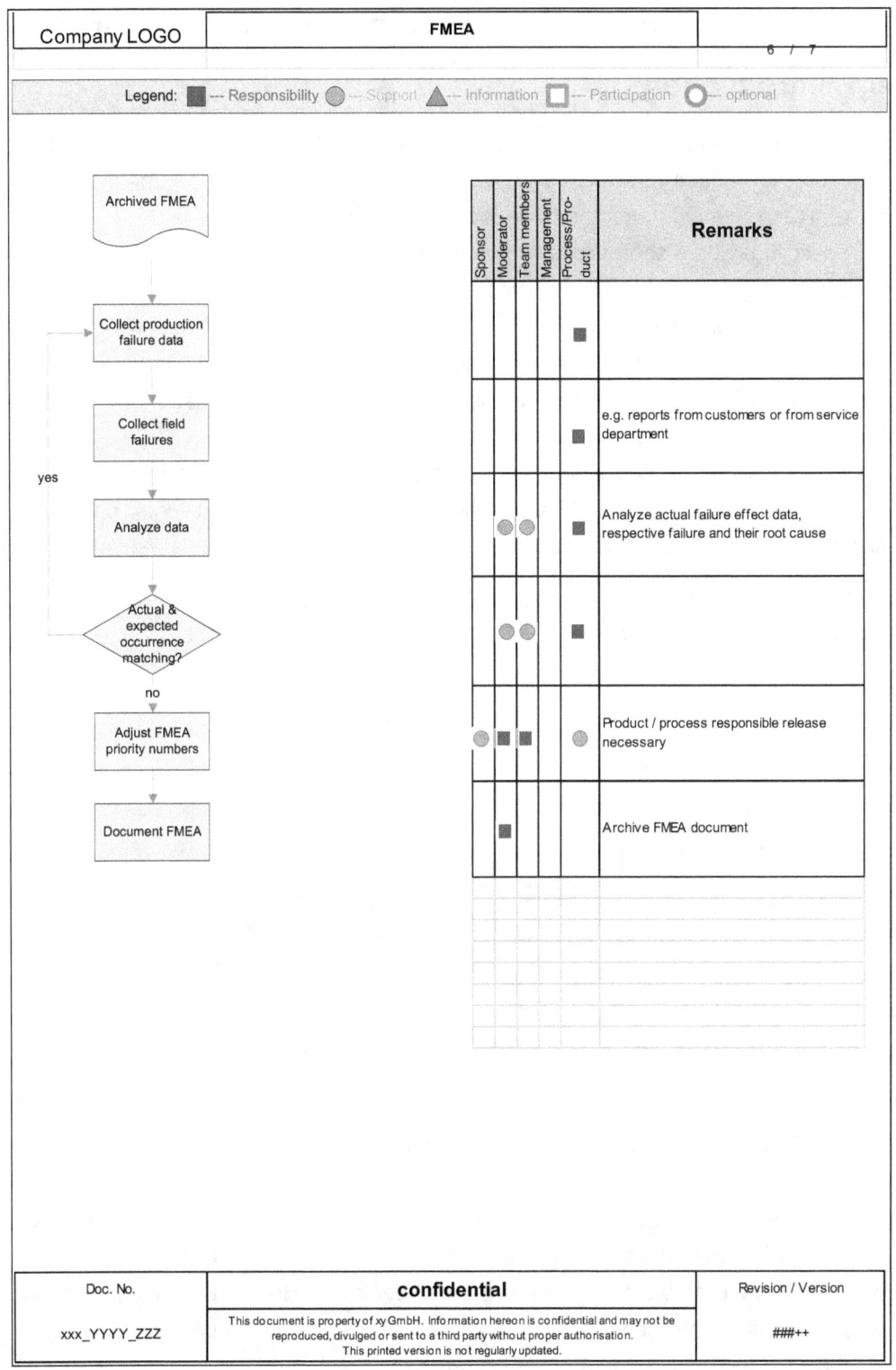

Fig. 5.19 Generic process "FMEA field data input"

# 6. Process and Product Development & Qualification

## 6.1. Introduction

Product development is a major cost factor in the manufacturing industry without a direct and short term cash return. Due to this fact the development departments frequently have to face cost pressure and budget cuts.

This policy may be a short term solution to reduce cost, but is detrimental to the overall company long term competitive situation. It is well known, that potential product failures found in the early phase of the product life cycle using the respective prevention methods can be avoided with relatively low effort, whereas failures detected very late, e.g. at the customer, generate huge cost for repair or recall activities (fig. 6.1, also see the rules of 10 mentioned earlier).

Thus, cutting development resources which very often result in reducing resources for error prevention, such as FMEA and other quality management tools, can result in significantly higher cost and loss of customer trust.

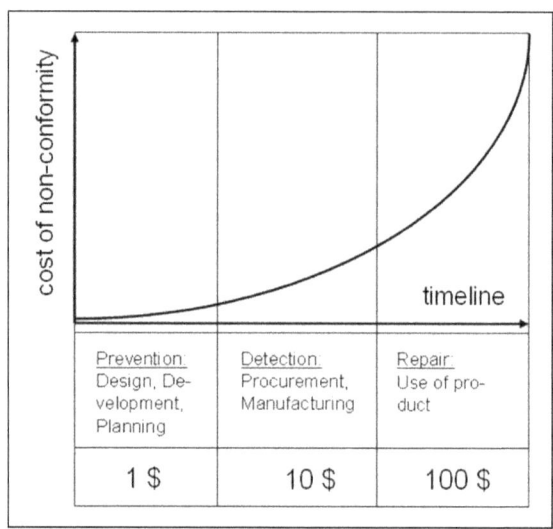

Fig. 6.1 Relative cost of product non-conformity[1]

In particular when it comes to recall actions in a high volume industry like car manufacturing, the number of affected products can be extremely large, i.e. millions of pieces2, and the problem might be even a safety issue. Hence it is not only a financial burden, but could also lead to injuries or death of the user of the product and may even put the existence of the company at jeopardy.

---

[1] Sondermann, J.P. (2007), p. 390
[2] Harloff, T., Salavati, N. (2014)

A similar line of argumentation is found when looking at when product failures emerge in the life cycle of a product and when product failures are eliminated. Fig. 6.2 shows a typical situation where the majority of failure causes arise in the development phase and the majority of correction activities take place after volume production release.

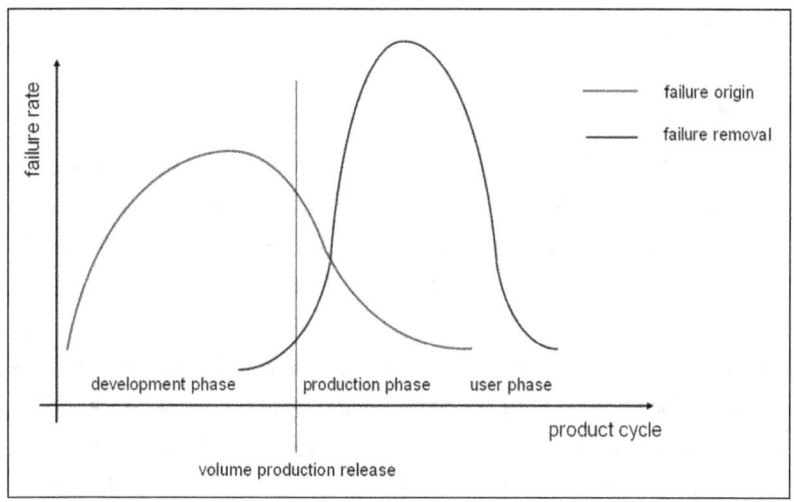

Fig. 6.2 Origin and removal of failures in the product life time[1]

Again, the use of preventive methods during development has the potential to reduce the number of failures in a product significantly, which eventually leads to a substantial cost reduction.
The consequence is that the development process should always be a central part of quality management activities in order to avoid product failures from the very beginning.

## 6.2. Requirements of the relevant standards

The certification of the quality management system according to DIN ISO 9001 is basically mandatory for every major producer of technical parts for the industry. In case the manufacturer also sells his products to a company in the automotive industry segment the certification according ISO/TS 16949 is mandatory. In this section the requirements of the standards regarding product development as to be implemented in a QM system are summarized.

[1] Prefi, T. (2007)

Chapter 8.3 of E DIN EN ISO 9001:2014-08 describes requirements for „design and development of products and services"[1]. In general, it states that an organization has to define and implement a development process for a physical product in case this particular product has not already been established. Additional requirements of ISO/TS 16949:2002[2] are shown separately.

Besides this general requirement various sub-chapters provide a more detailed view on development activities.

## 6.2.1. Planning

ISO9001: Planning of development activities shall include various phases. Typically, a particular development phase ends with the fulfillment of defined requirements. Often this is based on verification and validation of results, which are evaluated in a development review in order to achieve a predefined milestone (e.g. a "milestone review meeting").

The development process needs to be adequate for the product as well as for the company and needs to include the definition of responsibilities. In the product development it is more useful to apply a simple structured development process rather than to elaborate a comprehensive process with all more or less likely features, which in the end becomes too complex to be managed properly.

Planning of development activities shall also include an evaluation if customers have to be involved and it shall include the way of documentation of development results.

ISO/TS16949: The development process shall use a cross functional approach including the definition of special characteristics (also called "key control characteristics") in order to focus on the most important areas. Failure modes and effects analysis (FMEA) and control plans are mandatory.

## 6.2.2. Inputs

ISO9001: Inputs for the development process shall be adequate and include requirements regarding or originating from e.g.

- ➢ Functionality
- ➢ Performance
- ➢ Consortium or international Standards
- ➢ Resources
- ➢ Potential failure effects

---

[1] E DIN EN ISO 9001:2014-08
[2] ISO/TS 16949:2009

- Customer expectation regarding development process control

ISO/TS16949: Inputs for the development process shall include

- customer requirements, e.g. traceability, transport…
- competitor product analysis results
- results from previous development projects
- information from suppliers or from the field
- and quality targets, e.g. life time, reliability, maintenance
- production targets, e.g. process capability targets, cost targets

## 6.2.3. Development Control Process

ISO9001: The purpose of development control process is to ensure that the following objectives are met. The outcome of the development process has to be specified precisely and accurately and needs to be reviewed as defined in the development planning phase. Verification and validation are used to show that the output of the development process matches the predefined input requirements and that the product is capable for the intended use.

ISO/TS16949. In addition to the checks regarding the product this international standard also focuses on the development project itself and asks for metrics for the critical path and cost of the project. The objective here is to keep the scheduled targets and timelines.

In addition, the customer may request prototypes for validation and the process- and product release process has to be approved by the customer. This also applies to the development process carried out at the suppliers.

## 6.2.4. Outputs

ISO9001: In addition to the fact, that input requirements have to be satisfied, the development outputs, which also include acceptance criteria and requirements for product checks, have to be suitable to ensure a functional product. Eventually the development output shall ensure that the product is safe and suitable for the intended use.

ISO/TS16949: Development outputs shall also include e.g.

- Design- and process FMEA
- Information about reliability, maintenance measures, quality

- Special Characteristics, i.e. characteristics of the product or of the manufacturing process, which have an impact on safety, compliance to standards and regulatory requirements, performance etc.
- Product definition
- Information about the production process down to process flows and standard operating procedures and acceptance criteria for the process

### 6.2.5. Change Management during Development

<u>ISO9001</u>: If during or after the development phase changes to the development output are identified, those changes have to be reviewed and controlled. The organization has to ensure conformity to requirements. Activities and results have to be documented.

**<u>ISO/TS16949</u>: No additional requirements.**

Additional ISO/TS16949 requirements for the semiconductor industry are listed in the document "Customer Specific Requirements (ISO/TS-16949) Semiconductor Commodity"[1]. In addition to extra requirements like customer approval for qualification tests, technology & product roadmaps and quality roadmaps, due to the nature of the semiconductor manufacturing process, it allows for a family approach when it comes to Design FMEAs. In this context a family is a group of products with many similar or even identical parts with each having the same or similar behavior.

## 6.3. Generic product development process

### 6.3.1. Market and customer requirements

The purpose of product development is to establish a new or a modified product. Product development is defined as:

> "The creation of products with new or different characteristics that offer new or additional benefits to the customer."

Nowadays the product development process has to satisfy several to some extent conflicting requirements:
First of all, the outcome of the product development process, the new or modified product has to be

---
[1] ISO/TS16949 Semiconductor Commodity (2004)
[2] BusinessDictionary (2015)

1. functional, i.e. the product performs as defined and all modes of operations are working
2. reliable, i.e. the product remains functional over a defined period of time
3. manufacturable, i.e. the product can be produced in high volume with stable and high yielding processes at reasonable cost

These product requirements may be driven by the customer or originate from competitor product analysis or market and cost analysis. In addition to these product related requirements, the product development process also has to cope with other challenges:

1. increasing product variety[1,2] leading to an increase in product development and qualification efforts
2. decreasing product development cycle times[3] resulting in "first time right" expectations by management or customers and time pressure
3. high cost pressure
4. an expected very steep volume production ramp up
5. and last but not least the obligation to meet the respective ISO9001 and ISO/TS16949 requirements

The generic development process described below addresses all those requirements.

## 6.3.2. Approach

First of all, the semiconductor development process has two major branches: technology development and product development. The objective of the technology development process is to generate new technology features, e.g. smaller devices and metal lines, new technology integration schemes or use of new materials etc. as compared to the previous technology. The initial steps of technology development are executed using test structures instead of actual products.

As soon as the technology development process demonstrated basic functionality on those test structures a pilot product is used to qualify the technology and the pilot product. After completion of a successful technology qualification additional product development projects may be started on the qualified technology platform.

Outputs of the technology process are e.g.

➢ process of records (POR)

---

[1] ElMaraghy, H., et. al. (2013)
[2] Hu SJ, et. al., (2011)
[3] Chesbrough, H.W. (2007)

- control and monitoring plan
- process FMEA
- operational working SPC with defined control limits

Hence product development is based on an existing technology targeting e.g. for new applications, different chip size (memory) etc. Frequently new products are using an existing technology platform, but require some additional technology features like special layers for sensors or special devices for power applications.

| Technology | product | complexity / maturity |
|---|---|---|
| New | pilot | ++ / -- |
| still new | follower | + / - |
| new features | pilot | + / - |
| new features | follower | + / o |
| Known | follower | o / o |

Table 6.1 Qualitative description of the technology/product complexity (+ : high, o : medium, - : low)

In these cases, a partial technology development / qualification in combination with a new product becomes necessary. Eventually various combinations of technology and product innovations are conceivable with differing degrees of complexity and maturity.

Another important aspect to keep in mind is that the semiconductor device / component typically is part of a major system with various levels of system integration up to the final product.

Each level undergoes a development process starting with system relevant requirements, documented in the respective requirements specification, and ending in the verification and/or validation of the results of the development process. Figure 6.3 shows the various levels as described for automotive electrical / electronic modules.

In order to meet the above mentioned requirements companies define a product development process in due consideration of product and market specific requirements. The actual execution of product development, however, is typically performed applying a project approach. The product development process together with the respective development project execution plan has to satisfy the ISO 9001 or other relevant standard requirements regarding development planning and execution. Very general the development process is part of the product life cycle[1]:

---

[1] Liu, B. (2003)

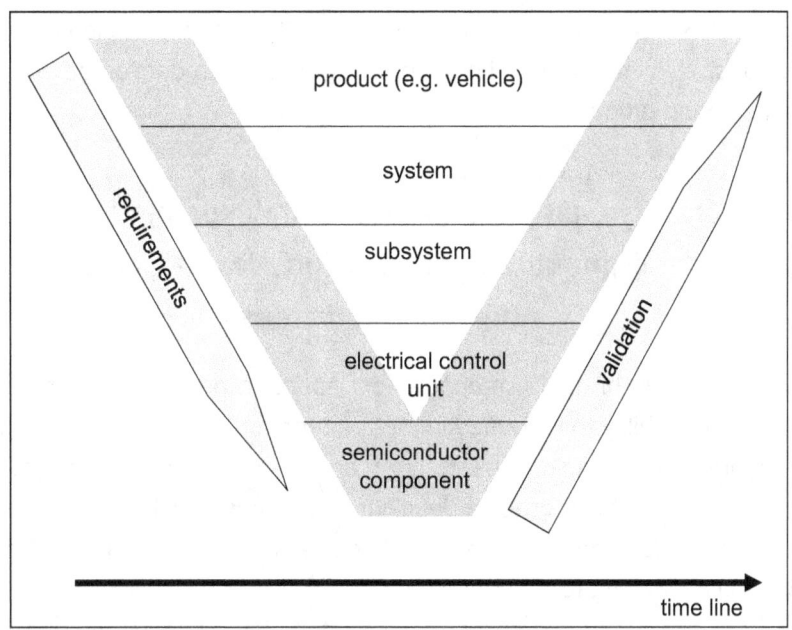

Fig 6.3 V-model with various levels of requirements[1]

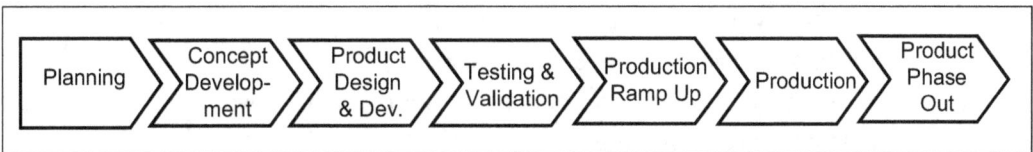

Fig. 6.4 Generic product life cycle

This chapter gives an example for a generic product development process. It is derived from the semiconductor industry, but is basically also applicable to other technology driven products.

### 6.3.3. Phases and milestones[2]

A development process is segmented into a number of phases (fig. 6.5). Each of these phases aims to provide deliverables and ends at a quality gate, i.e. a milestone, as soon as the gate requirements are fulfilled.

The actual phase description specifies the activities within a phase in order to achieve the deliverables. Eventually the fulfillment of gate requirements is reviewed using predefined checklists and the milestone achievement is approved by defined functions.

---

[1] ZVEI (2007)
[2] Auner, M. (2008)

An example for an approver list is given in table 6.2. Since production is the receiver of a new product, it is important to implement the head of production as approver for the qualification of the product.

> Quality Gates are objective checkpoints for the fulfillment of defined criteria for particular development steps.

In addition, the head of quality management needs to be an approver for this phase. This way the probability to introduce immature products to production is reduced.
Prior to starting the actual development activities, i.e. circuit design, technology development, layout etc., a product definition phase has to be executed.
The product definition phase results in a milestone which can be called M0 and contains several activities leading to

- Product idea
- Competitor, market & application analysis
- Initial feasibility studies & timelines
- Make-or-buy decision & production strategy
- Preliminary business case & business plan
- Customer specific requirements, in some cases even customer specific milestones, quality requirements
- Qualification strategy & plan

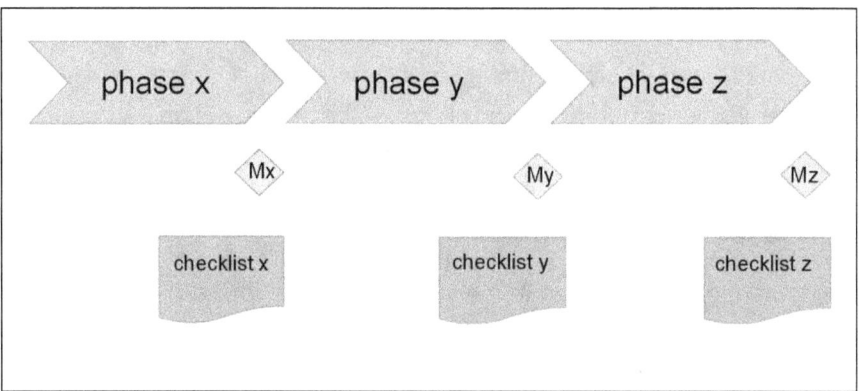

Fig. 6.5 Generic development process

Based on those results a product development project may or may not be initiated, i.e. the above mentioned items are content of the M0 checklist.

In addition to the mandatory project milestones additional optional technical milestones might be part of the process, e.g. engineering sample release for early product investigation, customer sample release in order to get early technical feedback from selected customers and various levels of volume release depending on customer strategy or customer quality requirements.

The development process shown in fig. 6.6 is applicable to the first product using a new technology node. For follower products or transfer of products to other manufacturing sites phases like product definition, design and layout are not necessary or are reduced, e.g. the qualification phase, due to reduced requirements (see "Product Qualification"). Along with the project execution another way to reduce the risk for the customer or production to receive a high failure rate product is to make a Failure Mode and Effects Analysis (FMEA) mandatory for particular milestone achievements. I.e. an FMEA has to be performed e.g. in the design phase or prior start of qualification hardware (HW). All risks for the product or the project should be identified. High risk topics need to be addressed and must not have an impact on the project. When producing for the automotive industry, FMEAs are mandatory according to ISO/TS 16949 anyway.

| Function \ Milestone | M0 project release | M1 product specification | M2 circuit design | M3 layout | M4 engineering HW | M5 qualifiCation HW | M6 qualification |
|---|---|---|---|---|---|---|---|
| General Mgr. or BU Mgr. | X | | | | | | |
| Financial head of unit | X | | | | | | |
| Project Program Management Head | X | | | X | | X | X |
| Marketing head | X | | | | | | X |
| Quality head | X | | | | | X | X |
| Project Manager | X | X | X | X | X | X | X |
| Production head | | | | | | | X |

Table 6.2 Example for an approver list for product development milestones

### 6.3.4. Phases and deliverables

In this section potential deliverables of various phases are described, which might be an input for a milestone achievement checklist. Milestone M0 (project release) is the starting point of the development process; it's deliverables were described in 6.3.3.

M1 product specification

In this phase product requirements are determined and frozen. Changes of product requirements after this point have to run through a defined change process. The main requirements and deliverables are:

- Documented specification of product requirements available
- Plan for verification of product requirements & test plan available
- Plan for product qualification available

M2 circuit design

In this phase product requirements and functions are transferred to a electrical circuit solution, i.e. design is created[1]. This design is verified, e.g. by the use of simulation methods. The main requirements and deliverables of this phase are:

- Geometrical overview of the chip (chip size, kerf location etc.)
- Pad location is frozen (enables ordering of probe cards)
- Arrangement of chips on the wafer ("wafermap")

M3 layout

In this phase the electrical circuit design is transformed into geometrical patterns on various mask levels[2]. Subsequently, in the manufacturing process the devices are built level by level using those masks. The main requirements and deliverables of this phase are:

- Layout data prepared
- Tape out, i.e. data for photomask production available
- Simulation model available

M4 engineering hardware processing

After meeting M3 requirements and having photomasks available developmental wafers can be started in production or in developmental process lines in order to confirm the process and the product functionality. The main requirements and deliverables of this phase are:

- Product functionality verified
- Design & process improvement potentials identified
- Product verified in the final product (application)

M5 qualification hardware processing

After M4, with product functionality proven and design, process or mask weaknesses addressed, process and product are supposed to meet all requirements. Again wafers,

---

[1] Allen, P.E. (2010)
[2] Allen, P.E. (2010)

so called qualification hardware, are started in production in order to provide a sufficient number of chips to the subsequent qualification process. The main requirements and deliverables of this phase are:

- Productive test program & test specification available
- Qualification hardware available (i.e. sufficient yield level achieved)
- Customer sampling plan available

M6 qualification

In this phase a sufficient number of chips has to go through the qualification tests defined at milestone M0. The main requirements and deliverables are:

- Process and product qualified according quality plan, i.e. qualification tests passed
- Production yield is at an acceptable level

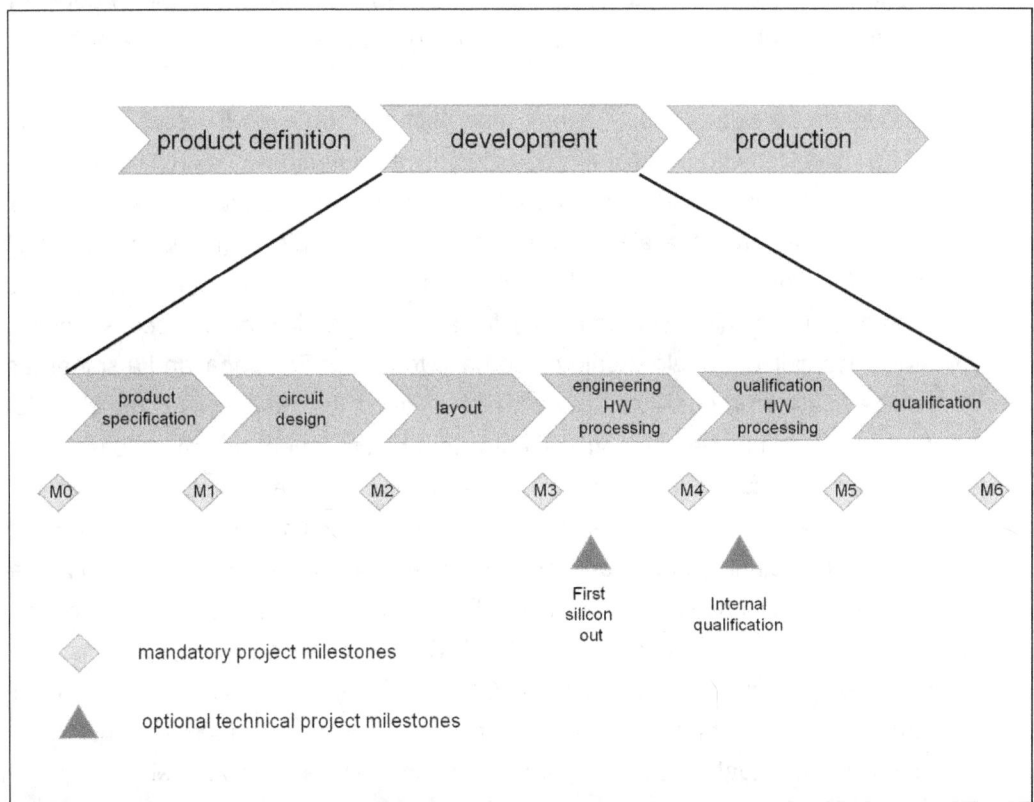

Fig. 6.6 Exemplary details of a development process (HW: hardware = silicon wafers)

## 6.4. Qualification & Market Introduction

> "The entire process by which products or production technologies are obtained, examined and tested, and then identified as qualified" [1]

### 6.4.1. Qualification Standards

The product qualification process intends to confirm manufacturability, functionality and reliability of a product. In order to make a sound statement about the maturity of a new product a defined sufficiently high number of parts have to go through functionality tests and reliability investigations. Depending on yield and reliability targets given by the customer the maximum allowable fail rate has to be defined.

Qualification tests have to produce information about how the product behaves over time. The design of those tests requires a deep knowledge of the product, the manufacturing process and the physical aging mechanisms, as explained in the preceeding chapters.

Semiconductor parts are subsequently manufactured in several separate production units. The initial step starts with the production of integrated devices on a substrate, e.g. a silicon or GaAs wafer and ends with a functional test on wafer level. Afterwards the integrated devices are separated and assembled to be single components. Those also undergo a functional test. Finally, depending on the product, several components are mounted on a printed circuit board (PCB), in case of dynamic random access memories (DRAMs) also called module. Again those have to pass a functional and a solder test.

In the qualification process only parts which passed the respective production level functional tests can be used for stress tests, i.e. on component and module level.

As for semiconductor components various qualification levels apply. International Rectifier, for example, distinguishes four levels with varying qualification requirements[2]: automotive, industrial, consumer and customer specific. For the automotive and industrial branch standards exist for the selection and execution of qualification tests, namely AEC-Q100 for automotive and JESD-47 for industrial applications. Other standards exist for discrete devices (AEC-Q101) and passive devices (AEC-Q200). For consumer products company internal regulations apply, whereas for customer specific applications the qualification requirements are agreed upon between the IC manufacturer and the customer. Companies selling into military applications have to

---
[1] ZVEI (2007)
[2] International Rectifier (2012)

follow the respective military standards. In addition to other test procedures typically the following tests are applied[1]:

➤ HTOL : High Temperature Operating Life

This stress test applies high temperature in combination with higher voltage in order to stress the integrated circuit. The purpose of these stress conditions is to provide information about the expected lifetime of the chip.

➤ HTSL : High Temperature Storage Life

These stress conditions accelerate physical and chemical processes like oxidation, diffusion or intermetallic growth. It provides information about the degradation of the chip package material.

➤ UHAST : Unbiased Temperature / Humidity

Galvanic as well as chemical corrosion processes are accelerated by these stress conditions.

➤ THB : Temperature and Humidity Bias

Temperature in combination with humidity accelerates chemical and physical processes like galvanic or electrochemical corrosion and ion diffusion.

➤ HAST : Highly Accelerated Temperature and Humidity Stress Test

These stress conditions are applied in order to achieve accelerated THB conditions.

➤ TC : Temperature Cycle

Different material exhibit different thermal expansion coefficients. The rise and fall of temperature causes thermal-mechanical stress within a stack of different materials.

➤ PTC : Power Temperature Cycle

In this case temperature cycling is combined with power on conditions at all temperatures.

➤ ELFR : Early Life Failure Rate

The purpose of these stress conditions is to identify the fail rate at the beginning of the product lifetime, e.g. several months.

---

[1] JESD47 I (2011)

Table 6.3 provides a selection of qualification test requirements for semiconductor components. For details and deeper information the author refers to AEC (http://www.aecouncil.com/) and Jedec (http://www.jedec.org/ ).

As mentioned above often components are mounted on a PCB. Hence the final product is not a single component, but a PCB (printed circuit board), e.g. the so called motherboard or a DRAM module in a PC.

Again a qualification process is required, in this case for the entire PCB, including e.g.[1]:

- Mechanical tests, e.g.
    - Drop test
    - Vibration test
    - Mechanical shock test
    - Transport simulation
    - Bend test
    - Stud pull test
    - etc.

- Reliability tests, e.g.
    - High Temperature Storage
    - High Temperature Humidity Bias
    - Temperature Cycling
    - etc.
    - Application tests in various applications.

The AEC-Q 100 flow is illustrated in fig. 6.7[2]. The qualification flow differentiates various test groups like accelerated environment stress test (group A), package assembly integrity test (group C) or die fabrication reliability tests (group D).

## 6.4.2. Family Concept

"A great challenge facing industry today is managing variety throughout the entire products life cycle"[3]. The product variety starts in the product definition and development phase and results in a huge effort when it comes to product qualification as described in the previous section. Hence it is very useful to use a family concept[4]

---
[1] Specht, J. (2008)
[2] Gruber, J. (2015)
[3] El Maraghy, H., et. al. (2013)
[4] El Maraghy, H., et. al. (2013)

from the beginning. A product family consist of a group of products with a leading parent product, with all members of the family being alike for the majority of features and differing in only a small number of features or at minor features (fig. 6.8).

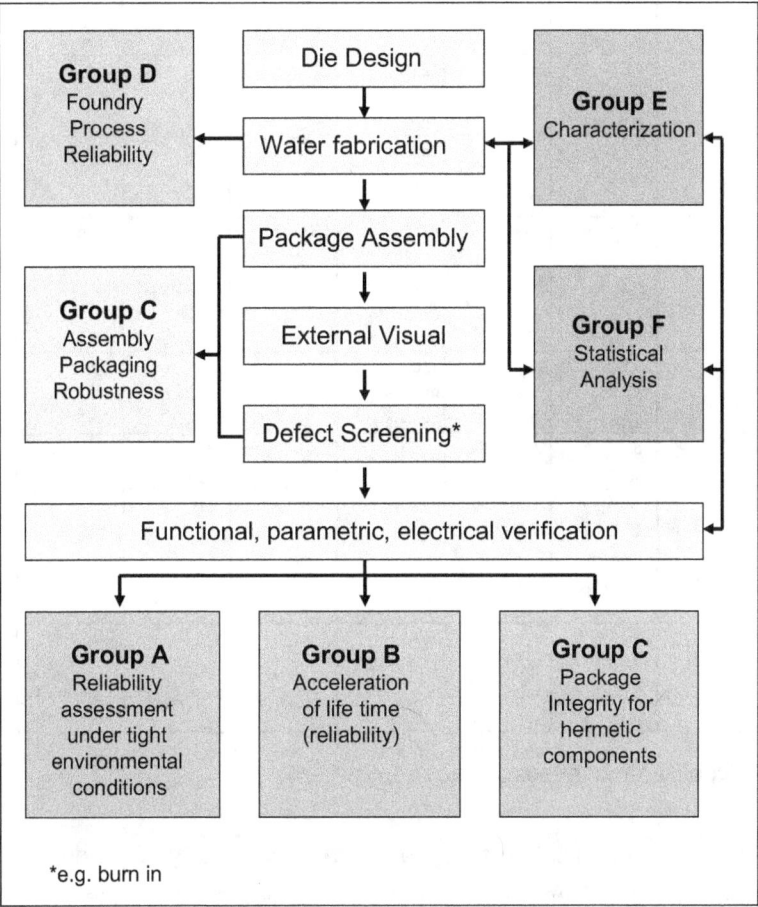

Fig. 6.7 Selection of qualification test requirements for Integrated Circuits using the example of International Rectifier ($T_j$ : Junction Temperature; $T_a$ : ambient Temperature; psia : pounds per square inch; RH : relative humidity)[1]

Depending on the nature of the product and its variety even the definition of subfamilies might be useful.
The technology (or "node") lead product has to go through a full qualification, the lead product of each family or sub-family needs to be qualified corresponding to all family attributes. Using the parents and family approach the efforts for qualification can be reduced significantly by referencing to already qualified family members.

---

[1] Gruber, J. (2015), p. 7

| | | Qualification level | | | | | | Customer specific |
|---|---|---|---|---|---|---|---|---|
| | | Automotive | | Industrial | | Consumer | | |
| Industry standard | | AEC-Q100 | | JESD-47 | | Company internal | | Customer guidelines |
| Test | Sample Size (# of lots x # of parts) | Condition | Duration | Condition | Duration | Condition | Duration [hrs] | |
| HTOL | 3 x 77 | Tj = 125°C | 1000 hrs | Tj = 125° | 1000 hrs | Tj = 125°C | 500 hrs | qualification tests agreed upon with the customer |
| HTSL | 3 x 77 | Ta = 150°C | 1000 hrs | Ta = 150°C | 1000 hrs | Ta = 150°C | 500 hrs | |
| AC | 3 x 77 | 121°C 29,7 psia 100% RH | 96 hrs | 121°C 29,7 psia 100% RH | 96 hrs | N/A | N/A | |
| UHAST | 3 x 77 | 130°C 85% RH 33 psia | 96 hrs | 130°C 85% RH 33 psia | 96 hrs | N/A | N/A | |
| THB | 3 x 77 | 85°C 85% RH biased up to 100% Vmax | 1000 hrs | 85°C 85% RH biased up to 100% Vmax | 1000 hrs | 85°C 85% RH biased up to 100% Vmax | 500 hrs | |
| HAST | 3 x 77 | 130°C 85% RH 33 psia | 96 hrs | 130°C 85% RH 33 psia | 96 hrs | 130°C 85% RH 33 psia | 96 hrs | |
| TC | 3 x 77 | -55°C / 150°C | 1000 cycles | -55°C / 150°C | 1000 cycles | -55°C / 150°C | 500 cycles | |
| PTC | 1 x 45 | -40°C to 125°C biased up to 100% Vmax | 1000 cycles | N/A | N/A | N/A | N/A | |
| ELFR | 3 x 800 | Ta = 125°C biased up to 100% Vmax | 48 hrs | N/A | N/A | N/A | N/A | |

Table 6.3 Qualification level specific requirement[1]

In addition to the complication that there is a large variety of products the customers might request to be delivered from various production sites. Hence each product has to be qualified several times, i.e. there is a separate qualification for each production site. In this case it is advisable to define a leading production site for each product with full qualification and define one or more follower sites with a reduced qualification.

Both JEDEC47 as well as AEC-Q100 use the family concept in order to decrease the efforts required for product qualification.

> "All products in the same product qualification family are qualified by association when one family member successfully completes qualification with the exception of the device specific requirements"[2]

---

[1] International Rectifier (2012)
[2] Automotive Electronics Council (2014)

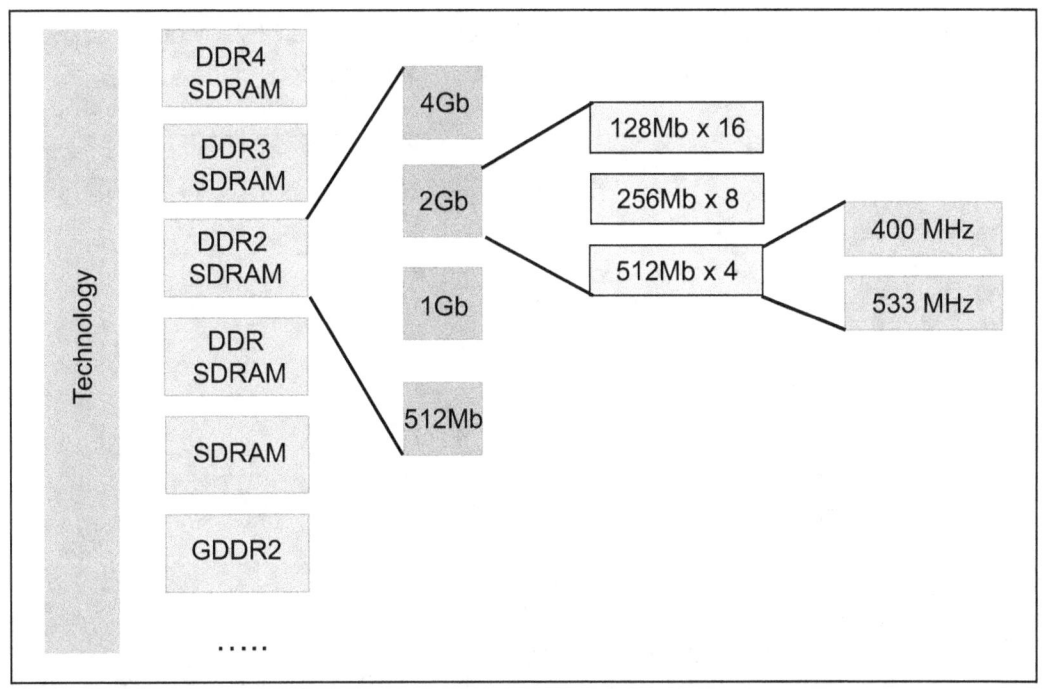

Fig. 6.8 Variety of products using the example of Micron DRAM components (http://www.micron.com/products/dram)

### 6.4.3. Customer Sampling & Market Introduction (non automotive)

For many products, e.g. a Dynamic Random Access Memory (DRAM) the technology qualification process ends with the customer sample qualification by the customer. For this purpose, a sufficiently high number of well performing dies or modules have to be delivered to the customer as early as possible.

As stated earlier a semiconductor manufacturing process is highly complex. After achieving the technology qualification requirements, the production process is not always meeting process stability targets of a high volume production. In addition to immature technology also new products might suffer from problems like design weaknesses or insufficient test coverage. Both will be detected and corrected in the early phase of market introduction ("learning"). However, this also means, that the customer may detects those failures in his application.

This state of affairs might lead to comparably low yields in production and high defect rates at the customer in the beginning (fig. 6.10). However, the customer expects to receive the new product with a similar defect rate as with the previous product. Here it is important to understand, that often from the customer point of view the new product is more or less identical to the old one, but from the manufacturer's point of view, a new, more advanced, technology was used. For example, a 4Gb DRAM could be produced on a 90 nm technology node as well as on a 65 nm technology node.

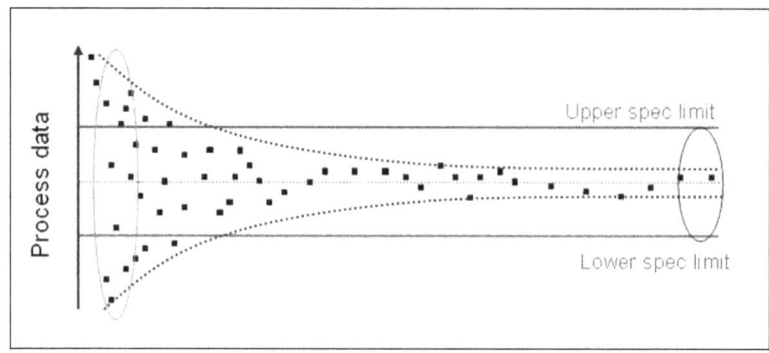

Fig. 6.9 Process stability improvement over time

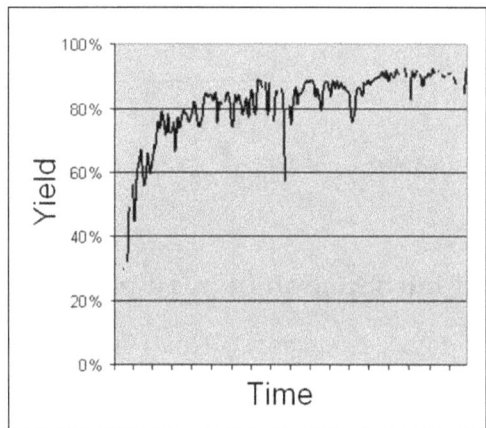

Fig. 6.10 Example for a yield learning curve over approximately 6 months[1]. Lower yield (as in the initial production phase) usually implies a higher rate of early fails / defect rate at the customer

For some products for sampling or low volume production in the early phase the defect rate at the customer can be modelled by testing the semiconductor devices in the final application ("application test"), e.g. a personal computer or laptop computer. Fails found with this method, which were not detected by the regular electrical test equipment, are then analyzed, in order to optimize the electrical functional test. This leads to a better test coverage and eventually to an improvement of the defect rate observed at the customer.

Unfortunately yield and defect learning requires volume production. Hence it is necessary to start production at yield and defects levels, which are not equivalent to the preceding product. In addition to extra application test there are basically two options to solve this problem:

---

[1] Bergholz, W. (2011)

1. A development partnership with a customer:

In this case volume production and yield and defect learning can start at an early phase, without alienation of the customer by higher defect rates. The benefit for this customer is to receive early products and the possibility to start his own learning phase ahead of his competitors.

2. Selling to less quality sensitive customers / applications

For instance, in the DRAM market there are various levels of quality requirements on the customer side. The highest level is typically found at server manufacturers, followed by personal systems manufacturers for business and finally manufacturers for private use.

Thus for customer sampling and shipping a phase model[1] can be used with a stepwise release of volume to the customers. Assuming customers with three different levels of quality requirements, three phases may be applied. Phase 1 customers have the lowest expectations, phase 3 customers have the highest level. The failure levels the product has to achieve based on the above described method for volume or sample delivery are e.g. 3x, 2x and 1x of the minimum defect rate of the preceding product (table 6.4) identified during application test.

In addition to the product defect rates there are several additional criteria customer samples need to fulfil in order to make a product qualification successful at the customer, e.g.

| defect rate of preceding product | volume delivery to customer | samples delivery to customer |
|---|---|---|
| 3x | Phase 1 customers | Phase 2 customers |
| 2x | Phase 2 customers | Phase 3 customers |
| 1x | Phase 3 customers | - |

Table 6.4 Shipping of samples and volumes to customer classes depending on defect rates

- ➢ The product shall not suffer from any excursion (e.g. material, process..)
- ➢ The product shall be tested for functionality with the latest release of the test program
- ➢ Yield and electrical parameter assessment show a healthy product (again: no excursions)

---

[1] Wittmann, J. (2008)

## 6.4.4. Market Introduction Automotive: Safe Launch

Due to high defect rates of new technologies as described above, it is not recommended to produce automotive or any other product with a high level of quality on an entirely new technology platform. Products with reduced quality requirements, e.g. consumer applications, may be used for technology learning first. With increasing maturity of the technology the phase of automotive products can start.

However, this might not be feasible for various reasons, for instance if the producer is focussing on the automotive market only, or if the automotive application requires additional technology features which were not required (and developed) for other applications.

Therefore, in the early phase of production of a new automotive product, it is necessary to take additional precautions in order to ensure a low failure rate in the customer application. Based on a risk assessment the so called Safe Launch Plan (also known as Product Special Characteristic Matrix (PSCM)). has to be defined which asks for additional activities like

- Higher inline sampling rate
- Tighter limits (inline / test)
- Additional parameters included
- Additional reliability monitoring

Those activities are an addition to the regular process and product control activities documented in the process flow. The safe launch phase is considered part of the design review process[1].

"Exit criteria for the Controlled - Safe Launch Plan is shipment of zero defect parts that meet either the defined period of time or number of lots. Any defect discovered during this period restarts the event to "0" pieces shipped."[2]

## 6.4.5. Robust Validation[3,4,5]

It is clear by elementary mathematics and statistics that for product specific defects, the limited number of qualification samples, e.g. zero fails from three lots with 77 pieces per lot (i.e. 0/231), does not provide any information about the expected dpm level of this product anyway. No fails in 231 pieces is actually statistically equivalent to up to 1% fails at 90% confidence level. In other words, the JEDEC / AEC-Q based qualification

---

[1] Magna (2016)
[2] Alcom (2013)
[3] ZVEI (2007)
[4] ZVEI (2010)
[5] ZVEI (2013)

concept only provides a more general evidence for 'no major problem' with the technology or product. It does not guarantee any reasonable level of defects later on. Hence, for the automotive market a somewhat different approach is taken, called "robust validation", to demonstrate sufficient product performance. Even though the main driver for robust validation seems to be the automotive semiconductor industry, of course this approach is applicable to other segments and industries, too.

> "Robustness Validation is a process to demonstrate that a product performs its intended function(s) with sufficient Robustness Margin under a defined Mission Profile for its specified lifetime" [1]

The purpose of the robustness validation is to validate "fitness for use" and to better understand the semiconductor component failure mechanisms, e.g. by end-of-life testing. The intention thereof is to identify the margin between the semiconductor component capability and the respective application requirements (fig. 6.11). The identification of the robustness margin is a process starting at product design and development throughout the product life cycle.

In addition to the failure mechanisms both the actual mission profile, i.e. the use conditions, as well as a sound understanding of the acceleration models are required for the validation process.

Sources of information are e.g.:

- Actual use data including environmental data (e.g. climate induced temperature cycles), i.e. a complete as possible mission profile
- Results and know-how from risk assessments (e.g. FMEA)
- Modelling /simulation results
- Failure analysis results from similar products
- Production data
- ....

Starting with the determination of the application robustness validation typically covers the following steps:

- Definition of the actual mission profile
  - Mission profile EEM (Electrical / Electronic Modules)
  - Mission profile component
- Determine requirements

---
[1] ZVEI (2013), p. 22

- Mission profile device
- Robustness requirements
- Run risk assessment (FMEA)
- Create qualification plan
> Qualification
- Run qualification tests
- Test according to plan
- Execute quantitative robustness assessment
- Component release
> Production
- Create Monitoring Plan using RIF figures (Robustness Indicator Figures)
- Regular sampling, monitoring & data review

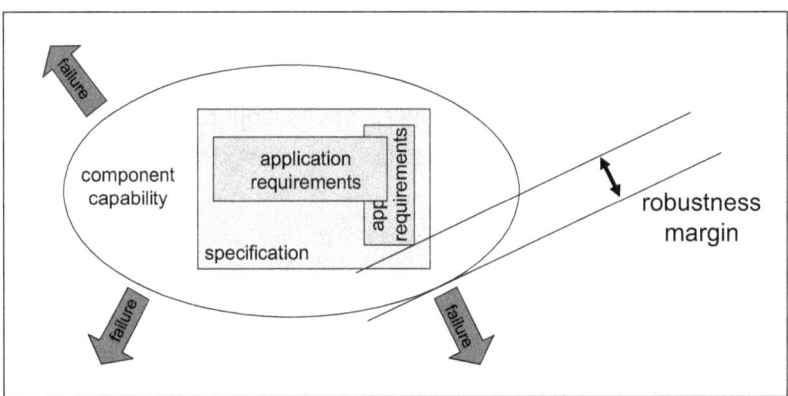

Fig. 6.11 Visualization of the margin between the edge of the application requirements and the actual product / component capability[1]

Robustness Indicator Figures (RIF)

In order to compare different products, designs, suppliers etc. in a meaningful manner it is necessary to develop a quantitative figure expressing the robustness margin. This figure is called Robustness Indicator Figure and is defined as follows:

$$RIF = \frac{estimated\_strength}{required\_spec}$$

It is useful to compare those at identical conditions (e.g. the same temperature). The calculation of the Robustness Indicator Figure depends on the acceleration model

---

[1] ZVEI (2007)

applied. Using the Arrhenius Model the RIF is calculated as follows, if test to fail data is available:

$$AF = \frac{\lambda}{\lambda_0} = \frac{R_0 \cdot e^{-\frac{E_a}{k \cdot T}}}{R_0 \cdot e^{-\frac{E_a}{k \cdot T_0}}}$$

$$AF = e^{\frac{E_a}{k}\left(\frac{1}{T_0} - \frac{1}{T}\right)}$$

with
- AF : acceleration factor
- Ea : activation energy
- k : Boltzmann's constant
- $T_0$ : max. specified temperature
- T : stress temperature

> "**estimated strength**
> = measured or calculated value of the item being considered, e.g. time to failure, $C_{pK}$, failure level, etc.
>
> **required spec**
> = requirement value based on Mission Profile or specification, which can also be associated with certain failure level criteria's (e.g. 10 years with 1% accumulated failure level)"  [1]

Example

Assumed the required test time at $T_0$ = 85°C is determined to be 2000h. The actual stress temperature is higher than that ($T_{Stress}$ = 125°C). A failure occurs under stress conditions after 2750 hours. Based on these results the acceleration factor as well as the RIF are calculated:

$$AF = e^{\frac{E_a}{k}\left(\frac{1}{T_0} - \frac{1}{T}\right)} = e^{\left(\frac{0.44eV}{8.617 \cdot 10^{-5} eV/K}\left(\frac{1}{358K} - \frac{1}{398K}\right)\right)} = e^{1.43} = 4.178$$

$$RIF = \frac{AF \cdot (time\_to\_fail)}{required\_test\_time} = \frac{1.43 \cdot 2750h}{2000h} = 5.75$$

---

[1] ZVEI (2013), p. 85

# 7. Supplier Quality Management

The depth of production varies from product to product and can be significantly less than 100%, i.e. a lot of components, materials and/or software are purchased from suitable suppliers. As a rule the quality of purchased parts and materials does have an impact on the final product's quality. Hence, in order to ensure the right quality on the procurement side, the supply and the supplier quality has to be managed with the goal of achieving stable high quality incoming parts or material.

Supplier quality management is dealing with procurement of „relevant to product quality" items like

- materials, components etc. the final product is made of and
- equipment, the final product is made with

Supplier management is normally not dealing with procurement of trivial items such as paper-clip, pens, etc. However, if pens and paper are relevant to product quality, e.g. clean room paper or clean room pens, they must be subject to supplier management.

## 7.1. Requirements of the standards[1,2]

In this section both the ISO9001 as well as ISO/TS16949 requirements regarding procurement are summarized.

### 7.1.1. General

ISO9001: Dealing with all external inputs, e.g. parts, materials, services, processes, the same way, the standard requires them to meet the same quality assurance requirements. In this section this is all summarized as "external product". They also need to be controlled as required, in particular, if

- the external product is integrated into the final product
- the external product is directly offered to the customer in the name of the organization
- a process is executed by an external provider

For selection, evaluation and re-evaluation of external providers' criteria have to be defined. The same applies to performance monitoring of those external providers.

---

[1] DIN EN ISO9001:2015
[2] ISO/TS 16949:2009

ISO/TS16949: Even though it is a matter of course ISO/TS16949 explicitly requires legal compliance of all parts and materials to be integrated into the product.
In addition, the supplier has to prove conformity to ISO9001 by a third party certification and has to show efforts to achieve the ISO/TS16949 certification. The customer has the right to impose the organization's suppliers.

## 7.1.2. Control of external supply

ISO9001: During the definition of activities to ensure the quality of external products the following has to be considered:

- the capability to meet customer's, authority's or legal requirements
- effectiveness of the external supplier's control activities

Verification activities regarding external supply to guarantee delivery of constant high quality have to be defined and implemented.
The organization's quality management system comprises outsourced processes, hence the quality assurance activities refer to the supplier and the supplied product.

ISO/TS16949: It's the organization's duty to ensure the quality of supplied goods. For these activities a process has to be defined and implemented, which includes e.g.

- evaluation of statistical data, i.e. supplier's production SPC data
- incoming inspection
- supplier production site audits
- part evaluation by a defined test laboratory

Additional techniques to ensure quality may be agreed upon with the customer.
The supplier performance needs to be monitored, too, using performance indicators like product conformity, field quality data and delivery performance.

## 7.1.3. Information

ISO9001: The organization has to provide information to the supplier, e.g. regarding

- the externally supplied part (material, service…) as a technical or procurement specification
- acceptance criteria for products in incoming inspection
- qualification and competence of the respective personnel
- the organization's quality control and verification activities towards the supplier

ISO/TS16949: There is no additional requirement in this section.

In addition to other requirements, one way to fulfil customer's, legal, authority's and standard requirements is to define and implement a supplier management process as well as an incoming inspection. For the handling of material, in particular incoming material for semiconductor material, a separate material management process may be necessary.

## 7.2. Supplier Management Process

The supplier management process is a crucial instrument to achieve a stable high incoming quality of purchased parts, materials or services. The supplier management process is embedded into the company's process landscape and starts with the company strategy, followed by supplier strategy, selection etc. (fig. 7.1) and eventually ends in the supplier phase out.

Fig. 7.1 Supplier Management Process

The main objectives of the supplier management process are:

- security of supply of material, parts and services
- at minimum total cost
- at consistently high quality

To some extent those targets are in conflict, i.e. lower failure rates may initially result in higher procurement cost. However, if the final product requires high incoming quality, in the long run there is a benefit for the company. The point is to balance these objectives in a way to achieve the optimum result for the company by defining suitable processes.

### 7.2.1. Supplier Strategy

The purpose of a supplier strategy is to define which supplier to cooperate with and to define the way how to cooperate with suppliers.[1] In addition important questions have to be answered like for e.g. which number of suppliers per product or product group is

---

[1] Wagner, S. (2007)

the optimum, is a partnership with suppliers required to achieve particular targets or what kind of supply risk is the company facing with the existing supply base.

The company or corporate strategy provides initial information in which way the supplier strategy needs to be adjusted. For instance, if the company is planning for a major growth strategy over the coming years, it is useful to choose a supplier, who is willing and capable to support this growth strategy.

| Supplier Strategy decision matrix | Weight | Supplier A | Result Supplier A | Supplier B | Result Supplier B | Supplier C | Result Supplier C | Supplier D | Result Supplier D | Supplier E | Result Supplier E |
|---|---|---|---|---|---|---|---|---|---|---|---|
| technological capability | 20 | 9 | 180 | 8 | 160 | 8 | 160 | 7 | 140 | 5 | 100 |
| sufficient capacity | 15 | 7 | 105 | 7 | 105 | 10 | 150 | 6 | 90 | 5 | 75 |
| financial strength & market share | 10 | 5 | 50 | 5 | 50 | 9 | 90 | 5 | 50 | 4 | 40 |
| quality of product in the field (yield, performance…) | 20 | 9 | 180 | 8 | 160 | 8 | 160 | 7 | 140 | 6 | 120 |
| pricing | 10 | 4 | 40 | 5 | 50 | 5 | 50 | 4 | 40 | 3 | 30 |
| quality of support (technology, quality, logistics) | 15 | 6 | 90 | 7 | 105 | 4 | 60 | 4 | 60 | 5 | 75 |
| …. | | | | | | | | | | | |
| …. | | | | | | | | | | | |
| **sum weighed** | | | 645 | | 630 | | 670 | | 520 | | 440 |

Grading  10 fully meets requirements
7 generally meets requirements
5 partially meets requirements
3 major improvements necessary
0 does not meet requirements

Table 7.1 Estimation of the degree of fulfilment of defined criteria by the supplier (arbitrary numbers)

In addition to the corporate strategy other inputs are necessary, to determine the most suitable suppliers, e.g.

- match of the company's and supplier's technology roadmaps
- match of the company's material roadmap with supplier's development roadmap
- market analysis & forecasts
- current supplier performance, i.e. regarding technology, capacity, quality, pricing, delivery performance etc.
- planned delivery from local or global supplier production sites

- supplier's strategy, i.e. willingness of the supplier to support the company
- supplier's openness to support cost reduction activities
- etc.

It is a reasonable approach to define the criteria which are relevant for the company, to assign a weight to each criterion them and to systematically evaluate each supplier's degree of fulfilment (table 7.1). The degree of fulfilment for each criterion may be estimated using another list of sub criteria or any other assessment method:

Example 1:

In order to estimate the supplier's technology capability and technology roadmap the present supplier's technologies are compared with the expected supplier's capabilities in 3 to 5 years.

| Production requirements today vs. in 5 years | | | | | | | | |
|---|---|---|---|---|---|---|---|---|
| | | Supplier A | Supplier B | Supplier C | Supplier D | Supplier E | Supplier F | Supplier G |
| today | COP* free material | P | P | P | P | D | D | R |
| today | 2mΩcm substrate | P | P | N | D | R | N | N |
| today | 200mm FZ** | P | N | D | D | ? | N | ? |
| | | | | | | | | |
| +5yrs | COP free material | P | P | P | P | P | P | D |
| +5yrs | 2mΩcm substrate | P | P | N | P | D | R | N |
| +5 yrs | 200mm FZ | P | N | D | D | ? | N | ? |

P production   R research         *Crystal Originated Pit
D development  N no activity      **Float Zone

Table 7.2: Generic Supplier Technology assessment & comparison for silicon wafer suppliers

The required information is gathered e.g. by the utilization of conventional market research activities, by attending specialized fairs and conventions or standardization activities on a consortium or international level, by direct exchange of information with a supplier during technical meetings or by alignment of technology roadmaps. It is clear that the reliability of the information is lowest for market research information and highest for technical meetings and technical roadmaps.

The comparison of the information gathered with the company technology roadmap then provides an initial overview of which supplier should be part of the supplier base (table 7.2) in the future.

Example 2:
Typically, security of supply is ensured by using dual or multiple sourcing. However, the larger the number of qualified suppliers the higher the effort to manage them and also the lower the share for each supplier with a direct negative impact on the price bargaining position of the company. The consequence is that also the supplier perceives the customer as minor and might not be willing to give support as required.
In order to optimize the efforts taken for the supplier management, frequently a simple ABC analysis is performed (fig. 7.2) in order to focus on the more important suppliers. Typically the annual turnover defines the supplier class[1]:

    A: large turnover and strategic importance
    B: medium turnover and marginal strategic importance
    C: small turnover and no strategic importance

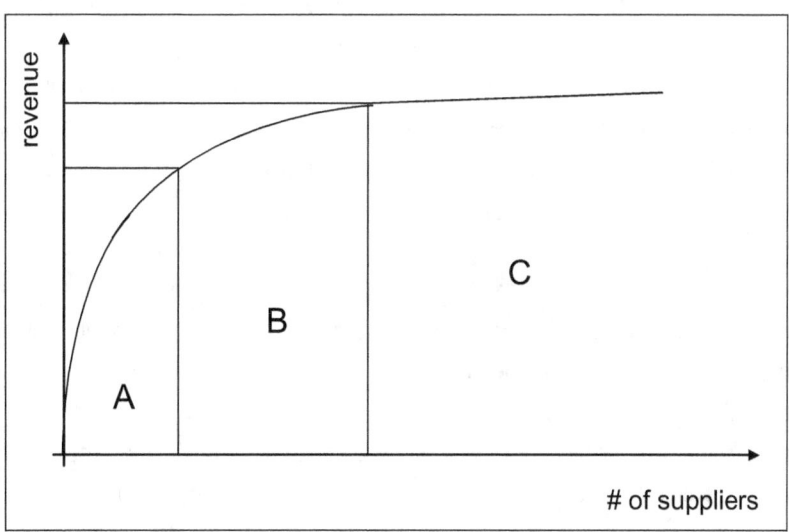

Fig. 7.2 ABC supplier distribution

For B suppliers often an individual case decision has to be made. Focussing on A suppliers significantly reduces efforts taken for supplier management and enables the company to even intensify the partnerships with A suppliers.

---

[1] Hartmann, H. (2004)

## 7.2.2. Supplier Selection & Qualification[1]

Driven by technology or product innovations, cost reduction activities, bad performance of existing suppliers or, for instance, environmental requirements, companies regularly need to reassess the supplier strategy and investigate alternatives to the existing supplier base. Typically, the need for an alternative is internally addressed to the procurement organization along with a set of information regarding quality system, product quality and volume and, if necessary, additional information, e.g. logistics aspects like special packaging and storage. In this context the term "procurement organization" or the "procurement team" always includes actual purchasing functions as well as technical functions and quality management.

Having these information available potential suppliers can be identified by the cross disciplinary procurement team. The comparison of market research results with the requirements of the organization leads to an initial list of potential suppliers. The initial supplier base is then to be reviewed based on a more technical and quality expertise in order to decrease the number of potential suppliers and, at the same time, in order to identify the most promising suppliers (table 7.5).

Before an order is placed with a new supplier, this new supplier has to be approved according to defined rules. Various levels of approvals should be defined according to the impact the supplier may have on the final product quality.

There are certainly many ways how to approach this topic, here we limit ourselves to a simple three level system. The levels basically reflect the focus on the technical impact the provided part / material may have on the final product quality.

Level 1

- those suppliers provide products and/or services which have no relevance for the process or have no financial impact, e.g. one-time suppliers, sample suppliers, minor local suppliers
- only basic information is required, e.g. address, bank account number, terms & conditions
- level 1 suppliers are not subject to supplier management

Level 2

- these suppliers may have a financial impact or a technical assessment is required for the supplied product, e.g. suppliers for production equipment, measurement equipment or spare parts

---

[1] Brunner, C. (2001)

- additional more detailed information in the form of a supplier assessment is needed
- level 2 suppliers are subject to supplier management

Level 3

- these suppliers have a potentially significant financial impact and their products are relevant for the final product quality, e.g. material which is integrated into the product (e.g. silicon substrates), parts which have a direct impact on the product quality (e.g. photo masks) or material which also has a strong impact on the product quality (e.g. photo resist)
- these suppliers have to undergo a full supplier qualification process in order to prove that their product is under control and does not have an detrimental impact on the final product quality
- level 3 suppliers are subject to supplier management

Once the exchange of information with the supplier starts, it should be supported by a Non Disclosure Agreement (NDA) between the two parties, in order to prevent technical know-how to escape to a third party, e.g. a competitor. Information is requested from the supplier in a formal way, for instance a supplier self assessment which may be part of the supplier profile later on.

Depending on the importance of the incoming part/material, a technical meeting and site inspection at a level 2 and level 3 supplier's site may be appropriate to gather more in-depth information.

Due to the fact that the level 1 approval process is simply a formal approval of minor importance in the following section we will focus on level 2 and level 3 supplier qualification.

Level 2 Supplier Qualification

The level 2 supplier qualification requires more detailed information about the way the supplier is doing business and the company's status. This includes more formal information and in particular more information about strategy, quality, technology and production capacity etc. in form of e.g. a supplier assessment sheet (fig. 7.3, 7.4).

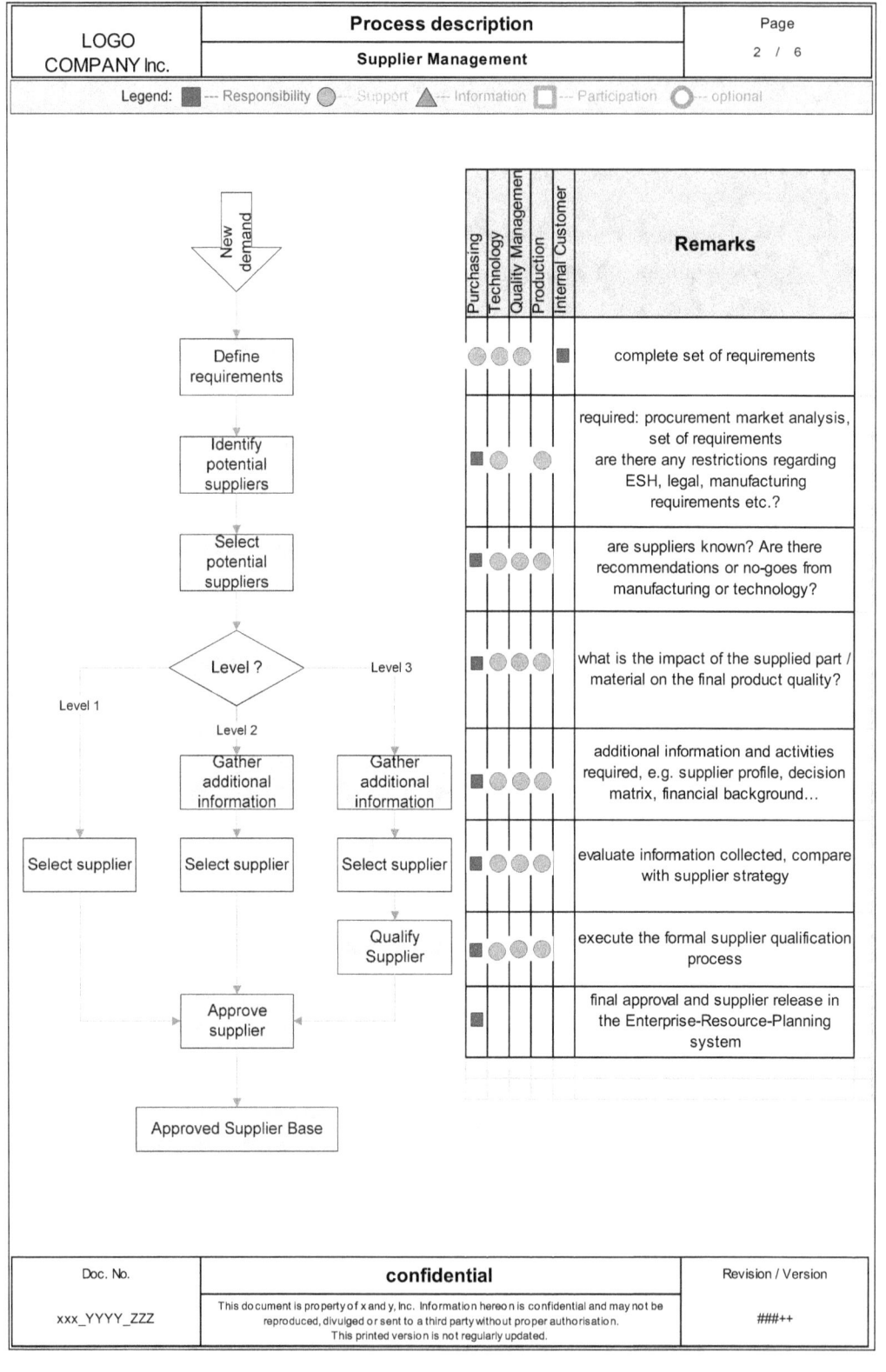

Fig. 7.3 Generic process supplier sourcing

| 1. | Address<br>Company name, division, street, P.O. box, Zip Code, Location, Country | | |
|---|---|---|---|
| 2. | Central company phone number<br>Web address | | |
| 3. | Contact persons | General Manager | |
| | | Sales | |
| | | Quality Management | |
| | | Person in Charge | |
| 4. | Company data: | Year of founding | |
| | | Legal form | |
| | | Organization | |
| | | DUNS number | |
| | | Turnover / earnings in Mio € | |
| | | Percent share of development cost for sold product | |
| | | Headcount | Total:<br>Administration:<br>Production<br>QM/QC:<br>Engineering: |

Table 7.3 Example for a supplier assessment sheet – general part

As mentioned earlier a decision matrix may be set up by selecting criteria based on the proposed assessment sheet, by assigning a weight to each of those criteria and by measuring the grade of fulfilment. This way a data based decision on gaps or fulfilment of the minimum assessment requirements can be made and several suppliers can be compared with each other.

Level 3 Supplier Qualification
Products supplied by level 3 suppliers are particularly critical to the final product quality. Hence the supplier qualification process needs to reduce incoming quality risks as much as possible, which requires significantly more efforts than level 1 or 2.
For the qualification of suppliers a set of minimum requirements has to be defined and summarized (table 7.5). In the subsequent step the real supplier conditions need to be assessed and reviewed based on the requirements list. There may be unacceptable or temporarily acceptable gaps. In the first case the qualification process is basically on

hold, whereas in the latter case the supplier can be qualified and delivery may start with some restrictions.

| 1.  | Company strategy 1 (Vision, Mission, Targets) | |
|-----|-----|-----|
| 2.  | Company strategy 2 (favoured markets, purchasing policy) | |
| 3.  | Company strategy 3 (alliances, R&D cooperations, long term agreements) | |
| 4.  | Supplier's market position | |
| 5.  | Supplier's major competitors | |
| 6.  | Manufacturing capacity | |
| 7.  | Manufacturing sites (adresses) | |
| 8.  | Delivery times | |
| 9.  | Contingency planning concept | |
| 10. | Quality management system (grade of certification) | |
| 11. | Sampling concept | |
| 12. | Traceability concept | |
| 13. | Production control concept (SPC, checklists, …) | |
| 14. | Logistics concept (JIT, Kann-ban, …) | |
| 15. | Service concept | |
|     | …. | |
|     | …. | |

Table 7.4 Example for a supplier assessment sheet – strategy & operational part

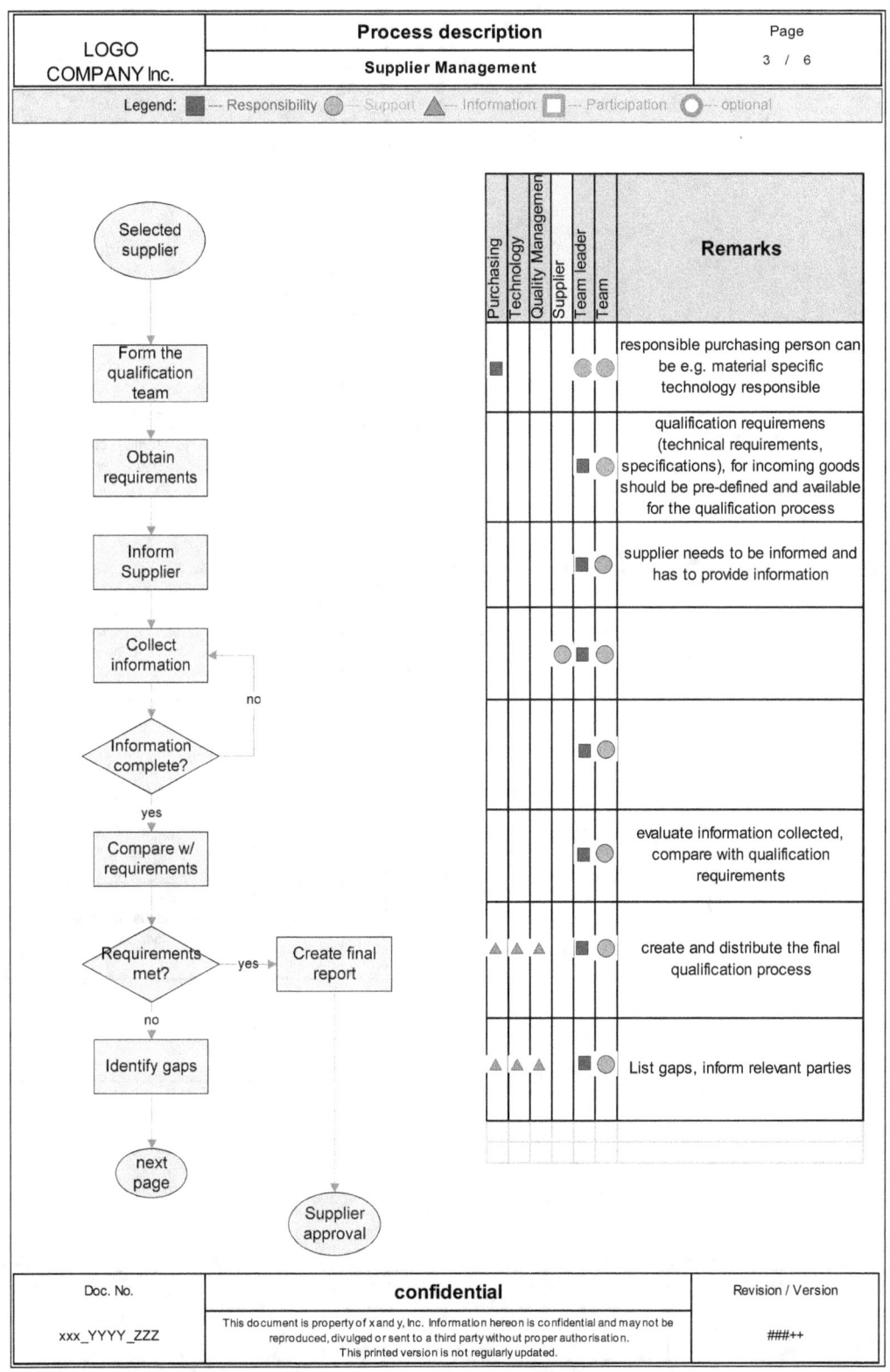

Fig. 7.4 Generic process supplier qualification part 1

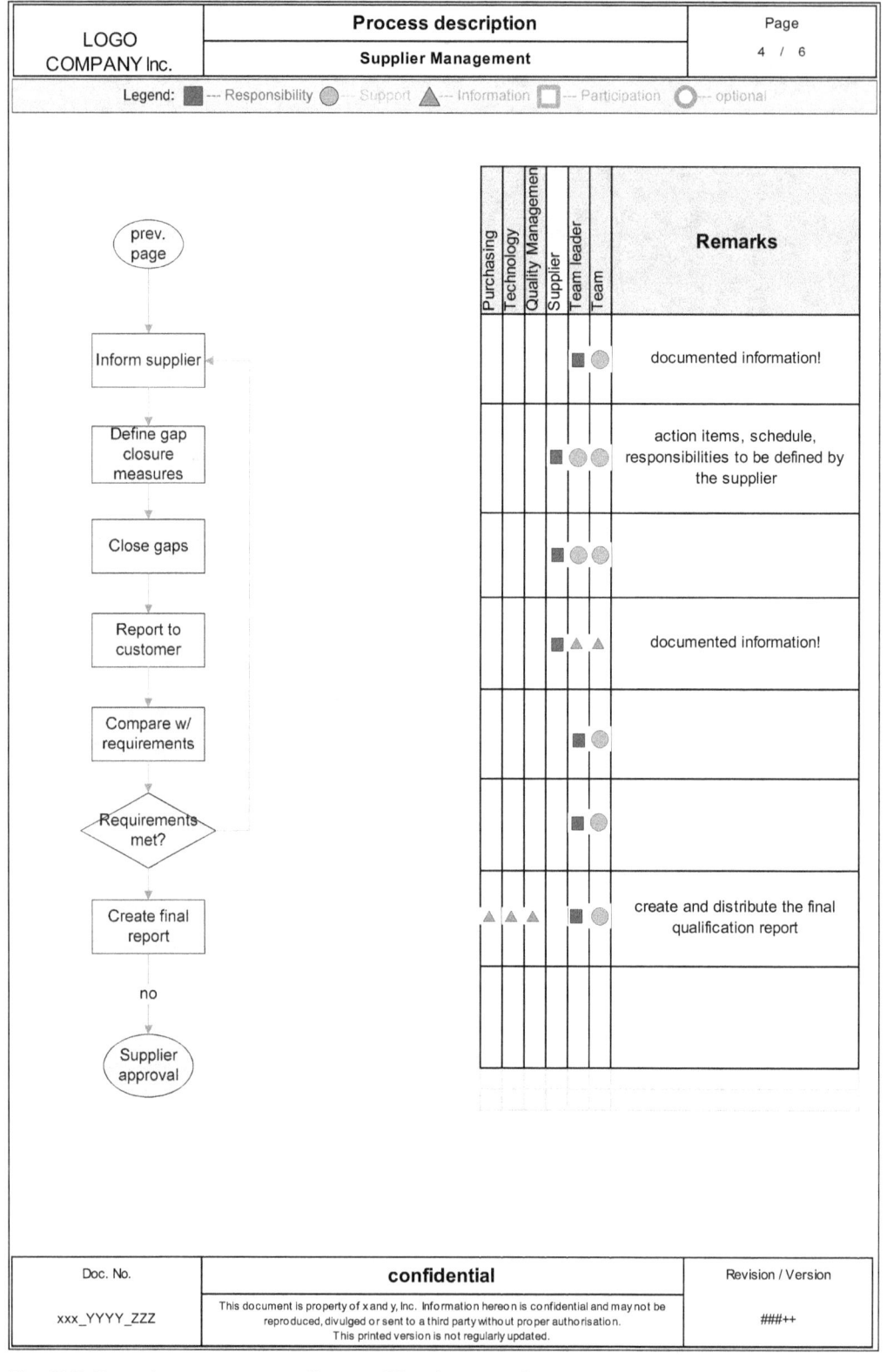

Fig. 7.5 Generic process supplier qualification part 2

| | Mandatory | Recommended | Supplier A | Supplier B | Supplier C |
|---|---|---|---|---|---|
| General supplier information complete | X | | complete | complete | complete |
| NDA | X | | signed | signed | signed |
| ISO9001 Certification | X | | expiry dd.mm.yy. | expiry dd.mm.yy. | expiry dd.mm.yy. |
| Compliance w/ required additional norms e.g. ISO/TS16949 | | X | certificate expiry dd.mm.yy. | certification planned mm.yy | no activities yet |
| ISO14001 Certification | | X | expiry dd.mm.yy. | expiry dd.mm.yy. | certification planned mm.yy. |
| Agreement to customer specific product related items (e.g. handling) | X | | signed | signed | signed |
| QAA agreed on and signed | X | | signed | signed | signed |
| Successful product qualification | X | | ongoing | done | ongoing |

Table 7.5 Generic status list of supplier qualification requirements fulfilment

Having the requirements defined and having suppliers pre-selected, the actual supplier qualification process can start. A generic process is shown in fig. 7.4. and 7.5.

### 7.2.3. Supplier Evaluation[1]

According to the relevant standards (see 7.1) the supplier performance has to be evaluated in order to ensure constant high incoming quality of materials. The typical approach to set up a supplier evaluation process is to define various areas of interest with each having several criteria and maybe one or two levels of sub-criteria and to weigh those according to their particular importance to the company. In the next step target fulfilment of each criterion for each supplier has to be evaluated and the overall target achievement can be calculated from those data.

The relevant areas are typically:

- Quality performance
- Logistics performance
- Technology performance
- Procurement & Cost performance
- Service & Commitment

---

[1] JESP146A

The weight distribution on this highest level depends, of course, on the company strategy. A technology driven company is expected to weigh "Technology performance" higher than a trading company.

For each of the aforementioned areas several sub-criteria are proposed in the following section as practical examples. Subcriteria may be split into several contribution measures. Those again, can be weighed. The following list is not meant to cover all possible criteria but provides an initial overview with different levels of detail and depth:

(1) Main Criterion 1: Quality

- Subcriterion 1.1.: Incoming Quality (e.g. number of events or rates or volume)
    - Rejection at incoming inspection e.g. due to[1],
        - Documentation
        - Packaging
        - Product
    - Field failure rate [dpm]
- Subcriterion 1.2.: Service Quality
    - Response times
    - Time to close a topic (including root cause analysis and permanent corrective action)
- Subcriterion 1.3.: Supplier Audit results
- Subcriterion 1.4.: Quality Assurance Agreement (QAA)
- Subcriterion 1.5.: Quality Improvement programs
- Subcriterion 1.6. Quality System
    - Certification according to ISO9001
    - Certification according to ISO/TS16949

(2) Main Criterion 2: Logistics & Security of supply

- Subcriterion 2.1.: Logistics Performance
    - On time delivery
    - Volume fulfilment
    - Flexibility regarding volatile order volumes
    - Early Warning System in case of late delivery
- Subcriterion 2.2.: Logistics Tools

---
[1] Melzer-Ridiger, R. (1995)

- Customer friendly tools, e.g. ship-to-line implemented, consignment stock...
- Electronic ordering system installed
- Subcriterion 2.3.: Production
    - Production capacity
    - Production stability

(3) Main Criterion 3: Technology

- Subcriterion 3.1.: Today's technology capability
    - Impact on product yield
    - Fulfilment of specified requirements
    - Openness & capability to fulfil new technical requirements
    - Technical assistance
    - Onsite support
- Subcriterion 3.2: Technology roadmap
    - Match with own technology roadmap
    - Development time for new products

(4) Main Criterion 4: Procurement & Cost

- Subcriterion 4.1.: Prices
    - Price level
    - Price trend[1]
- Subcriterion 4.2.: Openness to cost reduction
- Subcriterion 4.3.: Terms & conditions
- Subcriterion 4.4.: Financial stability of the supplier

(5) Main Criterion 5: Service & Commitment

- Subcriterion 5.1.: Support in daily business
- Subcriterion 5.2.: Availability of competent contact persons
- Subcriterion 5.3.: Willingness to cooperate
- Subcriterion 5.4.: Reaction time
- Subcriterion 5.5.: Market position

---

[1] Hartmann, H., (2004)

| | | | | | | |
|---|---|---|---|---|---|---|
| 1.5. Quality Improvement Program<br>    well established program: 100 %<br>    program initiated     : 80 %<br>    program planned     : 60 %<br>    no activities     : 0 % | 15 | | | 60 | 100 | 80 |
| 1.6. Quality System | 15 | | | 80 | 80 | 100 |
| 1.6.1. Certification acc. ISO9001<br>    certification achieved : 100 %<br>    certification planned : 80 %<br>    QMS available     : 60 %<br>    no activities     : 0 % | | 50 | | 100 | 100 | 100 |
| 1.6.2. Certification acc. ISO/TS 16949<br>    (if required)<br>    certification achieved : 100 %<br>    certification planned : 80 %<br>    QMS available     : 60 %<br>    no activities     : 0 % | | 50 | | 60 | 80 | 100 |
| | | | | | | |
| **2. Logistics & Security of Supply** | 20 | | | 62,45 | 91,04 | 85,62 |
| 2.1. … | | | | | | |
| | | | | | | |
| **3. Technology** | 15 | | | 94,01 | 79,93 | 85,38 |
| 3.1. … | | | | | | |
| | | | | | | |
| **4. Procurement & Cost** | 25 | | | 72,67 | 75,64 | 78,43 |
| 4.1. … | | | | | | |
| | | | | | | |
| **5. Service & Commitment** | 20 | | | 92,89 | 84,55 | 86,63 |
| 5.1. … | | | | | | |
| | | | | | | |

Table 7.6 Supplier Evaluation Sheet part 1

Typically, the results of the supplier evaluation are summarized in table form with explanations on how to rate a particular supplier performance in a way that a comparison of different suppliers is possible (tables 7.6 and 7.7 with arbitrary values for different suppliers).

The status of the supplier is derived from the evaluation results. There may be a defined minimum percentage for preferred suppliers and for standard suppliers, respectively. In case the minimum percentage is not achieved by a supplier the supplier's status may be set to "on probation", eventually resulting in a supplier phase out if improvement is not implemented. Preferred suppliers, on the other side, may have the opportunity for a larger purchase volume share, which motivates them to improve according the customer requirements.

| SUPPLIER EVALUATION RESULTS | weight | | | | Supplier 1 | Supplier 2 | Supplier 3 |
|---|---|---|---|---|---|---|---|
| level / final result | 1 | 2 | 3 | 4 | 77,88 | 82,59 | 84,41 |
| **1. Quality** | 20 | | | | 72,69 | 82,88 | 87,75 |
| 1.1 Incoming Quality | | 25 | | | 77,75 | 68,5 | 53 |
| 1.1.1 Rejection at incoming / quarter | | | 50 | | 75,5 | 57 | 46 |
| due to Documentation<br>= 0 : 100 %<br>< 1 : 80 %<br>< 2 : 60 %<br>< 3 : 30 %<br>> 3 : 0 % | | | | 20 | 60 | 30 | 100 |
| due to packaging<br>= 0 : 100 %<br>< 1 : 70 %<br>< 2 : 45 %<br>< 3 : 15 %<br>> 3 : 0 % | | | | 30 | 45 | 70 | 70 |
| due to product<br>= 0 : 100%<br>< 1 : 50%<br>< 2 : 20%<br>> 2 : 0 % | | | | 50 | 100 | 100 | 50 |
| 1.1.2. Field failure rate [dpm]<br>< 50 : 100 %<br>< 100 : 80 %<br>< 250 : 60 %<br>< 500 : 40 %<br>> 500 : 0 % | | | 50 | | 80 | 80 | 60 |
| 1.2. Service Quality | | 20 | | | 67,5 | 100 | 87,5 |
| 1.2.1. Response times<br>< 1 working day : 100 %<br>< 2 working days : 80 %<br>< 3 working days : 60 %<br>< 4 working days : 40 %<br>> 4 working days : 0 % | | | 50 | | 60 | 100 | 100 |
| 1.2.2. Time to close a topic (ave.)<br>< 5 working days : 100 %<br>< 10 working days : 75 %<br>< 15 working days : 50 %<br>> 15 working days : 0 % | | | 50 | | 75 | 100 | 75 |
| 1.3. Supplier Audit results<br>no deviations : 100 %<br>only minor deviations : 75 %<br>major deviations : 50 %<br>critical deviations : 0 % | | 15 | | | 75 | 75 | 100 |
| 1.4. Quality Assurance Agreement<br>QAA signed : 100%<br>QAA planned : 50 %<br>no activities : 0 % | | 10 | | | 50 | 50 | 100 |

Table 7.7 Supplier Evaluation Sheet part 2

All suppliers are informed about the evaluation result and are asked to initiate corrective or improvement actions. The entire supplier evaluation process is typically a periodical process, e.g. once per year, with a constant set of activities (fig. 7.6) according to a stable process and evaluation criteria to ensure comparability with earlier results.

## 7.2.4. Supplier Development

As a matter of fact, the performance of suppliers in the field of quality and/or cost and/or strategic orientation is typically never at an optimum for the customer. In other words, most of the time there is considerable improvement potential. Three options are possible in this situation[1]:

(1) Sourcing-In
The company decides to produce the part in house. This option can turn out to be a expensive approach due to limited know how and experience in this particular field

(2) Change of supplier
This is a simple and fast option if the procured product is a commodity product which can be replaced easily. For the decision to do that, of course, the company has to consider the life cycle of the product and the availability of alternative suppliers. A supplier can be available in general, but most of the time needs to undergo a lengthy qualification process.

(3) Supplier development
If the procured product is complex or has a strong interaction with the company's product performance and/or if the supplier's processes are highly integrated in the company's processes (e.g. Just In Time) or if it takes a lot of time to qualify another supplier within the company or, worst case, at the next customer level, then it makes sense to focus on supplier improvement rather than supplier substitution. In addition, typically supplier development is less expensive for the company than supplier substitution.
This is the reason for companies to motivate and to push the suppliers towards improvement. Many companies follow a proactive approach in terms of supplier development, which requires a continuous measurement of the supplier performance e.g. by incoming control, supplier audit or during price negotiations or technical meetings.

---

[1] Wagner, S. (2014), p. 566

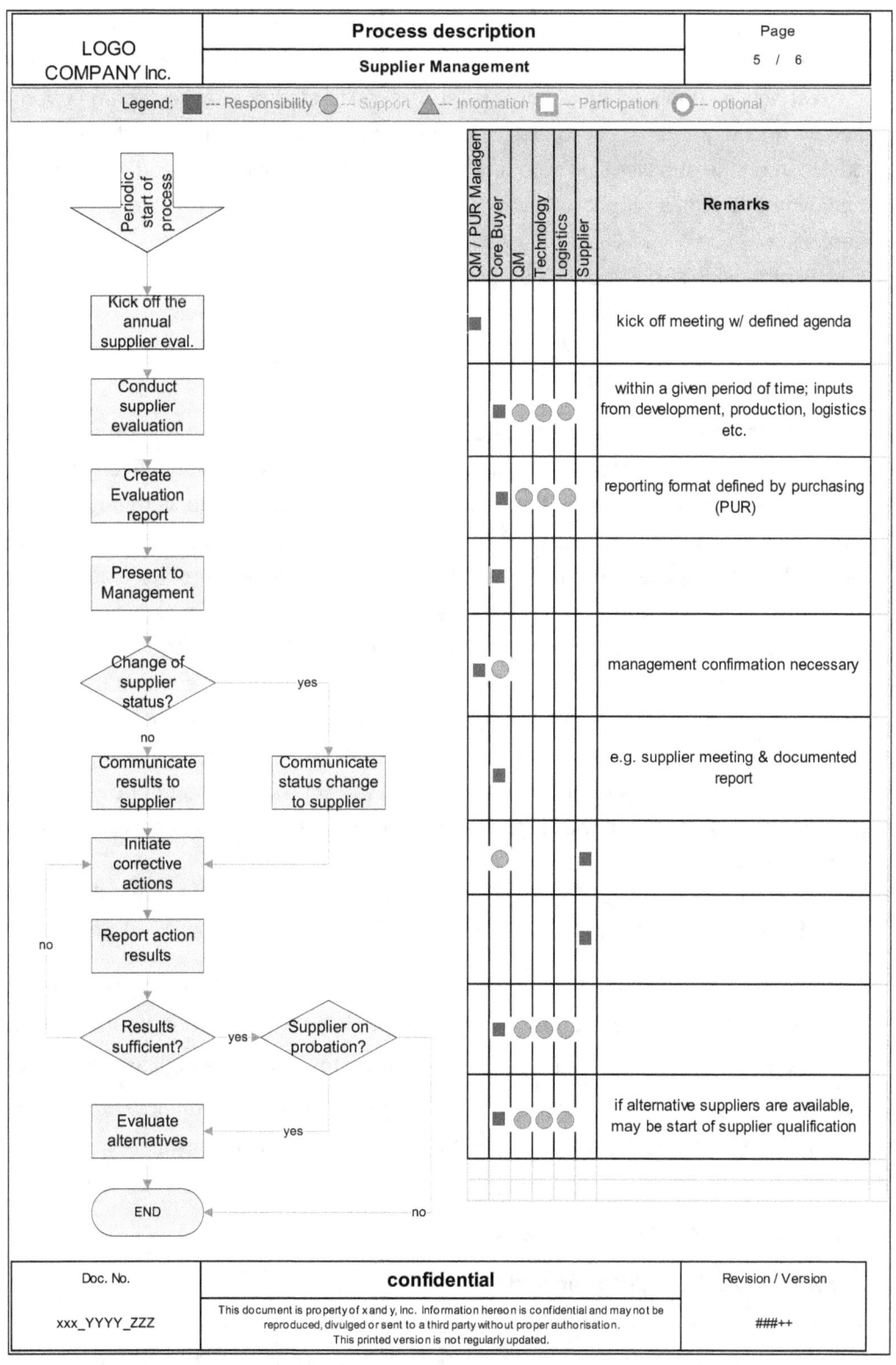

Fig. 7.6 Generic Supplier Evaluation process

The results of this supplier performance monitoring are then summarized in the regular supplier evaluation document or report (fig. 7.6) and those results are also shared with the respective supplier. The supplier is then asked to start corrective and improvement actions and is granted a reasonable amount of time to complete those actions. A major deviation on the supplier side also has the potential to trigger an immediate request for improvement with a stop of delivery until the effectiveness of a corrective action is verified.

On the other side supplier development can be used as a more strategic tool, e.g. when executing the strategy toward a higher level of production outsourcing.

### 7.2.5. Supplier Integration

Certainly the highest and most beneficial level of supplier development is supplier integration. Supplier integration means the supplier is actually involved in the customer's activities to conduct the business. The concept of supplier integration has evolved over the last twenty years in order to benefit from the suppliers' know how[1]. In particular if the supplied product has a strong impact on the final product's performance and if the respective interactions are highly complex it makes a lot of sense to cooperate and to commonly solve technical problems. This cooperation can range from occasional technical meetings to discuss current technical topics through standardized regular technical meetings using a defined agenda and cooperation in technology or product development[2] up to material and technology roadmap and strategy alignment. The higher the degree of supplier integration, however, the higher the efforts required for coordination of those activities. Supplier integration can be regarded as a tool to increase competitiveness[3].

## 7.3. Supplier Audit

So far, measures to ensure the quality of external products and reduce risks associated with using external suppliers has been based mainly on the exchange of information and practical cooperation. At a more stringent level to ensure the quality of the product the quality management system of the supplier should be evaluated on-site. This process is called "supplier audit", the details are explained in this section.

### 7.3.1. Definition and Motivation

According ISO19011:2011[4] an audit is a

---

[1] Krampf, P. (2012)
[2] Hartmann, H. (2004)
[3] Wagner, S. (2007)
[4] DIN EN ISO19011:2011

> "systematic, independent and documented process for obtaining audit evidence and evaluating it objectively to determine the extent to which the audit criteria are fulfilled."

In general, the motivation for supplier audits is two fold:

1. The supplier audit is used to evaluate a supplier's quality capability and
2. The supplier audit is used to find gaps in the supplier's quality management system und eventually serves to improve the supplier.

The supplier audit may be initiated for several reasons like

- as part of a regular supplier evaluation
- as part of the qualification of a new supplier
- in case an existing supplier
    - plans to deliver from a new site or
    - plans to change the production process
    - suffers from major or repeated quality problems

## 7.3.2. Supplier Audit Process

The supplier audit process consists of several steps starting from audit initiation up to final report and closure (fig. 7.7) and it should conform to the standard ISO19011.

Fig. 7.7 Generic supplier audit process

Proper execution of a supplier audit requires a qualified lead auditor and appropriate audit team members. In particular, the lead auditor should meet the following requirements:

- sufficient knowledge of the respective standards
- several years of experience in the field of activities, at least two of them in a quality management department function
- audit experience as audit team member in at least four audits
- sufficient management skills to lead an audit team

> clear and fluent oral and written expression and proficiency in the English language in case of an international setting
> objectivity, persistence and ability to work under pressure
> personal maturity

In more detail the audit process is described in figures 7.9 – 7.13:

Preparation

Typically, the audit team consists of the lead auditor, which often is a member of the quality management department, a responsible buyer and a technical expert who is able to judge the validity of the supplier's statements regarding technology or other relevant aspects in the provision of external products during the audit. The preparation of the questionnaires is described in "7.3.3. Questionnaires".

An audit agenda is mandatory and serves as guidance through the audit and should be followed in general, unless the audit team identified a serious issue which makes it necessary to deviate. It is the task of the lead auditor to make up the time in order to keep the schedule for the final audit result presentation. The reason behind is that typically also the supplier's upper management plans to attend this meeting which puts more emphasis on the auditor's conclusion.

Even though it is up to the lead auditor and also depends on the particular topic, the proposed agenda (fig. 7.9) turned out to be reasonable and accepted by suppliers. Figure 7.9 shows a one and half day audit agenda. It contains a major portion of presentations by the supplier, which gives the audit team the chance to learn what the supplier's quality management system should be like in general or specifically if selected areas have to be audited. In the following sections the audit team has the opportunity to compare reality with the theoretical model for the QM system presented. This presentation is meant to take only a minor share of the available audit time.

Execution

The actual execution of the audit should include an opening session which gives all participants the opportunity to get to know each other, to clarify open questions and to get all participants on the same level of knowledge regarding the audit and expectation of the customer. It helps to reduce unnecessary tensions.

This initial meeting is followed by supplier presentations on specific topics or on the supplier's quality management system in general in order to make those known to the audit team and the actual audit tour through the production line or other departments.

Product and process related audits typically follow the flow of the product from incoming inspection through production and ends at final test or even at the warehouse. It is helpful to ask the supplier before the audit to provide the production flow including relevant quality inspections to the audit team. The process flow is later one part of the audit report. During the audit tour the audit team is asking the supplier's employees for information and/or evidences. In this context it is important to receive answers from the person asked, not from the respective head of department or even the quality management responsible.

In case of finding non-conformances those need to be documented immediately. It may be difficult or even impossible later in the line tour to describe the non-conformance in a concise manner. In addition, a non-conformance immediately needs to be communicated to the attending member of the supplier's quality management team. This way the topic can be discussed on the shop floor with all persons and evidences available and a common understanding of the state of affairs can be discussed and developed. This approach turned out to be very useful to prevent misunderstandings and to avoid lengthy discussions and argumentation in the final session.

At the end of major phases of the audit, e.g. presentation session, line tour day 1 etc., open items should be reviewed together with the supplier. The gives the supplier the opportunity to supply additional information or evidence for identified non-conformances prior to the auditors' meeting.

After finishing the actual audit tour the audit team summarizes, discusses and evaluates the findings. A preliminary audit report may be filled in handwritten or using a PC (fig. 7.16), which will be signed by both parties at the end of the audit.

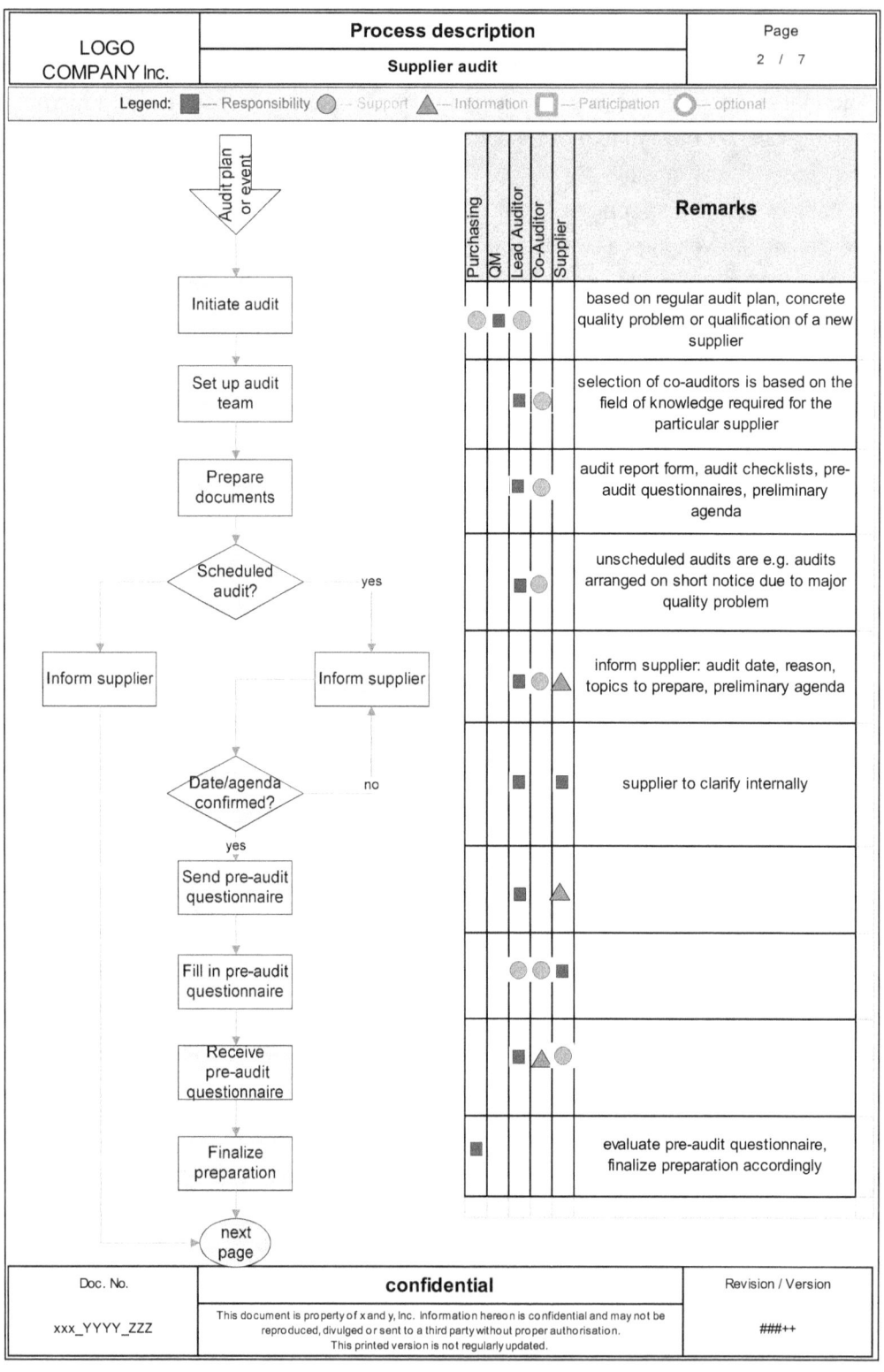

Fig. 7.8 Generic process description for supplier audit preparation

**Agenda**            J. Wittmann, 2006/03/24

Supplier May 04/05, 2006

| Time | Place | Item | Who |
|---|---|---|---|
| **Day I** | | | |
| 09:00 – 09:10 | | Welcome / Introduction | all |
| 09:10 – 09:30 | | Presentation Customer Organization / Customer Audit Procedure | Customer |
| 09:30 – 09:45 | | Presentation Organization / overview of Supplier | Supplier |
| 09:45 – 10:15 | | Presentation overview QM-System / Documentation | Supplier |
| 10:15 – 10:30 | | Presentation overview of the manufacturing process | Supplier |
| 10:30 – 10:55 | | Presentation Process control process (including SPC rules and data, measurement equipment control, non-conformance management, customer information…) | Supplier |
| 10:55 – 11:15 | | Presentation Change management | Supplier |
| 11:15 – 11:25 | | Presentation Preventive maintenance | Supplier |
| 11:25 – 11:35 | | Presentation Spec approval process and transfer into Supplier internal documentation down to production level | Supplier |
| 11:35 – 12:20 | | Review pre-audit questionnaire | Customer + Supplier |
| 12:20 – 13:05 | | Lunch on site | Customer + Supplier |
| 13:05 – 13:20 | | Review open items | Customer + Supplier |
| 13:20 – 17:30 | | Line tour day 1 | Customer + Supplier |
| 17:30 – 18:00 | | Open items day I | Customer + Supplier |
| **Day II** | | | |
| 09:00 – 09:30 | | Open items day I | Customer |
| 09:30 – 11:00 | | Line tour day II | Customer + Supplier |
| 11:00 – 11:30 | | Open items / final review of open items | Customer + Supplier |
| 11:30 – 12:15 | | Lunch on site | Customer + Supplier |
| 12:15 – 13:45 | | Internal meeting auditors | Customer |
| 13:45 – 14:00 | | Audit result presentation | Customer + Supplier |

Fig. 7.9 Example for a one and a half day supplier audit

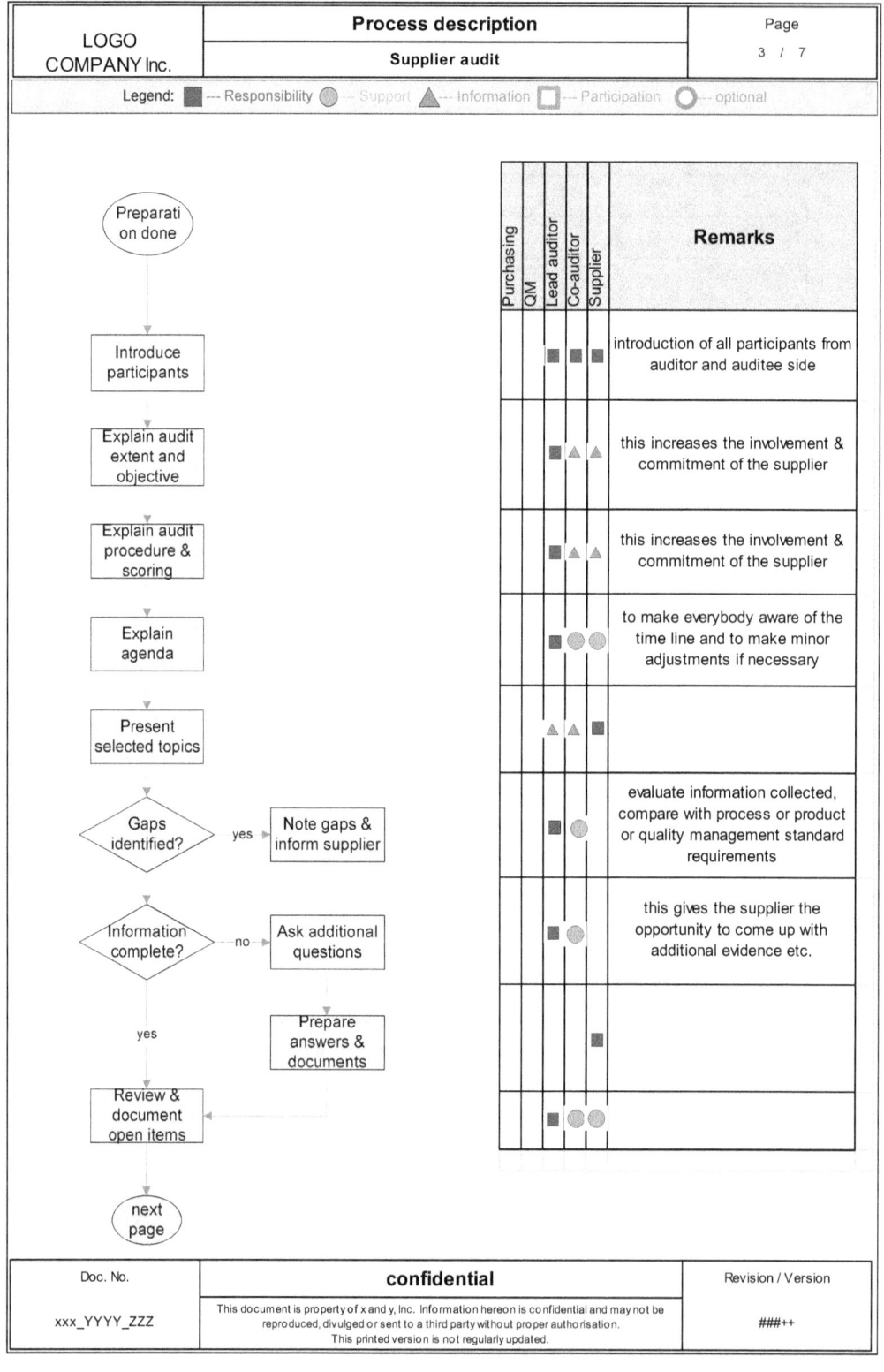

Fig. 7.10 Generic process description for audit execution – part I

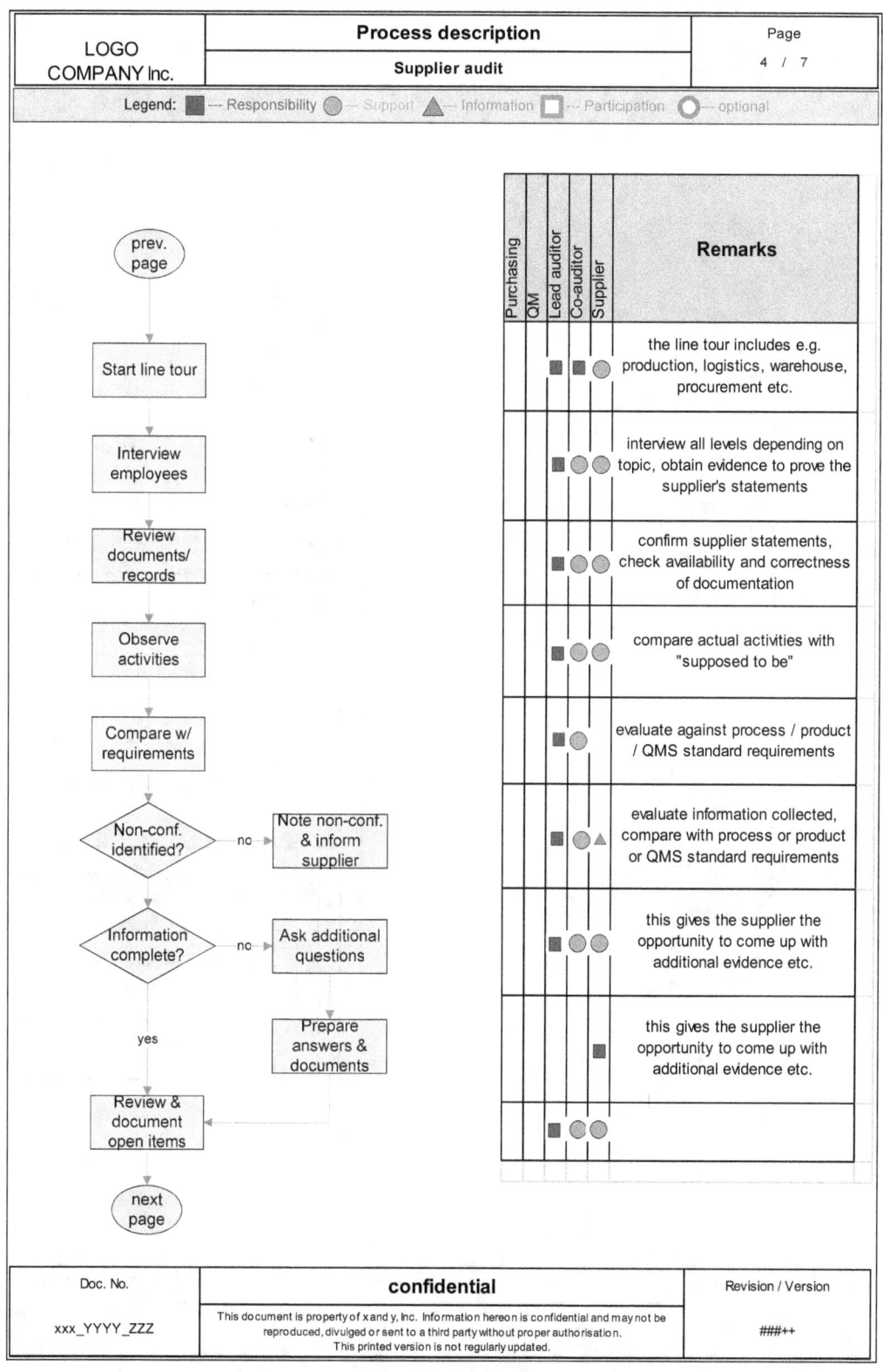

Fig. 7.11 Generic process for audit execution – part II

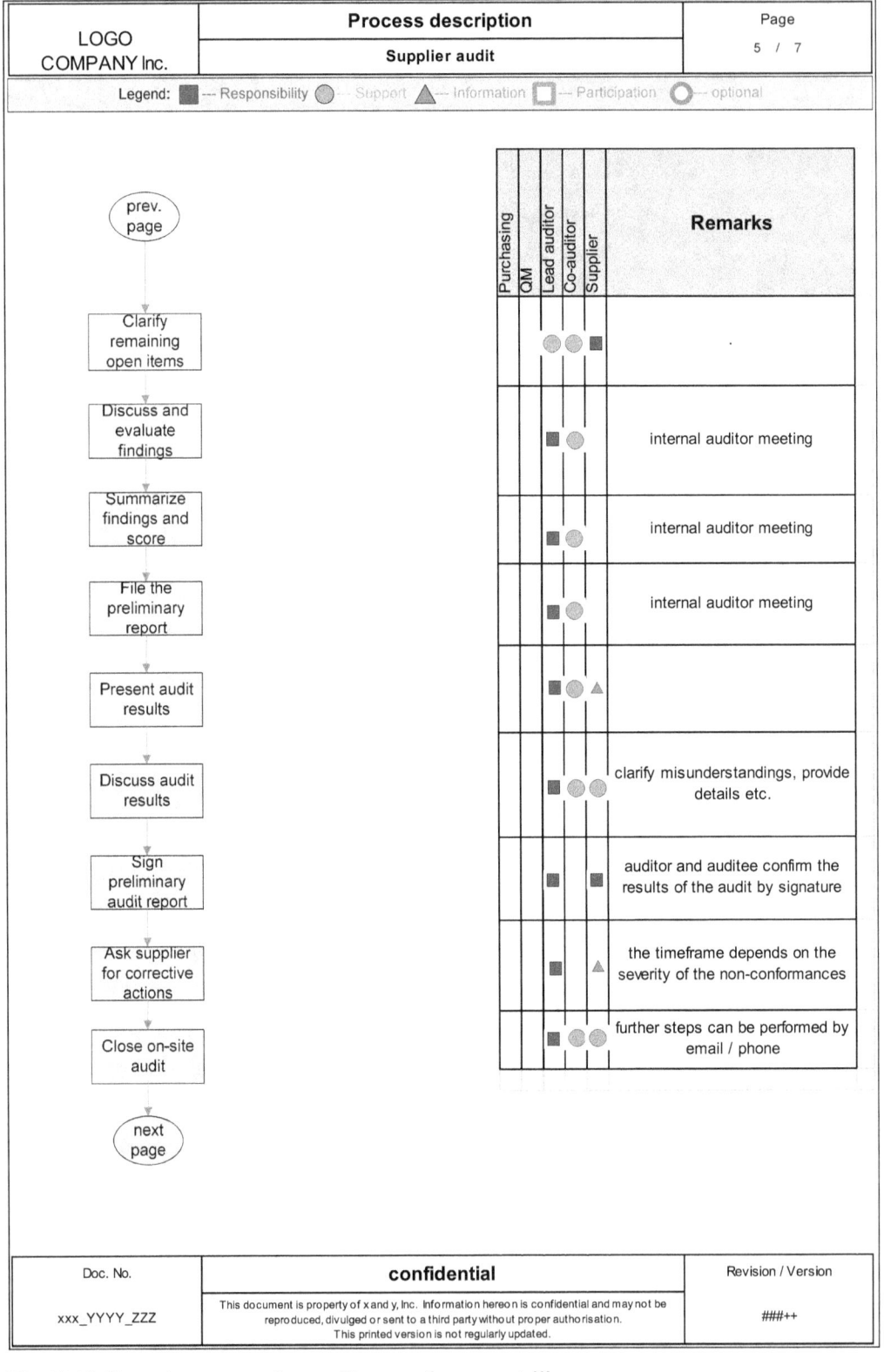

Fig. 7.12 Generic process for audit execution – part III

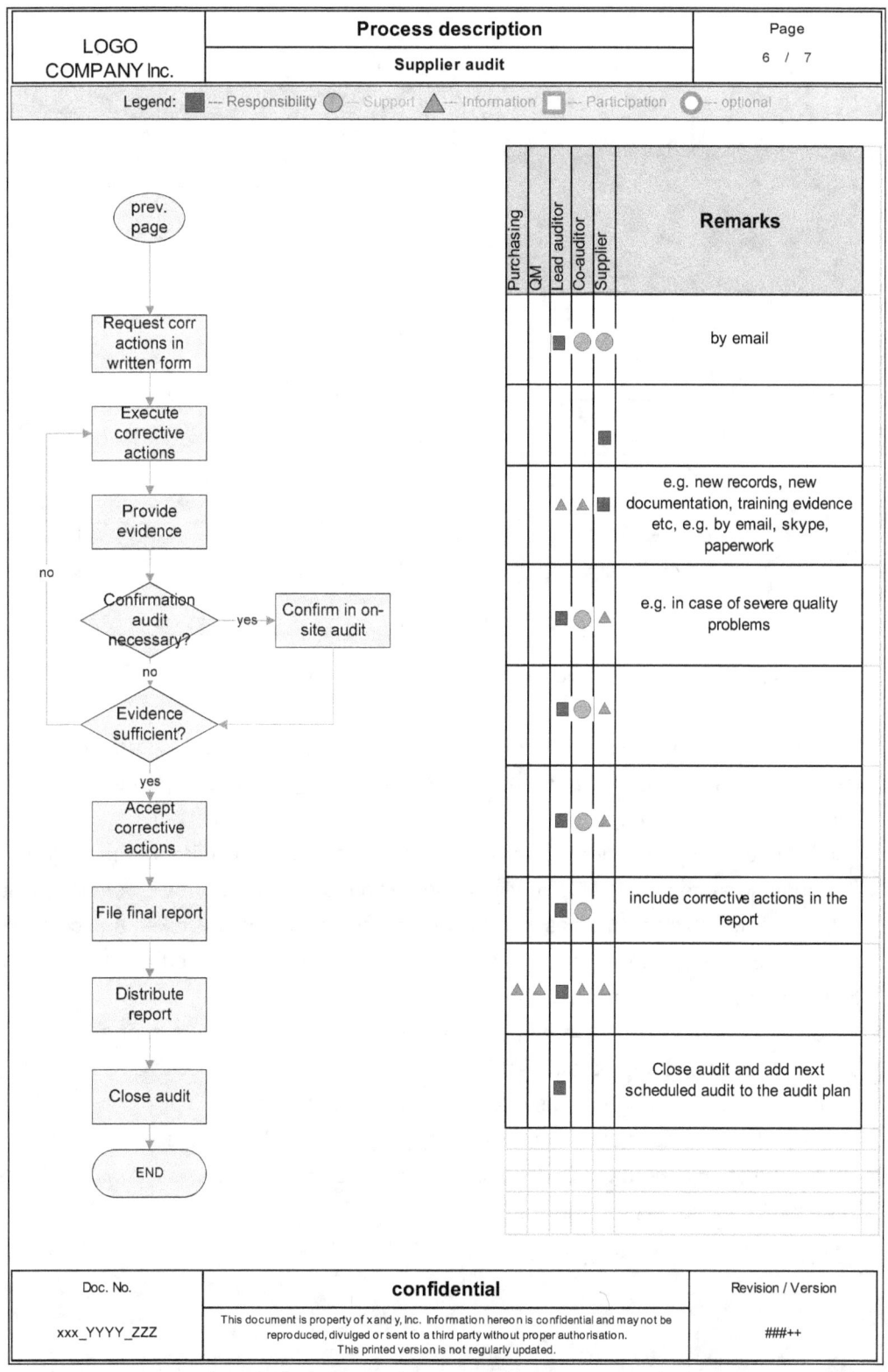

Fig. 7.13 Generic process flow for corrective actions and audit closure

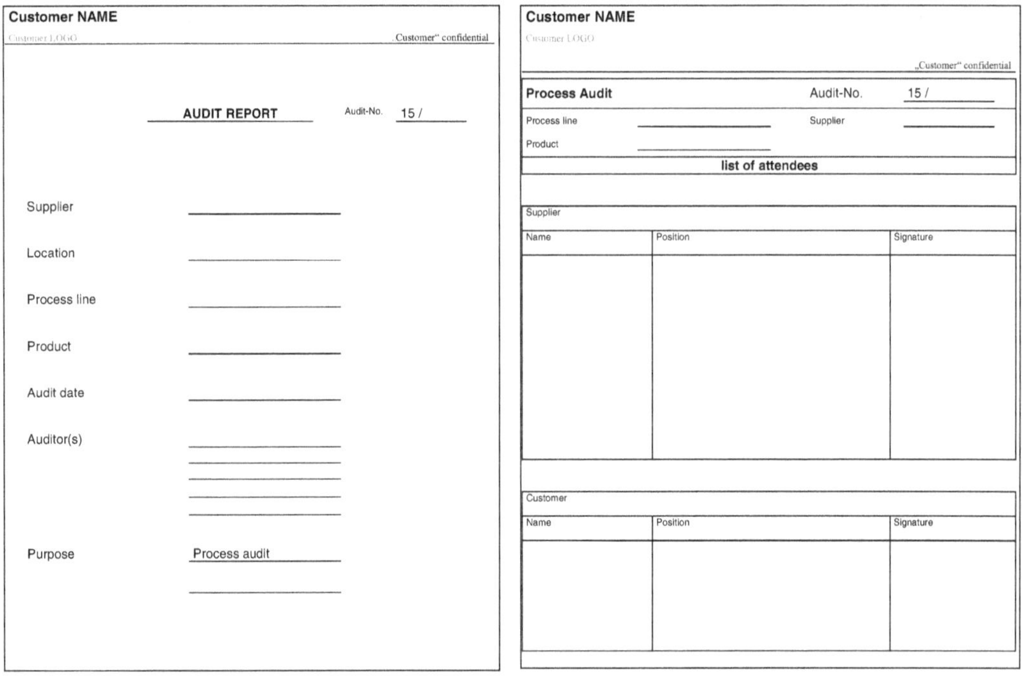
Customer NAME

Customer LOGO

Customer" confidential

AUDIT REPORT Audit-No. 15 /

Supplier

Location

Process line

Product

Audit date

Auditor(s)

Purpose Process audit

Customer NAME

Customer LOGO

„Customer" confidential

| Process Audit | | Audit-No. | 15 / |
|---|---|---|---|
| Process line | | Supplier | |
| Product | | | |

list of attendees

Supplier

| Name | Position | Signature |
|---|---|---|
| | | |

Customer

| Name | Position | Signature |
|---|---|---|
| | | |

Fig. 7.14 Audit report cover page and attendance sheet

With the list of attendees participation of auditors and auditees is documented. In case the audit lasts over more than one day it is advisable to have separate attendance sheets for the two days.

The list of concerns (i.e. non-conformances) provides a fast overview of the customer's findings, the more detailed Corrective Action Request (fig. 7.16) provides a description of the finding with signatures as well as fields for corrective actions taken and corrective actions accepted, hence comprises the complete controlled (feedback) cycle from finding to closure of a particular topic.

In conclusion, the audit report also contains the overall judgement and/or evaluation of the audit result and a separate part for remarks, which may be positive aspects of the supplier's quality management system or very minor findings.

At the end of the onsite audit a final meeting takes place which the lead auditor uses to inform everybody about the audit result. It is also used to clarify open items and to ask the supplier for corrective actions.

**Customer NAME**

Customer LOGO

„Customer" confidential

| **Process Audit** | | Audit-No. | 15 / |
|---|---|---|---|
| Process line | | Supplier | |
| Product | | | |
| Process step | | | |

**process description**

| | techn. evaluation infineon | audit-result |
|---|---|---|
| **1) method** | | |
| **2) material**<br>2.1) production material | | |
| 2.2) auxiliary material | | |
| **3) equipment** | | |
| **4) environment** | | |
| **5) quality control**<br>(characteristic - sample - frequency - specification) | | |
| **remarks** | | |

**Customer NAME**

Customer LOGO

„Customer" confidential

| **Process Audit** | | Audit-No. | 15 / |
|---|---|---|---|
| Process line | | Supplier | |
| Product | | | |

**process flow**

| No. | Process step | Blatt x.n | remarks |
|---|---|---|---|
| 1 | | | |
| 2 | | | |
| 3 | | | |
| 4 | | | |
| 5 | | | |
| 6 | | | |
| 7 | | | |
| 8 | | | |
| 9 | | | |
| 10 | | | |
| 11 | | | |
| 12 | | | |
| 13 | | | |
| 14 | | | |
| 15 | | | |
| 16 | | | |
| 17 | | | |
| 18 | | | |
| 19 | | | |
| 20 | | | |

Fig. 7.15 Process flow and process description as part of the audit report

**Customer NAME**

Customer LOGO

"Customer" confidential

| **Process Audit** | | Audit-No. | 15 / |
|---|---|---|---|
| Process line | | supplier | |
| Product | | | |

**List of Concerns**

| Severity C | M | m | process step + criteria | Blatt / page x.nn |
|---|---|---|---|---|
| | | | | |
| | | | | |
| | | | | |
| | | | | |
| | | | | |
| | | | | |
| | | | | |
| | | | | |
| | | | | |
| | | | | |
| | | | | |
| | | | | |
| | | | | |
| | | | | |

C = critical
M = major
m = minor

**Customer NAME**

Customer LOGO

"Customer" confidential

| **Process Audit** | | Audit-No. | 15 / |
|---|---|---|---|
| Process line | | supplier | |
| Product | | process module | |
| Process step | | | |
| Type: C / M / m | **Corrective Action Request** | | |

**1. Concern**

Supplier (Date/Signature) — Lead Auditor (Date/Signature)

date for corrective actions: ______

**2. Corrective action taken**

Supplier (Date/Signature)

**3. Completion**

Corrective action accepted? YES ☐ NO ☐

Follow-up audit necessary? YES ☐ Ddate: ______ NO ☐

Lead Auditor (Date/Signature)

Fig. 7.16 Summary and detailed overview of findings including corrective actions

**Customer NAME**
Customer LOGO

"Customer" confidential

| **Process Audit** | | Audit-No. | 15 / |
|---|---|---|---|
| Process line | | supplier | |
| Product | | | |

**Conclusion**

**1. Pro's**

**2. Concerns**

| | | |
|---|---|---|
| number of critical concerns | (C) | |
| number of major concerns | (M) | |
| number of minor concerns | (m) | |
| number of remarks | (R) | |

Gesamterfüllungsgrad / total grade of fulfilment (%)

**3. General assessment**

**4. Sign off**

Auditee (QA responsible) Date Auditor

**Customer NAME**
Customer LOGO

„Customer" confidential

| **Process Audit** | | Audit-No. | 15 / |
|---|---|---|---|
| Process line | | supplier | |
| Product | | | |

**Remarks & Corrective actions**

| No. | Remarks |
|---|---|
| | |

| No. | Corrective actions |
|---|---|
| | |

Fig. 7.17 Audit report conclusion and remarks

Corrective actions and final report

In order to close the audit evidence for successful implementation of corrective actions have to be provided to the lead auditor. Depending on the severity of the non-conformances, it is sufficient to assess documented evidence, e.g. a new Standard Operating Procedure, training records etc., or a re-audit may be necessary to convince the lead auditor from the improved status. The target date for finishing the corrective actions ranges from a few days for very severe deviations to weeks or months. Very minor issues could even be subject of another regular audit later on.

After closure of all open items the lead auditor files the final report and distributes it internally, e.g. to the management and to the procurement organization, and externally to the supplier's responsibles. Then the audit is officially closed.

## 7.3.3. Quantification of the Audit Result

The quanification of the audit result can be done in different ways, e.g. by enumerating the number of non-conformances found with or without distinguishing between severe or minor deviations, or by calculating an audit score or audit rating. Even though it is obvious that audits are always only yielding a scoring of the supplier's quality performance based on samples, hence the audit score should not be understood as precise and absolute measure, it turned out the audit score does actually reflect the supplier quality capability[1]. Being aware of this it is useful to have an audit score in order to be able to compare audit results from different suppliers. Here two exemplary methods of determining a quantitative score are shown basically:

(1) additive method[2]

For each single question of a defined questionnaire the additive method assigns points in order to rate the degree of fulfilment of requirements, e.g.

| | |
|---|---|
| 10 points | requirements fully satisfied |
| 8 points | requirements predominantly satisfied, minor deviations |
| 6 points | requirements partially satisfied, major deviations |
| 4 points | requirements insufficiently satisfied, severe deviations |
| 0 points | requirements not satisfied |

The degree of fulfilment E[%] is then calculated to be:

$$E[\%] = \frac{sum\_of\_achieved\_points}{maximum\_number\_of\_points} \cdot 100\%$$

The degree of fulfilment results in the audit rating:

| Degree of fulfilment [%] | Audit rating |
|---|---|
| E ≥ 90 | Capable |
| 80 ≤ E < 90 | Conditionally capable |
| E < 80 | Not capable |

Table 7.8 Audit rating

A "not capable" audit rating may result in the start of supplier phase out activities.

(2) subtractive method[3]

---
[1] Wittmann, J. (2006)
[2] VDA Prozessaudit (2010)
[3] Wittmann, J., Bergholz, W. (2006)

In the subtractive method the supplier starts with a rating of 100%. Findings are assessed regarding their severity and depending on their severity percentage points are subtracted from 100%.

The degree of fulfilment E[%] is then calculated to be:

$$E[\%] = 100\% - C \cdot 20\% - M \cdot 10\% - m \cdot 2.5\%$$

In this case C stands for the number of critical deviations and M and m stand for major and minor deviations, respectively. In this connection deviations are classified as:

Critical
means a severe deviation which will impact the product quality negatively, e.g. use of scrapped material, falsification of quality records, use of unreleased production equipment

Major
means a deviation which probably leads to a negative impact on product quality, e.g. volume production with improper SPC, production equipment with overdue maintenance, measurement equipment with overdue calibration

minor
means a minor deviation which probably does not lead to any negative impact on product quality, e.g. corrective action taken, but not documented, minor deviation from Standard Operating Procedure

Again, the degree of fulfilment results in the audit rating:

| Degree of fulfilment [%] | Audit rating |
|---|---|
| E ≥ 95 | Fully accepted |
| 80 ≤ E < 95 | Accepted with limitations |
| 60 ≤ E < 80 | Major improvements necessary |
| E < 60 | Not acceptable |

Table 7.9 Audit rating

As before the rating E < 60% will most likely result in the initiation of phase-out activities for the supplier.

### 7.3.4. Questionnaires

Questionnaires are a very useful tool for audit preparation and execution since they will improve the efficiency and effectiveness of the audit process. Depending on the purpose of the audit and the quality or certification level the questionnaire may be based on the respective applicable standard, e.g. ISO9001, or is more driven by branch and product specific requirements. Questions which are specific to a particular product are more likely to reveal relevant information than generic questions. In this section a few examples are shown what a questionnaire could look like for various fields:

| Nr. | Element | Question | defined | proven | score | Evidence |
|---|---|---|---|---|---|---|
| 8. Operation | | | | | | |
| 8.1. Operational planning and control | | | | | | |
| | 8.1.a. | What methods are used to determine product requirements | | | | |
| | 8.1.b. | What methods are used to determine criteria for product acceptance. What kind of criteria are set up for processes. | | | | |
| | 8.1.c. | What methods are used to determine resources required to achieve product conformity | | | | |
| | 8.1.d. | What system is in place for monitoring and controlling processes based on defined criteria | | | | |
| | 8.1.e. | What system is in place to ensure conformity of products can be demonstrated (archiving of documentation) | | | | |
| | 8.1. | How do you ensure that operational planning and control is adequate to the company's size | | | | |
| | 8.1. | How does the company ensure that planned changes and unintended changes are under control | | | | |
| 8.2. Determination of requirements for products | | | | | | |
| 8.2.1. Customer communication | | | | | | |
| | 8.2.1.a. | How does the company organize customer communication regarding product related information | | | | |
| | 8.2.1.b. | How does the company organize customer communication regarding contracts, orders etc. | | | | |
| | 8.2.1.c. | How does the company organize customer communication regarding customer feedback and complaints | | | | |
| | 8.2.1.d. | How does the company organize customer communication regarding customer property | | | | |
| | 8.2.1.e. | How does the company organize customer communication regarding contingency planning / actions | | | | |

Table 7.10 Example for ISO9001 related questionnaire in the area of element 8 "Operations"

| Nr. | Element | Question | defined | proven | score | Evidence |
|---|---|---|---|---|---|---|
| Production | | | | | | |
| | Production | | | | | |
| 1 | | What procedure is in place to ensure appropriate maintenance of production equipment | | | | |
| 2 | | What statistical methods are used (e.g. SPC) for monitoring and improvement of important product characteristics | | | | |
| 3 | | How do you determine important product or process characteristics | | | | |
| 4 | | What system is in place to regularly analyze the process capabilities | | | | |
| 5 | | What procedure is applied in case of nonconformities in production | | | | |
| 6 | | What methods are used to ensure that all relevant to product quality parameters are included in the control and production plans including measurement planning, sampling and execution | | | | |
| 7 | | Is there a quarantine storage area established for nonconforming material | | | | |
| 8 | | What documented procdure is available describing handling of nonconforming material | | | | |
| 9 | | How do you distinguish material prior and after quality inspection | | | | |
| 10 | | What method is used to determine in-process inspection to ensure product quality | | | | |
| 11 | | What system is used to document inspection results | | | | |
| 12 | | What method is in place to ensure traceability of pre-products up to the final product | | | | |

Table 7.11 Example for production related questionnaire

In addition to a regular production audit the audit may also be executed in an area with very area specific requirements, e.g. in the semiconductor industry (table 7.12) or in chemical plant engineering (table 13).

For instance questions in the semiconductor industry need to cover Statistical Process Control (SPC), stability of production processes, Electrostatic Discharge (ESD) and in particular clean room relevant topics.

Chemical plant engineering on the other hand has to ensure high performing welding processes with no weaknesses in the piping especially under pressure. So in this case questions have to focus on the training level of the welders, on weld inspection using xray etc., traceability of piping elements etc.

| Nr. | Element | Question | defined | proven | score | Evidence |
|---|---|---|---|---|---|---|
| | Methods | | | | | |
| 1 | | What methods are used to identify unstable processes | | | | |
| 2 | | What methods are used to manage unstable processes | | | | |
| 3 | | To what extent statistical methods are used for production control | | | | |
| 4 | | What documentation and reporting system is in place for SPC data (e.g. cp, cpk) | | | | |
| 5 | | What procedure is in place for root cause analysis and follow up of nonconforming or unstable situations | | | | |
| 6 | | What procedures are in place defining introduction and/or qualification of new or changed manufacturing processes | | | | |
| 7 | | What procedures are in place defining introduction and/or qualification of new or changed equipment | | | | |
| 8 | | What way does the company ensure that the customer is informed about changes (if required) | | | | |
| 9 | | What methods and inputs are used to define a process control strategy | | | | |
| | Environment | | | | | |
| 10 | | What methods are used to ensure clean and well organized work environment | | | | |
| 11 | | How do you ensure that all employees wear correct and adequate clothes (e.g. clean room clothes) | | | | |
| 12 | | What way does the company ensure cleanness of clothes | | | | |
| 13 | | What procedures are in place controlling employee's clean room behaviour | | | | |
| 14 | | What methods are used to keep the clean room under control regarding particles | | | | |
| 15 | | What strategy is in place for defect reduction | | | | |
| 16 | | What methods are used to keep the clean room under control regarding metals | | | | |
| 17 | | What procedures are in place to avoid ESD related damage of the product | | | | |

Table 7.12 Example for audit questionnaire in the semiconductor industry

In addition to company audits which cover 100% of all relevant company sections selected departments or projects can be subject for a supplier audit. Typically this would be a major project over several years. In this special case again the questionnaire may be adjusted in order to meet the overall project requirements. One important topic in a large project is proper documentation which covers project specific manuals, SOPs, checklists etc. as well as project specific records management (fig. 7.14).

| Nr. | Element | Question | defined | proven | score | Evidence |
|---|---|---|---|---|---|---|
| **Welding** | | | | | | |
| 1 | | Are all welding procedures required for the project established and approved | | | | |
| 2 | | Are all welding procedures distributed to the welders | | | | |
| 3 | | Is an updated welding procedure list available | | | | |
| 4 | | Are welders tests carried out according to specified standards | | | | |
| 5 | | Do welding inspectors have required qualification and all equipment and gauges required | | | | |
| 6 | | Is a procedure for handling of welding filler material existing and distributed to the personnel concerned | | | | |
| 7 | | Is there a system for rebaking electrodes | | | | |
| 8 | | Are the ovens calibrated | | | | |
| 9 | | Are the hot boxes returned to the stores every night | | | | |
| 10 | | Is access to filler material given to authorized personnel only | | | | |
| 11 | | Are different types of electrodes segregated in the store and ovens | | | | |
| 12 | | Is a defined system used of issuing filler materials | | | | |
| 13 | | Is an updated welders list available | | | | |
| 14 | | Are welders qualification certificates available and filed | | | | |

| Nr. | Element | Question | defined | proven | score | Evidence |
|---|---|---|---|---|---|---|
| **NDE** | | | | | | |
| 27 | | Are NDE procedures existing and available on site | | | | |
| 28 | | How do you control the performance of the subcontractors | | | | |
| 29 | | Is storing of films adequate | | | | |
| 30 | | Are NDE reports available for all NDE performed | | | | |
| 31 | | Are repair films checked against originals | | | | |
| 32 | | How is it ensured that all ordered NDE is carried out | | | | |
| 33 | | Is guaranteed that the first xx welds of each welder are 100% radiographed. | | | | |
| 34 | | Where is the NDE-scope defined. Is guaranteed that NDE is performed and documented (Steel Structure) | | | | |
| 35 | | Are welds which have to be tested or repaired clearly marked in field | | | | |

Table 7.13 Example for questionnaire in chemical plant engineering in welding and non-destructive examination (NDE)

| Nr. | Element | Question | defined | proven | score | Evidence |
|---|---|---|---|---|---|---|
| | Documentation | | | | | |
| | General | | | | | |
| 11 | | Do you have a list of all site relevant documents (manuals, SOPs, Forms, Checklists, method statements, ...) | | | | |
| 12 | | Is there a documented procedure defining access rights to documents and records on site | | | | |
| 13 | | Is the position of document control described by a approved job description | | | | |
| | Drawings | | | | | |
| 14 | | Is a drawing register available? What system is used? | | | | |
| 15 | | How do you ensure the register is kept and updated by nominated personnel? | | | | |
| 16 | | How do you ensure the revision status is indicated? | | | | |
| 17 | | What system is in place to ensure a distribution list for drawings is available? | | | | |
| 18 | | How do you ensure old drawings are withdrawn from workshop and properly marked "Cancelled", "Superseded" or "Obsolete"? | | | | |
| 19 | | What procedure is in place to ensure only latest revisions of drawings are in workshop, etc.? | | | | |
| 20 | | How do you define limitation of access to the document centre to authorized personnel only? | | | | |
| 21 | | What system is in place for proper storage of all relevant documents? | | | | |
| 22 | | How do you ensure revised status is indicated on drawings and material lists? (red line mark up) | | | | |
| 23 | | Is the documentation, e.g. "as build", recorded progressively? | | | | |
| | Records | | | | | |
| 24 | | Is there a process/SOP "Records Management" | | | | |
| 25 | | How do you determine required records? | | | | |
| 26 | | What records are required for final documentation? | | | | |
| 27 | | What records are managed on site / in in the office? | | | | |
| 28 | | Where are records archived? | | | | |
| 29 | | Who has access to records? | | | | |
| 30 | | How do you identify the required retention time for the records? | | | | |

Table 7.14 Example for project documentation related audit questionnaire

# 7.4. Quality Assurance Agreement

## 7.4.1. Motivation and Purpose

Quality Assurance Agreements (QAA) are contracts between supplier and customer which regulate quality related issues which detail or reach beyond the basic standards requirements.

The purpose to have an additional supplier-customer-contract[1] is to

- ensure product quality by defining product and/or production process specific rules and procedures
- increase the supplier's general quality capability
- define areas of responsibilities and accountabilities on the supplier and the customer side
- to ensure a consistent quality assurance system which avoids redundant quality inspections, hence reduces cost overall

In some cases, the purpose of the QAA also includes a general improvement program for the supplier quality management system. However, since the QAA is a contract, the content of the QAA is subject to negotiations. Eventually this may result in content wise differing QAAs even for the same kind of supplier.

## 7.4.2. Contents of a QAA

The contents of a Quality Assurance Agreement depend on the particular product, but in many cases cover the following items[2,3,4]:

- General

This item typically includes requirements regarding quality management system and/or company certification levels (e.g. DIN EN ISO9001, ISO/TS 16949). Based on the customer's quality management philosophy "Zero Defects" can be expressed as a target for the supplier.

- Documents & records

The documents section covers format, language and content of required documents. As for records, for instance, it needs to be defined which records are required and what is the necessary minimum storage time. Records have to be available for the customer, if required.

---

[1] Ensthaler, J. (2007)
[2] Kroonder, M. (2007)
[3] ZVEI (2006)
[4] Bosch (2011)

➢ Communication

An uninterrupted continuous flow of material is important for the customer to guarantee a smooth and continuous production. Therefore, the supplier is requested to provide a product termination notice (PTN) to the customer well ahead of time. Also, if the supplier intends to change the product or it's production process, the customer needs to be informed using a process change notification (PCN). Both help the customer to keep his own production running without any interference caused by the supplier. It is necessary to define the kind of process/product change which initiates as PCN (major change list), for silicon wafers in the semiconductor industry e.g.

- ➢ Relocation of a production to a different site
- ➢ Production change to a new technology
- ➢ Change of a raw material supplier
- ➢ Change of the crystal pulling process
- ➢ …

Communication includes also logistical information like delivery dates and volumes and production performance data like SPC (Statistical Process Control) data.

➢ Product realization

In addition to requirements like machine and process capability targets (i.e. $c_p$, $c_{pk}$ values) and accepted failure rates (e.g. dpm levels), the customer uses the QAA to establish and confirm the right to execute supplier process audits in order to verify product realization activities.

➢ Product requirements

This section often covers the way new products are implemented and introduced to the customer, i.e. the requirement for an Advanced Product Quality Planning (APQP) and the execution of the Production Part Approval Process (PPAP) prior to volume production.

➢ Procurement

The supplier has to establish a system to ensure the quality of purchased parts (parts, materials, services) from his suppliers. The customer is allowed to reduce incoming inspection. Hence the supplier hast to ensure incoming quality on the customer side by suitable QA measures on his side.

➢ Traceability

In order to minimize the consequences of a quality issue in many cases the requirement of traceability is part of a QAA. The level of detail depends on the product and the application and may include part number, production lot number, manufacturing date codes, equipment used etc.

> Corrective action request

Typically, the supplier is requested to set up a failure analysis request (FAR) / corrective action request (CAR) process. The customer's intention hereby is to have fast and sound response from the supplier in case of the receipt of a non-conforming product. The supplier has to ensure maximum response times e.g.

| | Response after: |
|---|---|
| Receipt of returned material | 1 working day |
| Initial problem verification on supplier side complete | 2 working days |
| Containment plan available | 3 working days |
| Problem analysis completed | 9 working days |
| Corrective actions implemented | according plan |

This list is not complete and not detailed enough for an actual Quality Assurance Agreement, but intends to give an idea of the most important contents.

## 7.5. Technical Material Management

First of all material in this section always means direct material

> "All items such as raw materials, standard and specialized parts, and sub-assemblies required to assemble or manufacture a complete product." [1]

Direct material can have a major impact on the quality of the final product, i.e. performance, yield and/or reliability. Therefore, in addition to supplier management, an active management of technical material related topics is an obligation in order to achieve constantly high quality. The approach described in this section is taken from the semiconductor industry, but applies also to other areas, e.g. the photovoltaic industry.

### 7.5.1. Material Specification

As described earlier the quality management standards require the organization to provide information to the supplier regarding the externally supplied part (material,

---
[1] BusinessDictionary (2015)

service…). This includes general information relevant for all kind of materials, more precise information on a group of materials, e.g. gases, wafers, CMP slurry etc., and very specific information on a particular unique material.

General Information

This general part typically lists material related requirements on the supplier level like standards or legal requirements. As for materials this may include quality information on the applicable standards, but focuses on environmental standards, e.g. ISO14001, EMAS (Eco-Management and audit scheme) or OHSAS 18001 (Occupational Health and Safety Assessment Series).

On the legal side the suppliers of potentially hazardous or polluting materials may be requested to comply with the respective law, e.g. 67/548/EEC (EU), OSHA 29 CFR 1910.1200 (US) or RoHS requirements.

In this context it is relevant to note, that a QM system can form the basis to ensure compliance, and if a compliance management system exists, there are strong links and synergies between these two systems.

Material Group Information

Material Group information is valid for all products within a particular material group. It is more convenient to summarize common material group requirements in one document rather than listing the same requirements on each material specific document. It is used to describe general technical requirements regarding supply of a material group.

If the general technical requirements which apply to all items in a material group are covered in a separate document, the administrative effort in the procurement process is significantly reduced.

Typically, this covers at least:

➢ Packaging and shipping

Packaging and shipping conditions have to be controlled in order to protect the material from e.g. temperature, humidity, particles…

➢ Delivery conditions

Depending on the way of transport, e.g. air freight or sea freight, the supplier has to treat the material in a different way. In general the supplier needs to be informed about conditions like the requested charge size, special packing units, the packing list,

Incoterms 2000[1], delivery address, net weight, net volume, net value, delivery note number...

➤ Marking

Has to guarantee an undelible and unique identification marker and traceability of the materials and includes e.g. production lot number, quantity, date, number of respective documentation, technical information ...

➤ Test procedures

Testing is necessary to ensure the required level of quality. The customer may stipulate specific test methods or only refer to standards.

➤ Process changes

Changes in the manufacturing process of the incoming material, e.g. process changes, relocation of the production site or change of the subsupplier can have a detrimental impact on the final product quality. Hence the customer and the material supplier have to agree on a "major change" list, which requires the material supplier to inform the customer about changes and give the customer the opportunity to test samples of the changed material in his production process prior to implementation in production.

➤ Customer complaints

Depending on the impact the material might have on the final product's quality, the handling of customer complaints has a different level of priority. For material with direct and significant impact on the product quality, the supplier has to ensure a fast execution of problem solving within a few days. The customer, on the other hand, is obliged to report complaints to the supplier in written form.

Specific Material Information

This part of the material related information to the supplier comprises specific information regarding the product and the respective quality level.

This covers e.g.:

➤ Technical parameters

Those include e.g. geometrical specifications, resistivity, surface roughness etc. including the respective target value and upper and lower specification limits. In addition this may be complemented by a defined particular production site or a defined particular production process and preferred test methods for relevant material parameters.

---

[1] Berlin Partner GmbH (2007)

- Required quality level

If necessary the accepted quality level (AQL), the accepted defects per million (dpm) level or the level of contamination in or on the surface of the supplied material has to be defined by the customer.

- Additional data like drawings, failure catalogue etc.

All of the above mentioned information has to be documented as controlled documents.

## 7.5.2. Introduction of a New Material

Prior using a new material for production this material has to go through a qualification process in order to prove that there is no negative impact on product quality, performance or reliability. Typically, a material change process is part of a production change process with risk assessments, trial runs, pilot production, yield and reliability assessment etc. The general production change process will be discussed in "Production Quality Management" later.

In this context "new material" can stand for:

- New type of material

For instance, the material supplier comes up with a new material production method resulting in a material with significantly altered properties and an entirely new specification.

- Changed material

For instance, driven by cost or performance optimization, the material supplier offers a slightly changed material, but keeps basically the existing specification. This case also includes the case where the material supplier changes the production site.

- New material supplier

For instance, again driven by cost or performance optimization or by the need for additional sources, material from a so far unknown supplier needs to be used. From a quality management perspective this is a critical case, since the supplier's methods and standards (e.g. monitoring level, qualification procedures) are unknown and have to be considered during the qualification process.

- Change of material supplier

For instance, as above, another source is required, but the supplier is providing other materials already.

Of course the above mentioned examples for material changes come with a different level of risk for the production as well as the product and will be treated accordingly. In addition to the actual change process the qualification of a new material has to follow a defined procedure including formal items like PCN (process change notification) and PPAP (production part approval process). In case the product is produced in more than one production site, it is useful to introduce a new material in one site first applying a complete qualification process. After a successful introduction of the material in the first site, a reduced qualification process is sufficient to introduce the now known material in other production sites. This way the efforts taken to qualify a new material in the entire production of a company are reduced and the risks regarding quality are not increased at the same time.

A brief generic process flow for the introduction of new material is shown in fig. 7.18.

### 7.5.3. Material Evaluation

In addition to executing a controlled process for the introduction of material into production, the material performance regarding yield and product performance have to be subject to continuous monitoring and control.

Material performance monitoring may include

- Number of rejects in incoming control
- Monitoring of incoming control results via $c_p$ / $c_{pk}$ values monitoring per delivery of production lot
- Supplier's capability data ($c_p$ & $c_{pk}$) of the material production process
- Regular material detailed yield and performance analysis (e.g. electrical parameters)
- Regular material detailed reliability monitoring
- Regular direct comparison of material from different suppliers regarding the entire set of material parameters and yield, performance and reliability monitoring

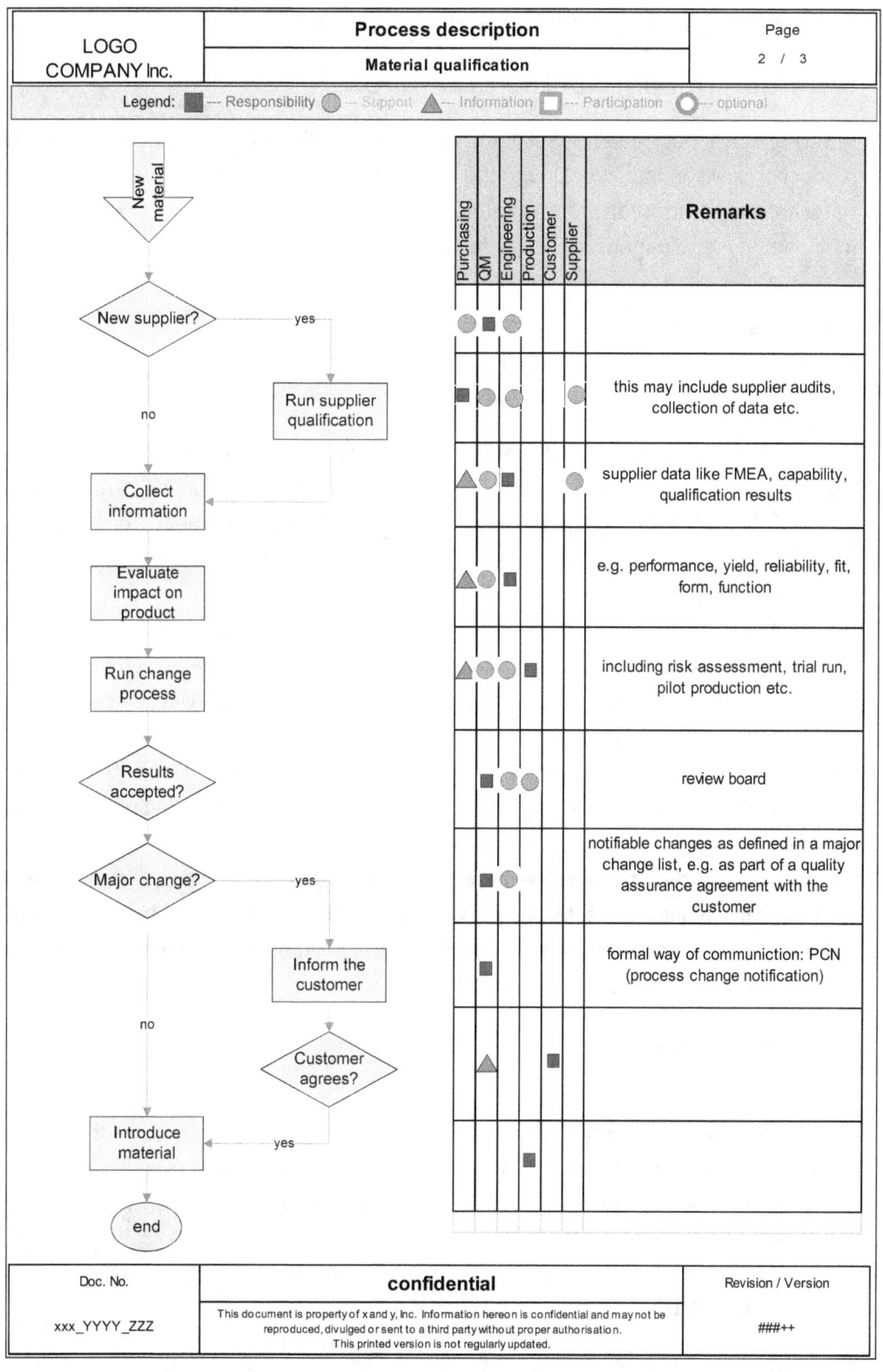

Fig. 7.18 Generic process flow for material qualification

# 8. Production Quality Management

## 8.1. Requirements of the Standards[1,2]

In this section both the ISO9001 as well as ISO/TS16949 requirements regarding production are summarized. Due to the fact that measuring activities are often found to be within a production environment the respective standard for measurement methodology requirements are included in this section.

### 8.1.1. Measuring

#### 8.1.1.1. General

ISO9001

The verification of product conformity has to be based on reliable and valid measuring results. For this purpose the measuring equipment has to be maintained and controlled accordingly. Subject, method and frequency of monitoring and measurement have to be defined.

ISO/TS16949

Appropriate statistical methods have to be determined and used for monitoring the measurement equipment and the measurement. Also, the capability of measurement systems has to be analyzed applying statistical methods.

Measurement equipment management has to include documentation of e.g. gauge labelling and traceability to national measurement standards including a capability assessment.

Internal laboratories have to be capable for the intended scope of measurement, external laboratories have to show evidence for their capability. The standard ISO TS16949 requires that for measurements which ensure the fulfilment of key control characteristics need to conform to ISO17025.

#### 8.1.1.2. Traceability

ISO9001

If required, the measurement equipment needs to be verified by the use of secondary standards which are traceable to national standards. It has to be protected from damage or false adjustment. In case of false adjustments the previous measurements have to be assessed.

### 8.1.2. Production

---

[1] DIN EN ISO9001:2015
[2] ISO/TS 16949:2009

### 8.1.2.1. Control of Production

ISO9001: The main objective of the norm is to execute production under well defined and controlled conditions. Those conditions include e.g.

- existence of documented information regarding the product characteristics
- documented information regarding details of the production process
- product and process monitoring as well as measuring activities
- appropriate environment and personnel
- validation of process and product in case verification is not possible
- avoidance of human error
- …

ISO/TS16949: ISO/TS16949 contains quite a large number of additional requirements for keeping a production under control, e.g.

- up to date control plans considering design and process FMEA results have to be available for pilot and volume production
- customer information as well as special characteristics (key control characteristics) have to be considered, too
- contingency planning is required for the case of uncontrolled conditions
- standard operating procedures have to in line with other quality related documents and have to be available on the shop floor
- verification of equipment set up is required as well as preventive and predictive maintenance activities have to be implemented and documented
- production equipment needs to be actively managed
- customer information about product problems have to be fed back to production

### 8.1.2.2. Marking and Back Tracing

ISO9001: The company has to ensure adequate capability to identify the product as well as its status. For back tracing this information needs to be documented.

ISO/TS16949: Back tracing is a general requirement. The requirement for back tracing is not to be defined by the supplier, but may be requested by the customer.

### 8.1.2.3. External Property

ISO9001: The company has to pay attention to the customer's property with regards to identification, protection etc. Reporting to the customer is required in case of loss or damage.

ISO/TS16949: Customer owned tools need a permanent label.

### 8.1.2.4. Preservation of Conformity

ISO9001: The product needs to be protected from any damage etc. in order to keep it compliant to requirements.

ISO/TS16949: In this standard products in the warehouse are included. First in first out (FiFo) is mandatory.

### 8.1.2.5. Post Delivery

ISO9001: Post delivery activities may be necessary. Those depend on legal conditions, information from customers, expected or guaranteed product characteristics etc.

ISO/TS16949: The pathway of information from the customer to production requires a defined process. Effectiveness of after sales service activities has to be verified.

### 8.1.2.6. Changes

ISO9001: Changes in production have to be controlled in order to protect the customer from non-conforming products. Content of and authorizing person for changes need to be documented.

ISO/TS16949: No additional requirements

## 8.1.3. Product Release

ISO9001: The standard requests to verify product requirements. A product release is to be granted only after meeting those requirements. Respective proof and authorizing person of releases need to be documented.

ISO/TS16949: Dimensional, visual and functional checks are to be included

## 8.1.4. Nonconformances

ISO9001
Unintentional use of non-conforming products has to be avoided prior and after delivery. Potential actions upon detection of a non-conformance are rework, containment, request for return of material from the customer and / or request for customer approval for delivery. Non-conformances, respective activities and approvals have to be described and documented.

ISO/TS16949: Products with no labelling or with non-conformance probability have to be treated like nonconforming products. Rework has to follow documented procedures. The customer may, upon reporting of delivery of nonconforming material, grant a special release. Handling and activities regarding nonconforming products have to be well documented.

## 8.2. Process and Product Control

### 8.2.1 Clean Room

In general the standards are asking for an appropriate environment for production. Due to the nature of the semiconductor manufacturing process the main objective here is to execute the production in a clean room and to keep the clean room clean and under suitable environmental conditions, in particular temperature and relative humidity.
In this context cleanliness essentially means absence of particles as well as absence of metal contamination, because both have a detrimental impact on either product functionality, i.e. yield, or product reliability.

#### 8.2.1.1 Particles

Clean Room Class
„Normal air" contains roughly 100.000 – 10.000.000 particles / ft$^3$. Depending on the technology requirements clean room classes are defined which allow a maximum number of particles of a particular size within a defined volume[1]. For instance, in order to achieve ISO Class 1 the clean room air must not exceed ten 0.1 µm particles / m$^3$ and two 0.2 µm particles / m$^3$ at the same time. Nine clean room levels are defined on a logarithmic scale, i.e. the maximum number of particles of a particular size allowed within on class multiplies tenfold when moving from ISO Class 1 to ISO Class 2.

Impact of particles on the product
Depending on what a particle is composed of and where a particle is located on an integrated circuit during the production process, for instance the following problems may occur:

➢ structuring

A particle may block parts of a metal line during structuring. The reduction of the metal width may result in a malfunctioning chip immediately or in an increase of electromigration. The latter situation will lead to a reliability issue at the customer.

➢ structuring

---
[1] DIN EN ISO14644-1:1999

A conductive particle placed in between two metal lines and touching both metal lines leads to a electrical short. This results in a malfunctioning circuit.

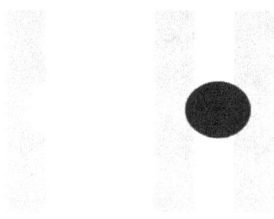

Fig. 8.1 Schematics of a narrowed metal line (left) and shorted metal lines (right)

> human particle with contact to sweat

Particles containing mobile ions (e.g. Na) are prone to lead to metal contamination. If they diffuse to a transistor this may lead to a $V_t$ shift of the transistors

> metal containing particle

At higher temperatures metal atoms diffuse very fast through silicon. At cool down those metal atoms either precipitate as conductive metal silicides (e.g. Cu, Ni) within or on the surface of silicon or the atoms are individually distributed within the silicon lattice and might act as generation centre (e.g. Fe). Both could eventually lead to a leakage problem later on, i.e. a reliability problem.

Driven by technology and feature size reduction the criticality of particles is constantly increasing:

"Leading IC manufacturers require 2X defect reduction every technology node" [1]

The size of the particles as well as the number of particles has to be reduced continuously in order to guarantee defined levels of quality and reliability.

### 8.2.1.2. Clean Room Management

Particle Sources

Typically, 80 – 90% of the particles in clean room air are human caused (flakes, hair). A person which is not even moving generates 100.000 to 1.000.000 particles per minute of various sizes. For instance, tobacco smoke particle diameters are found to be in a range between 10 nm up to 500 nm[2].

---

[1] Song-Moon, S. (2013)
[2] Van Zant, P. (1997)

In addition, other sources of particles are active, such as contamination of incoming materials, e.g. silicon wafers, gas, chemicals, particles are generated within the production chambers themselves.

The latter may be caused by the friction of moving parts like a handling system within a process chamber and mainly consist metals.

Particle prevention and reduction

In consequence of the above mentioned detrimental effects of particles and metals on product functionality and reliability a clean room has to be well controlled and in addition people working there have to comply with various behavioural patterns.

Clean room monitoring includes activities like regular temperature, humidity and particle density control of the clean room air and to monitor those parameters using statistical process control methods. In case of a particle counts increase a root cause analysis has to be executed and corrective actions have to be implemented ("defect engineering").

In addition to air control also particle generation as well as metal contamination of production equipment has to be monitored via regular monitoring activities or special activities prior restart of production.

The high fraction of particles caused by humans drives the motivation to keep the particle generation rate as low as possible by applying various simple rules for people working there as well as for managing the clean room itself:

- use of clean room garments with or without clean room gloves
- use of air lock showers: prior entering the clean room an air shower is used to remove particles from the clean room suit
- no smoking prior to entering the clean room – too many particles would be released from the lungs
- no jewelry, particles might stick underneath them
- no make up – it consists of particles
- slow motion – fast motion might increase the particle generation
- no talking or manipulation above open cassettes with wafers in them
- avoidance of big groups of people
- use of clean room pens and paper
- sit-over in order to separate the "clean" side of the gowning area from the incoming side of that area (also a psychological barrier!)
- air locks in order to prevent particle loaded external air to intrude into the clean room
- …

## 8.2.2. Inline Process Control

The semiconductor manufacturing process is characterized by a long sequence of single consecutive process steps. Today's most advanced semiconductor manufacturing processes consist of many hundreds up to 1000 steps in front end production[1]. This results in cycle times of up to several months. Hence, instead of applying only one final test at the end of the process, it is useful to have many inline tests in between (fig. 8.2). Only then fast reactions upon process and or material deviations are possible. In addition short feedback loop support root cause analysis in case of yield and/or reliability problems.

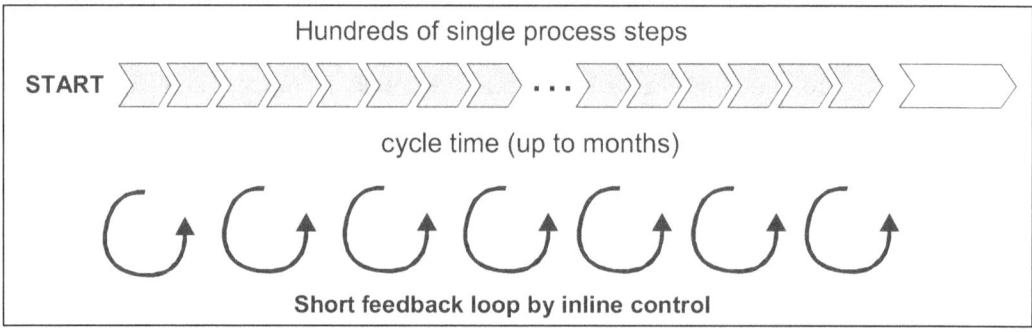

Fig. 8.2 Illustration of short feedback loops in semiconductor processing

Initial steps of a generic 250 nm DRAM process are shown in table 8.1[2] (extract of the full process without cleaning steps). Each of these steps requires more or less intense testing and monitoring activities (statistical process control).
The definition of the concrete measuring instruction for a particular characteristic is based on the criticality of the process step regarding product quality and reliability as well as on the nature of the process step. For instance lithography processes have to be monitored using various dimensions of the structures as well as alignment of different mask levels (see fig. 8.3).
Table 8.2 provides an overview of inline parameters to be monitored[3]. The frequency of sampling and the number of measurement points depends, among other things, on the stability of the process and could include measurements of:

> chip centre and chip edge
> centre and edge of the mask (which typically contains more than one circuit)

---

[1] Puffer, W. (2007)
[2] Widmann, D. et al. (1996)
[3] May, G. S., Spanos C. J. (2006)

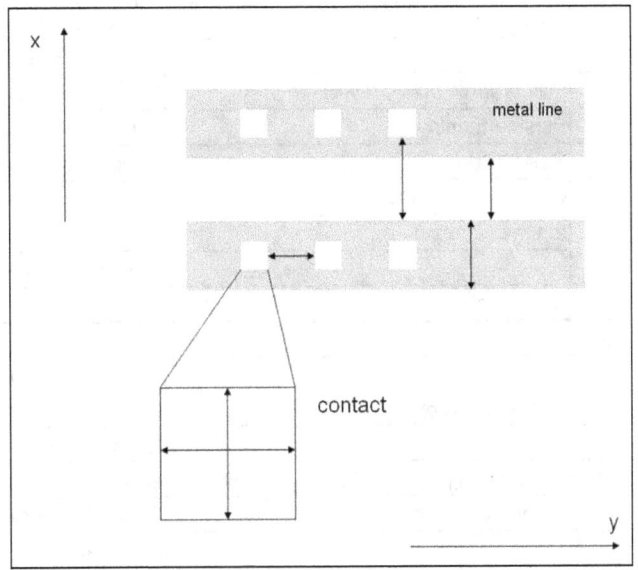

Fig. 8.3 Schematic of lithography process characteristics

- nested and isolated structures on a chip
- various exposure fields on a wafer
- wafer centre and wafer edge
- wafer to wafer & lot to lot & run to run
- production chamber to production chamber
- ...

| No. | Process Step | Characteristics |
|---|---|---|
| 1 | Oxidation | Layer Thickness |
| 2 | Lithography (buried layer) | Critical Dimension |
| 3 | Plasma Etching | Critical Dimension |
| 4 | Resist Strip | Optical Control |
| 5 | Oxidation (sacrificial oxide for implant) | Layer Thickness |
| 6 | Ion Implantation | Junction depth |
|   |   | Dopant concentration |
| 7 | Oxide etch |   |
| 8 | Annealing |   |
| 9 | Oxide etch |   |
| 10 | Epitaxial Si growth | Layer Thickness |
|   |   | Dopant concentration |
| 11 | Oxidation (hard mask for ion implant) | Layer Thickness |
| 12 | Lithography (n-well) | Critical Dimension |
|   |   | Alignment |
| 13 | Oxide etch | Critical Dimension |
| 14 | Resist Strip |   |
| 15 | Oxidation (sacrificial oxide for implant) |   |
| 16 | Ion Implantation | Junction depth |
|   |   | Dopant concentration |
| 17 | Oxide etch |   |
| 18 | Annealing |   |
| 19 | Oxide etch |   |
| 20 | Oxidation | Layer Thickness |
| 21 | $Si_3N_4$ Deposition | Layer Thickness |
| 22 | Lithography (active area) | Critical Dimension |
|   |   | Alignment |
| 23 | $Si_3N_4$ etch | Critical Dimension |
| 24 | Resist Strip |   |
| 25 | Local oxidation (LOCOS) |   |
| 26 | ... | ... |

Table 8.1 Initial steps of a 0.25 μm DRAM process[1]

This way a large amount of data is collected and needs to be evaluated using statistical methods.

---

[1] Widmann, D., et al. (1996)

| Area | Parameters |
|---|---|
| Lithography | critical dimension (line width) |
| | overlay (alignment) |
| | resist profile |
| Etch | critical dimension (line width) |
| | etched profile |
| | etch rate |
| | uniformity |
| | selectivity |
| Litho/Etch | bias |
| Insulator Deposition | layer thickness |
| | layer uniformity |
| | dielectric constant |
| | refractive index |
| Conductor Deposition (incl. Epitaxy) | layer thickness |
| | layer uniformity |
| | sheet resistance |
| | surface concentration |
| Doping | sheet resistance |
| | junction depth |
| | surface concentration |

Table 8.2 Examples for inline monitoring parameters

Eventually the process control activities are summarized in a control plan.

| Part Name / Process Flow Number DRAM 0.25 µm / 1234 | | Control Plan Number 4567 | Date 2015-11-19 | | | Prepared by N.N.4 | | |
|---|---|---|---|---|---|---|---|---|
| Team Members N.N.1, N.N.2, N.N.3 | | | Supplier Company XY | | | Factory Site A | | |
| Reference to process flow | | Parameter | Test | | | Evaluation | | |
| No. | Process Step | Process Characteristics | Prozess Specification | Measurement Tool | Sample Size | Sample Frequency | Method | Reaction Plan |
| .. | | | | | | | | |
| .. | | | | | | | | |
| .. | | | | | | | | |
| 10.1 | Layer Thickness Measurement | Oxide Thickness | ABC123 | Optiprobe | 1 wfr / 9 pts | Sampling plan | SPC | N14 |
| 10.2 | Resistivity Measurement | Resistivity | ABC124 | Four Point Probe | 1 wafer | Sampling plan | SPC | N15 |
| 11.1 | Layer Thickness Measurement | Oxide Thickness | ABC125 | Optiprobe | 1 wfr / 5 pts | Sampling plan | SPC | N16 |
| 12.1 | Geometry Measurement | Critical Dimension | ABC126 | SEM | 1 wafer | Sampling plan | SPC | N17 |
| 12.2 | Control of overlay | Alignment | ABC127 | SEM | 1 wafer | Sampling plan | SPC | N18 |
| 13.1 | Geometry Measurement | Critical Dimension | ABC128 | SEM | 1 wafer | Sampling plan | SPC | N19 |
| 14.1 | Inspection SEM | Inspection | ABC129 | SEM | 1-3 wafers | Sampling plan | visual check | N20 |
| 15.1 | Layer Thickness Measurement | Oxide Thickness | ABC130 | Optiprobe | 1 wfr / 9 pts | Sampling plan | SPC | N21 |
| 161 | Resistivity Measurement | Resistivity | ABC131 | Four Point Probe | 1 wafer | Sampling plan | SPC | N22 |
| .. | | | | | | | | |
| .. | | | | | | | | |
| .. | | | | | | | | |

Table 8.3 Example for a control plan for the above described DRAM process

## 8.2.3. Advanced Process Control (APC)

The conventional inline process control activities as described in previous chapters used to be sufficient for process control over many years. Over the last years, however, with constantly shrinking feature sizes it became more and more difficult to meet the tightening tolerance requirements[1].

Advanced Process Control had to be introduced to keep the manufacturing processes under control. The basic idea of APC is the online monitoring of equipment and sensor data during processing and making use of this data via online fault detection and classification (FDC).

> "integrated computer-controlled wafer fabrication is playing an increasingly important role in the semiconductor industry" [2]

Eventually the purpose of FDC is the implementation of online reactions, e.g. lot hold, process stop or simply notification of a process engineer by email.

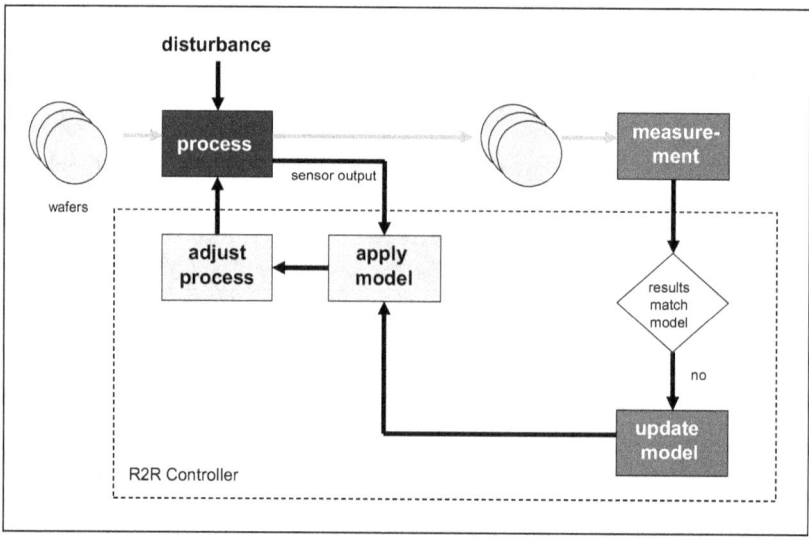

Fig 8.4 Simplified APC R2R flow[3]

In addition to FDC the Run-to-Run control (R2R) is meanwhile one of the mostly used techniques in semiconductor manufacturing[4]. Run to run control could be both wafer to wafer or batch to batch control.

---

[1] Emami-Naeimi, A. (2008)
[2] Emami-Naeimi, A. (2004)
[3] Boning, D. (2003)
[4] Susto, G. A. (2013)

In order to set up an Advanced Process Control the respective process engineer needs to have sound knowledge of the physics and chemistry of the process, e.g. plasma etch or CMP process, what sensors and actuators are implemented and relevant interference factors to the process. In addition a process model needs to be available, which should be validated using real process data. The purpose of the model is, based on sensor inputs, e.g. pressure or gas flow, to predict process results, e.g. layer thickness and homogeneity. From there the recipe for the consecutive run is adjusted (fig. 8.4). This way the altered recipe accounts for unpredictable disturbances.

The model itself is continuously with real process results. In case the model results deviate from actual measurement values, the model may be adjusted and/or fine-tuned.

## 8.2.4. Equipment Maintenance

In addition to a reasonable process control also production equipment needs to be maintained in order to keep a production process stable. In particular maintenance activities help keep a production tool available for production and failure free regarding the product. Various approaches for equipment maintenance are in place:

➢ Breakdown Maintenance

I.e. the production tool is running until it fails. Only then repair and service activities start. Within a semiconductor production environment this kind of maintenance is not the usual approach. Products, production time and production equipment are much too valuable to take this kind of risk.

➢ Preventive Maintenance (PM):

The system or parts of it are subject to service or replacement before a failure is likely to occur. This way an uniplanned interruption of production is avoided and maintenance resources can be used in a planned and scheduled way. Typically preventive maintenance focuses on the "weak" parts of the equipment and is executed on a regular schedule, e.g. monthly, yearly.

➢ Predictive Maintenance

There are two approaches to predictive maintenance. In case of "condition-based" predictive maintenance the parameters of the equipment are monitored on-line or periodically (e.g. vibration, temperature, pressure, resistance) in order to predict likelihood of failure. Those equipment parameters may be monitored using the aforementioned APC sensors or the similar methods.

The other case, the "statistical-based" predictive maintenance, uses data of previous fails of the equipment or parts of it for statistically analysis in order to predict future failures.

Maintenance activities are supposed to start well before the parts are likely to fail.

## 8.2.5. Electrical Parameter Control

An integrated circuit, i.e. a chip, consists of a large number of individual active and passive devices like transistors, metal lines, capacitors, diodes etc. The device characteristics have to meet the respective tolerances, e.g. threshold voltage, sheet resistance and capacity.

Hence, in addition to inline process control and equipment maintenance quality and performance of the product devices are also controlled and monitored using SPC for electrical parameters.

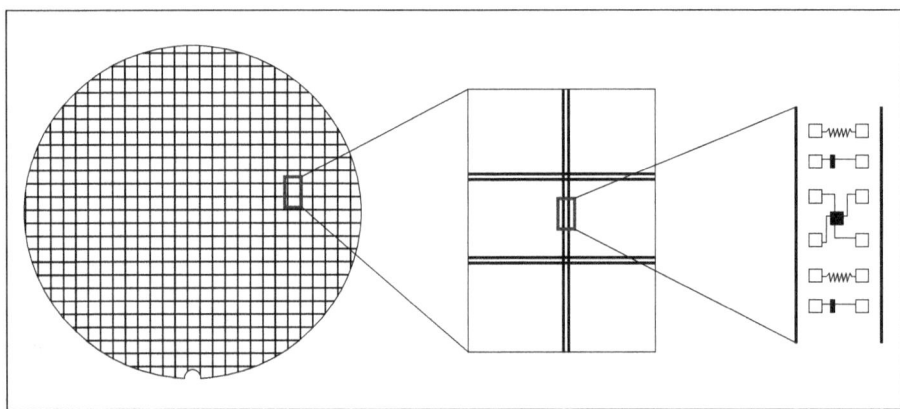

Fig. 8.5 Test structures for electrical parameter test are placed in the sawing street in between productive chips

They may include e.g.:
- metal lines
- contact and via chains
- resistors
- capacitors
- gate oxide test structures
- transistor test structures

with various dimensions and representing various levels of the process.

For this purpose, test structures are placed in the sawing street in between the product dies (integrated circuits) (fig. 8.5). Typically, electrical parameter tests are performed after finishing the front end production process and before the actual functional test. Results are often summarized using a wafer map (fig. 8.6), which provides information about the device performance across the wafer, and in form of $c_p$ and $c_{pk}$ data.

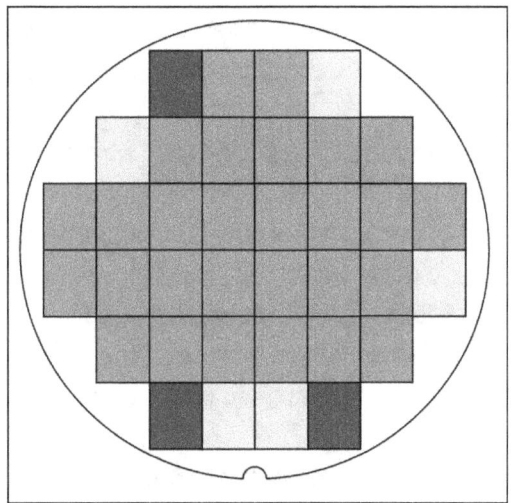

Fig. 8.6 Wafer map with visualization of performance

## 8.2.6. Yield

> "Yield is simply the percentage of "good" product in a production batch " [1]

Cost saving and increase of profit are the most obvious reasons for a company to invest in a lot of effort in increasing yield and to keep the product at a stable high yield level. High and stable product yield also indicates a stable production process and the absence of maverick wafers or lots.

However, there are many circumstances which lead to functional yield reduction in a semiconductor environment, e.g.

- particles falling on the surface of the chip during or in between process steps
- mask defects
- process parameters not perfectly centered (low $c_{pk}$)
- process parameters with bad uniformity across the wafer or wafer to wafer (low $c_p$)
- …

Due to the complexity and the length of the semiconductor manufacturing process, which, in addition, includes various separate production segments (front end production, front end test area, back end production, back end test area) and due to the

---

[1] Spanos, C. J. (2005)

fact that yield could refer to functional (i.e. electrical) yield as well as loss of wafers there are different kinds of "yield" in use, describing the loss of chips, parts or wafers:

| | |
|---|---|
| Line yield | : loss of wafers in the front end process, e.g. wafer breakage |
| Parametric test yield | : loss of wafers due to electrical parameters out of specification |
| Chip yield | : loss of chips based on the results of the front end functional test prior separation |
| Die yield | : loss of dies based on back end functional test |

Actually the view on yield is much more detailed, but this would lead beyond the scope of a general introduction. With the help of a suitable test program in functional test, malfunctioning circuits are identified, i.e. scrapping of the chip by inking. Those defects lead to yield reduction. Defects which cause performance very near to the specification limit (but still are "pass"), however, may lead to reliability problems at the customer later on.

> "Reliability defects are proportional to yield defects (typically 1 % - 2 %)" [1]

For this reason, low yielding lots or wafers are scrapped when not meeting a defined minimum yield level. This so called scrap limit could be either a fixed limit or could be based on statistics, e.g. a 6σ – distance from the mean value. In addition often rules for inking "good" chips in the neighbourhood of failing chips apply.

Microelectronics manufacturers are trying to predict the yield level of new products using various yield models. A sound understanding of the impact factors of yield is also useful to address manufacturing process issues, eventually to increase product quality, since, as explained, the lower the yield, the larger the percentage of product with potential reliability issues.

> "Semiconductor yield modeling is essential to identifying processing issues, improving quality, and meeting customer demand in the industry." [2]

---

[1] ON Semiconductor (2015)
[2] Krueger, D. C. (2011)

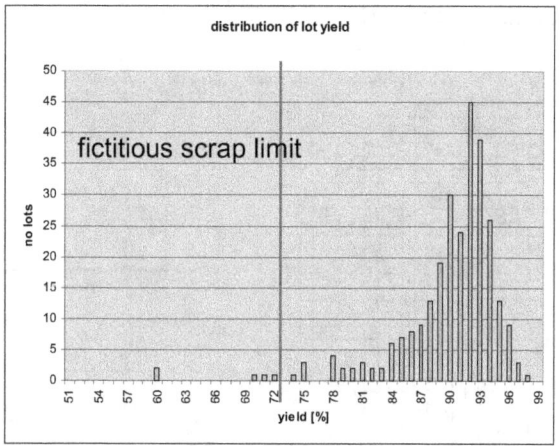

Fig. 8.7 arbitrary production lot yield distribution with fictitious lot scrap limit

There are several yield models in use. Typically they consider the defect density ($D_0$) of the manufacturing line and the chip size (A) as the main impact factors to yield. Some of the models are summarized in the following list[1]:

- Classic Poisson Model : $Y = e^{-D_0 A}$

- Murphy's Yield Model : $Y = \left[ \dfrac{1 - e^{-D_0 A}}{D_0 A} \right]^2$

- Seed's Yield Model : $Y = \dfrac{1}{1 + D_0 A}$

- Dingwall's Yield Model : $Y = \left[ 1 + \dfrac{D_0 A}{3} \right]^{-3}$

- Moore's Yield Model : $Y = e^{-\sqrt{D_0 A}}$

The effect of particles and other electrically effective defects is obvious, the effect of the chip size is lead back to a geometrical effect (fig. 8.8). Other models also include the

---

[1] Krueger, D. C. (2011)

radius of the wafer, the number of layers of a chip etc.[1] or clustered distributions of defects[2] in order to make the yield model closer to reality.

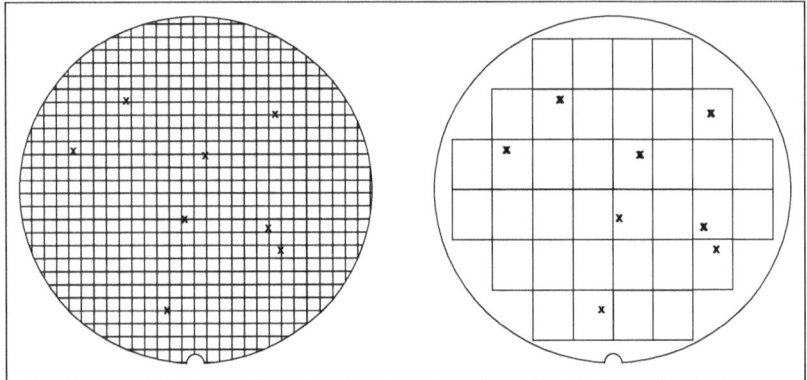

Fig. 8.8 Same number and distribution of defects leading to a different percentage of chips affected

## 8.3. Reliability in Production

The primary proof of product reliability is demonstrated during the product qualification. There the product and technology durability is tested under a variety of stress conditions like elevated temperature, humidity, voltage, by intense temperature cycles and the like. However there are two tasks regarding product reliability which have to be performed within the production environment. One is to monitor the expected product reliability performance using the wafer level reliability approach, the other one is to sort out "weak" parts (i.e. the circuits which marginally passed the functional test) by applying the so called "burn in".

### 8.3.1. Wafer Level Reliability

> "Wafer Level Reliability uses mechanism-specific test structures under accelerated conditions to assess the reliability impacts of process and tool changes" [3]

In most cases the aforementioned proof of reliability is based on a relatively low number of devices from three different lots with high stress applied over a few thousand hours. Wafer level reliability testing on the other hand is based on several test structures on a wafer, possibly on several wafers per lot over many production lots.

---

[1] van Zant (1997)
[2] Lin, J.-S. (2012)
[3] Fairchild Semiconductor (2015)

|  | Lot to lot variation | Wafer to wafer variation | Across wafer variation |
|---|---|---|---|
| Sample Plan | 5 sites per wafer | 2 sites per wafer | 50 sites per wafer |
|  | 5 wafers per lot | 25 wafers per lot | 1 wafer per lot |
|  | 100 lots | 20 lots | 50 lots |
| Total sample size | 2500 | 2500 | 2500 |

Table 8.4 WLR sampling plan example[1]

These test structures are placed in the sawing street in between productive dies (see also fig. 8.5). Hence thousands of devices are subject to wafer level reliability depending on the respective sample plan[2] (table 8.4). For regular production monitoring at least one lot per week should be used for WLR across key technologies[3].

The benefit of WLR is obvious. In contrast to conventional reliability stress tests the chips don't have to be packaged and assembled[4], which could take several weeks:

- No need for production time and resources regarding packaging
- Early feedback about reliability issues enables process engineers to act early, hence to avoid scrapping of material later on
- Typical stress times are significantly reduced as compared to conventional reliability stress[5]

In order to further reduce stress times fast Wafer Level Reliability (fWLR) has been developed. By applying more intense stress conditions the duration of a fWLR stress test ranges from fractions of seconds up to a few seconds[6] (fig. 8.9).

However, for the application of fWLR a very good understanding of the failure mechanisms is required to avoid stress related failures which do not occur in real product use. Due to self heating, of stress structures additional heating via the wafer chuck is not required. This approach protects the productive chips on the wafer. The fWLR sampling frequency is subject to some controversy and ranges from one lot per week up to each wafer from each lot[7]

---

[1] Turner, T. (2001)
[2] Turner, T. (2001)
[3] Fairchild Semiconductor (2015)
[4] Rubin, D. (2005)
[5] Fairchild Semiconductor (2015)
[6] Martin, A. (2007)
[7] Martin, A. (2007)

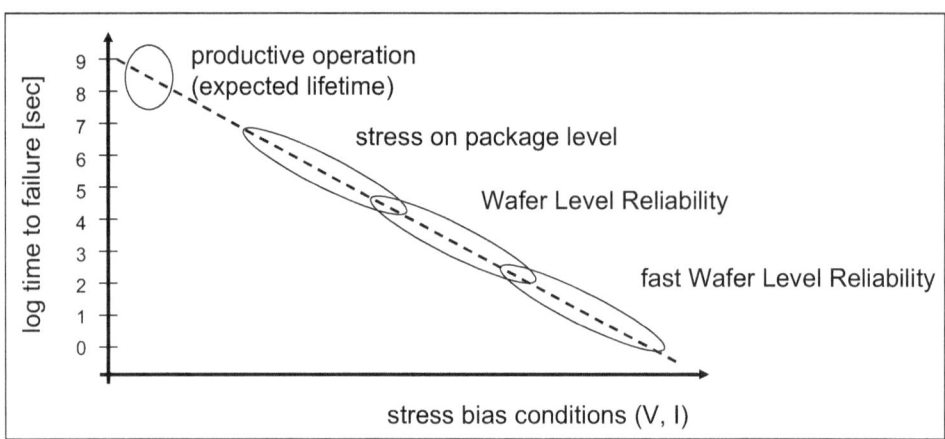

Fig. 8.9 stress time comparison depending on stress level[1]; ("stress on package level" represents the normal reliability tests to predict early fail rates and lifetime)

## 8.5.2. Burn In

The fail rate of semiconductors over their lifetime can be divided in three phases (fig. 8.10)[2]:

➢ Phase one: "early fails"

Represents fails due to process or material weaknesses and problems. The fail rate in this phase is comparably high.

➢ Phase two: "random fails", also known as "useful life"

In this phase random fails occur e.g. based on random stress situations. The fail rate in this phase is very low and may go down to practically zero[3]

➢ Phase three: "wear out fails"

During "useful life" devices, insolators and conductors are exposed to elevated temperature, temperature cycles, electric current etc. leading to fatigue and wear out of material. In this phase the fail rate increases significantly. All parts of the product have to be designed and manufactured in a way, that the wear out phase starts only after the contractually agreed lifetime of the product.

Such bath tub curves for failure of manufactured products are general, and can be applied to almost all technologies. The platted fail rate is the conditional fail rate h(t), i.e. the failure rate is calculated on the basis of the number of devices that have survived

---

[1] Aal, A. (2010)
[2] Renasas (2013)
[3] SemeLab (2002)

until the given point in time. By contrast the absolute fail rate f(t) is calculated with respect to the total number of tests which initially went into the test. Therefore, h(t) > f(t), only for t = 0 : h(t) = f(t).

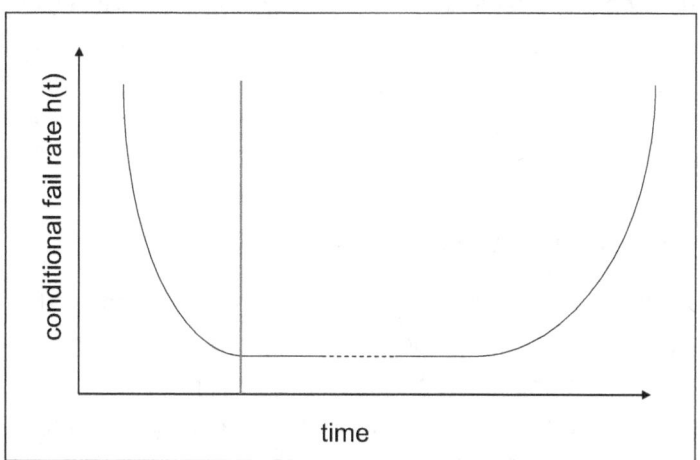

Fig. 8.10 Bathtub curve representing semiconductor fail rates over time; red line represents product life after burn in

For large and complex integrated circuits the fail rate of phase one typically exceeds the customer's expectation and need significantly. Hence, the purpose of the burn in is to artificially age the product and to make the early fails fail during the burn in phase. This way the customer only experiences a stable phase at a very low fail rate. The basic idea behind this strategy is to run the product under accelerating conditions[1] by applying voltage and temperature (typically at 125°C[2]) in order to provoke early fails. The voltage applied may be static or dynamic depending on the targeted failure mechanism. Another purpose of burn in is the early detection of reliability problems: a decreasing burn in yield signals reliability issues.

Of course stable products using stable technologies may also be manufactured at low early fail rates without burn in.

## 8.4. Deviation Management

Unintentional use of non-conforming products has to be avoided. In order to ensure this standards requirement and to protect the customer form receiving questionable products suitable procedures and rules are defined and followed in production. Those procedures describe identification and handling of deviations in detail.

---

[1] Marshal, J. (2011)
[2] Bozturk, C. (2015)

> "The specification details the requirements for review and disposition of nonconforming product or raw materials " [1]

Typically, deviation management consists of several subprocesses as described in the following sections. However, the level of customer involvement regarding decision making or release of material very much depends on the contractual situation as well as the respective quality standards requirements. Therefore the process described is one option only.

## 8.4.1. Identification and Risk Assessment

"The process begins when discrepant material is identified"[2]. The identification of deviations, i.e. excursions in production, is simple in case of obvious fails and rather complex if data mining and statistical analysis is required[3].

In this context, for instance, excursions may be detected in the form of

- Equipment failure
- Inline parameter not meeting SPC requirements (e.g. layer thickness, layer uniformity, line width…)
- Electrical parameters not meeting SPC requirements
- WLR / fWLR parameters not meeting requirements
- Yield below defined limits
- Defect density above defined limits
- …

Immediately after identification of deviating material the affected production lots are marked accordingly and put under quarantine. Then the material has to undergo a formal risk assessment including assessment of:

- Potential functional fails (expected yield loss)
- Potential reliability issues (much more problematic)
- Volume of affected material
- Potential customers affected
- Product delivery status (potential shortage / late deliveries)
- …

---

[1] Intel (2014)
[2] Intel (2014)
[3] Frankwicz, P.S. et al. (2009)

Formal risk assessments are executed by experts and reported to a review board, e.g. a Material Review Board (MRB) which consists of participants from various departments (e.g. manufacturing, engineering, quality, reliability). The set up often depends on the nature of the problem as well as on the manufacturing environment.

Starting with the risk assessment further steps regarding material evaluation, root cause analysis and corrective actions have to be initiated. Material evaluation could include additional testing in functional test, but could also include reliability tests using the final product. In these cases a final decision can only be made after all relevant results are available.

### 8.4.2. Analysis and Disposition

It is necessary to identify the root cause of the deviation. Only with a reasonable root cause at hand valid corrective actions as well as meaningful preventive actions can be initiated.

Based on the results of the risk assessment and of the root cause analysis different actions are possible, which need to be determined using clear and reproducible criteria[1]:

➤ Release of the material

In case fit, form and function of the product are not affected a release of the material may be justified. Depending on the contractual situation with the customer a release by the customer may be required in addition. Any release of material has to be based on defined criteria.

➤ Rework of material

Rework of material intends to establish the original status. However, it is a deviation from the regular process flow and, depending on the quality standard or the customer, a customer release. It is helpful, in particular for well known topics, to qualify the rework flow from the beginning.

➤ Downgrade of material

Depending on the customer's accepted quality level a downgrade of the product to a lower quality level, i.e. a higher failure rate, is possible. Typically this includes a price reduction. Delivery of downgraded material has to be agreed by the customer.

➤ Scrap of material

---

[1] Sandhu, J. S. (2015)

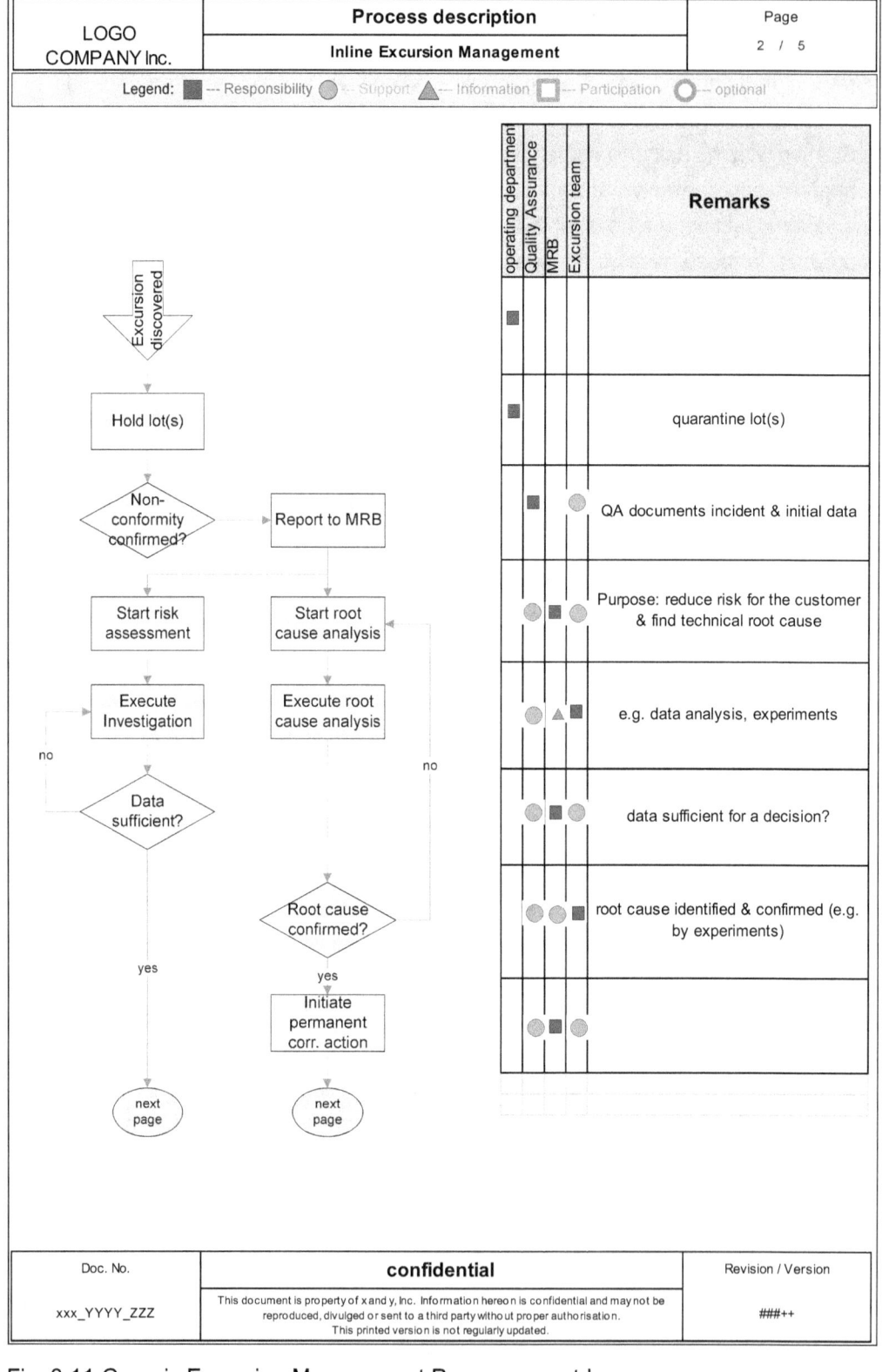

Fig. 8.11 Generic Excursion Management Process – part I

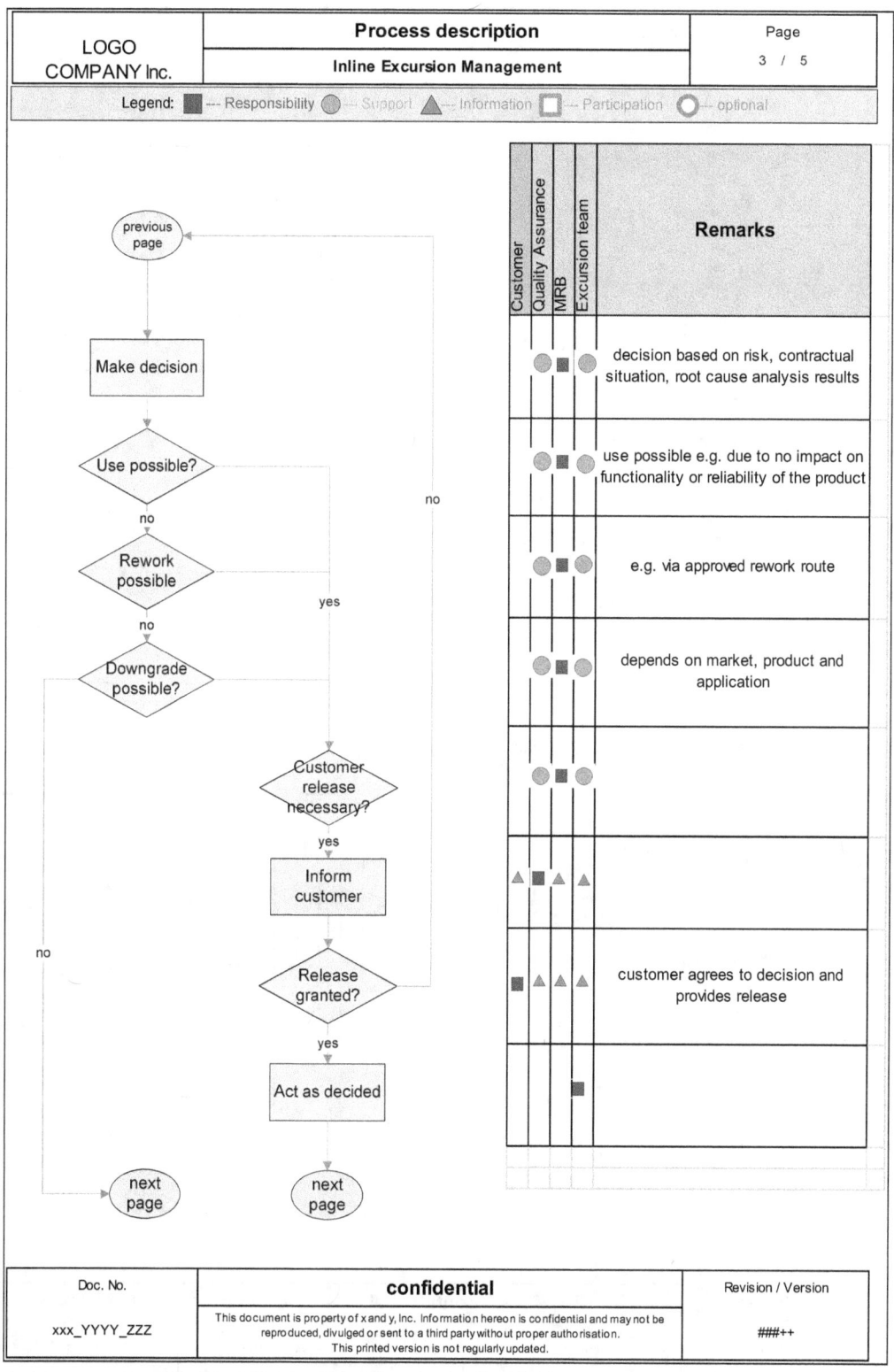

Fig. 8.12 Generic Excursion Management Process – part II

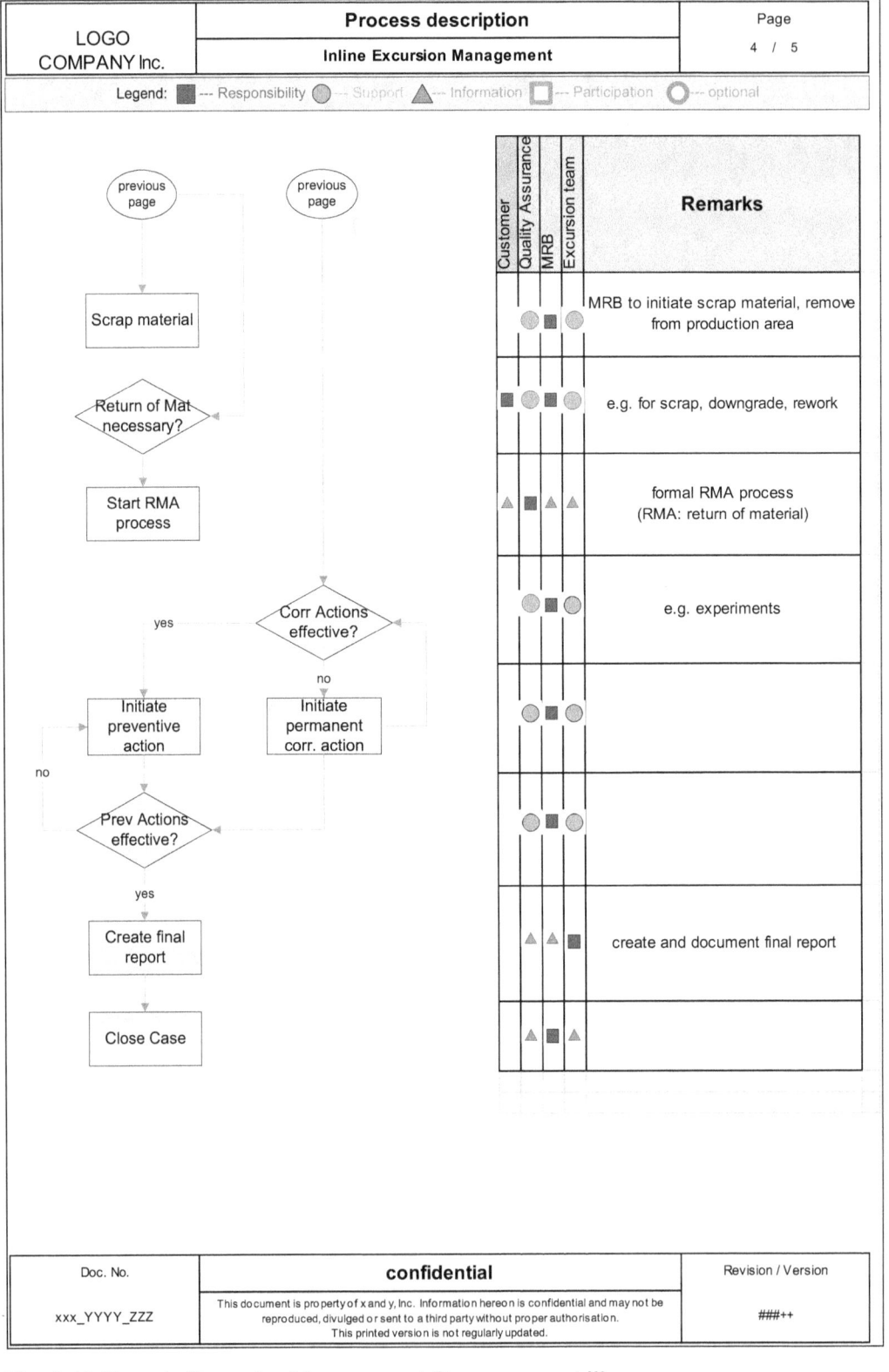

Fig. 8.13 Generic Excursion Management Process – part III

If the risk for the customer, in particular regarding reliability reduction, the material has to be removed from production, i.e. has to be scrapped. This is the safest way to ensure high product quality, but the most costly one.

➢ or, in case of incoming material as the root cause: return to supplier

### 8.4.3. Preventive Actions and Closure

After implementing short term corrective actions and decision making regarding disposition of material the customer has to be informed or asked for release. If customer approval is required for delivery of reworked material all relevant information has to be provided to the customer. This includes e.g.
➢ detailed problem description (when, what, how much, ...); support by visual material is often helpful
➢ detailed description of the rework process (kind and number of corrective actions)
➢ inline data
➢ electrical results
➢ reliability results.

In addition a preventive action plan needs to be developed and implemented in order to prevent recurrence of the deviation.

All actions defined are executed and completed and their effectiveness is assessed. The results are summarized in the form of a report and presented to the review board. An internal release has to be based on defined criteria. After approval by the review board the deviation is closed, i.e. the end of the deviation management process is reached.

Using the aforementioned procedure and including information of excursion and non-conformance handling from various semiconductor manufacturers Figs. 8.11 – 8.13 show a generic process flow for excursions.[1,2,3]

## 8.5. Process Change Management

"Process improvements are performed on a daily basis in the mass production phase to achieve even better product quality and reliability and to improve productivity"[4]. In other words, production process changes are a must and cannot be avoided. On the other hand, a qualified production process is "frozen", i.e. must not be changed. The

---

[1] ST Microelectronics (2013)
[2] Furtaw, R. (2015)
[3] Fujitsu Semiconductor Limited (2013)
[4] Seiko Epson Corporation (2009)

way out of this seemingly contradictory requirements is a structured process change management.

The purpose of the process change management is to ensure that upon a change, there are no detrimental effects on the product regarding quality and/or reliability which could affect the customer. Whether or not the customer needs to be informed is part of the "Customer Quality Management" session. Typically the process change process consist of the following main steps[1]:

- Proposal for a change (of process or product)
- Change assessment by a review board
- Classification of the change as to whether it is to be regarded as a minor or major
- change
- Qualification, Approval & Implementation
- Customer Notification

Change management covers both process and product changes. The following section, however, focuses on process changes.

### 8.5.1. Proposal for a Change

Changes of the manufacturing process may have a lot of different reasons. They could be motivated by

- Potential yield or reliability improvement
- Reduction of cost
- Higher throughput
- Shorter cycle time

and could affect materials or manufacturing equipment or the process itself, for instance by using a different or shorter process sequence.

In any case the responsible process engineer needs to compile data and is obliged to report this to a review board for approval. This set of data should at least contain[2]

- An assessment of the benefits of the change
- Initial results from experiments
- A feasibility study
- A risk analysis (e.g. FMEA) including the potential impact on the product

### 8.5.2. Change Assessment by a Review Board

---

[1] ST Microelectronics (2013)
[2] ST Microelectronics (2013)

After compiling the aforementioned data, these initial results are presented to a review board. The review board could be named PCRB (process change review board), PCCB[1] (process change control board) or simply PCB (process change board). Its task is to assess and to evaluate change proposals, risks and results and, based on that, to release productive lots for further evaluation and, eventually, to release a change.

The formal way of changing processes is often helpful to ensure proper documentation of the change.

> "This board is comprised of a multidisciplinary team of experts, including senior engineers and management, who evaluates then classifies, approves or rejects the change request. " [2]

The PCRB consists of a cross functional team with participants from Quality Management, Product Engineering, Process Engineering, Production, depending on the nature of the change also equipment and/or maintenance engineering or others. Additional participants are invited as needed. For instance this could be reliability engineers or device specialists.

## 8.5.3. Classification

Changes are classified in minor and major changes. Major changes are changes with impact on form, fit and function of the product or with detrimental impact on the quality and/or reliability of the product[3].

| | |
|---|---|
| "Form | the visual appearance of a device's documented external package color, mark and surface finish |
| „Fit | a device's external dimensional and associated tolerances" |
| „Function | a device's electrical, mechanical, thermal, quality or reliability performance characteristics, in accordance with specified tests and limits" |
| „Quality | results of tests or procedures that indicate conformance to documented requirements" |
| „Reliability | the probability of continued operation over time" [4] |

---

[1] Intel (2014)
[2] ST Microelectronics (2013)
[3] JEDEC 46
[4] Pericom Semiconductor Corporation (2007)

Major changes require the customer's approval prior implementation. Table 8.5 provides examples for major changes[1]:

| Item | Examples |
|---|---|
| Product | Chip shrink, i.e. change of design rule (e.g. from 90nm to 65 nm) |
| Front End Production | Manufacturing site, significant process flow change, silicon substrate diameter, gate oxide layer thickness, type of dopant, metallization layer thickness or material… |
| Back End Production | Manufacturing site, material (lead frame, die attach, plating, wire bond, molding compound), outsourcing to an external assembly house |
| Functional Test | Test reduction, widening of test conditions or outsourcing to an external test house |

Table 8.5 Examples for major changes which require customer approval

### 8.5.4. Qualification, Approval and Implementation

The main purpose of qualification is to guarantee that the planned change does not affect the process or product performance. Depending on the results of the risk assessment, the nature of the change and the initial results the qualification of a change may be executed stepwise. For high risk changes or if the expected difference in yield is small, for instance, a small number of split lots may be useful. The small number of lots reduces the risk for the company, the split lot approach provides the chance to identify small differences without interference from typical lot to lot fluctuations. For the set up of the split plan equipment specific set up needs to be considered, i.e. if a two chambers equipment is used with alternating use of the two chamber then an odd/even split should be avoided. Otherwise the impact of the two chambers might mask the change.

Afterwards, assuming positive results, or for low risk changes pilot runs over a defined number of lots or a defined share of production volume is used to evaluate the change. Inline parameters, electrical parameters, functional yield, WLR results, die yield, die reliability data may need to be considered for qualification. All parameters have to meet the defined acceptance criteria.

---

[1] Pericom Semiconductor Corporation (2007)

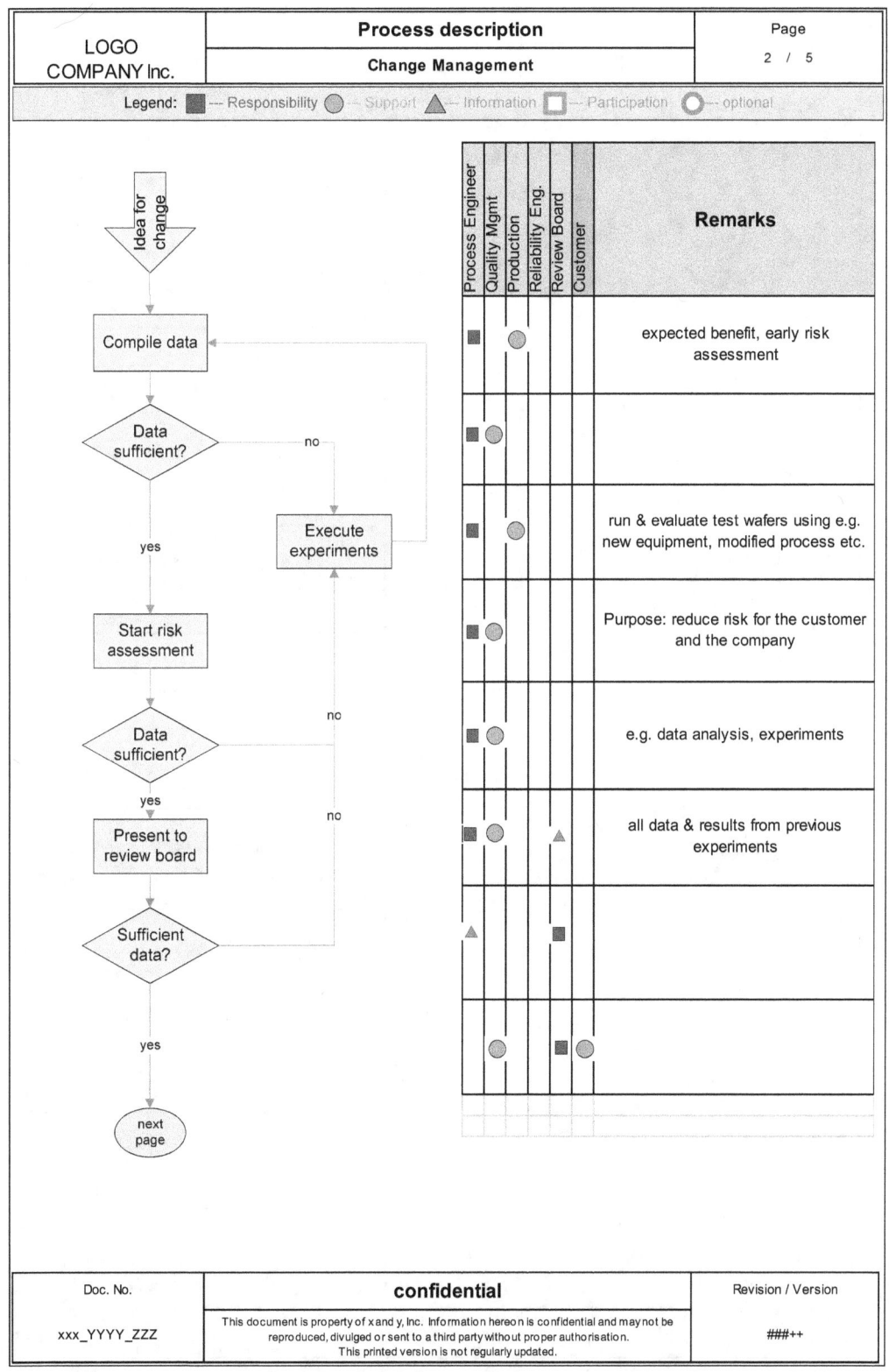

Fig. 8.14 Generic Process Front End Change Management – Experimental Part

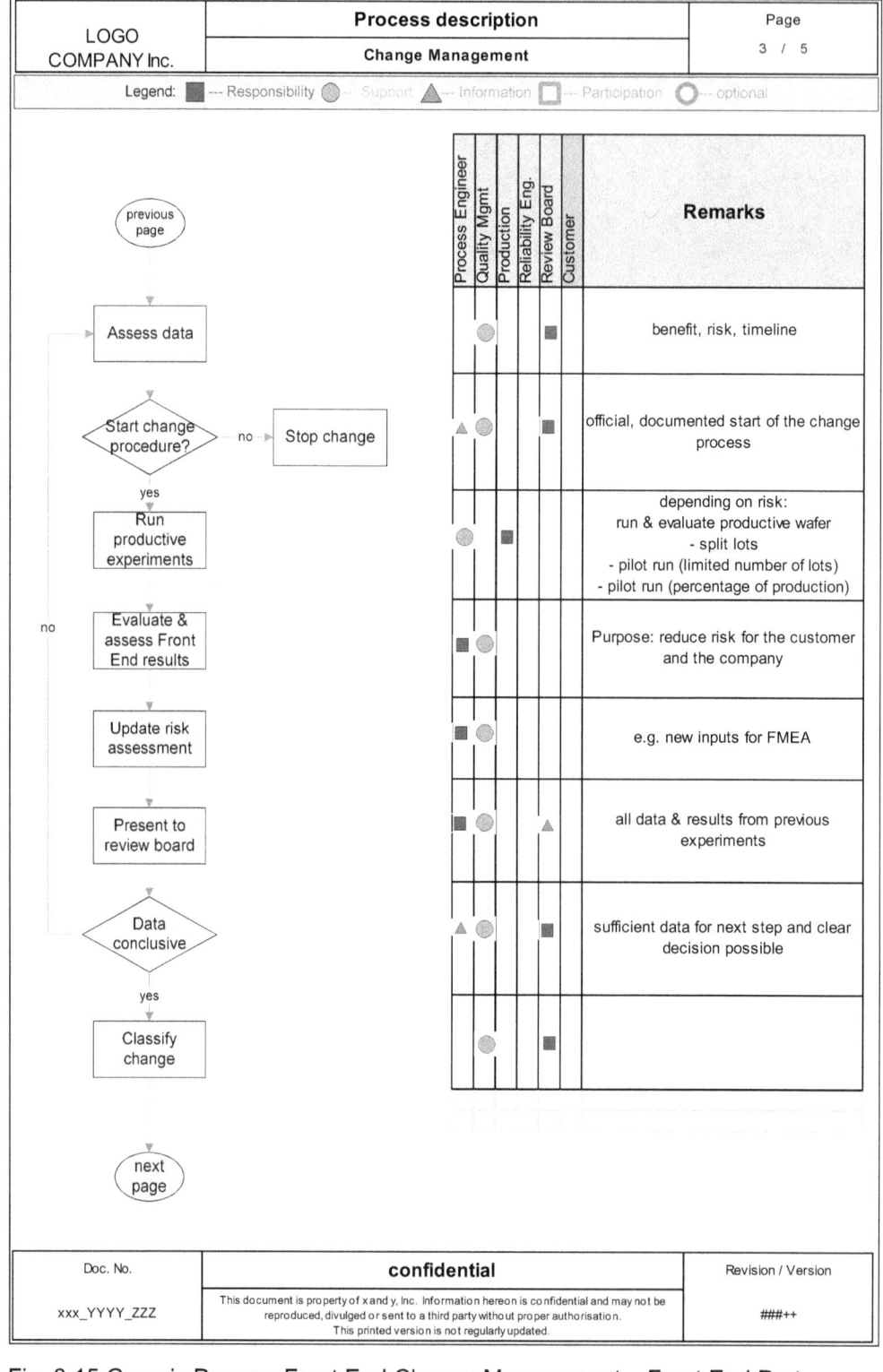

Fig. 8.15 Generic Process Front End Change Management – Front End Part

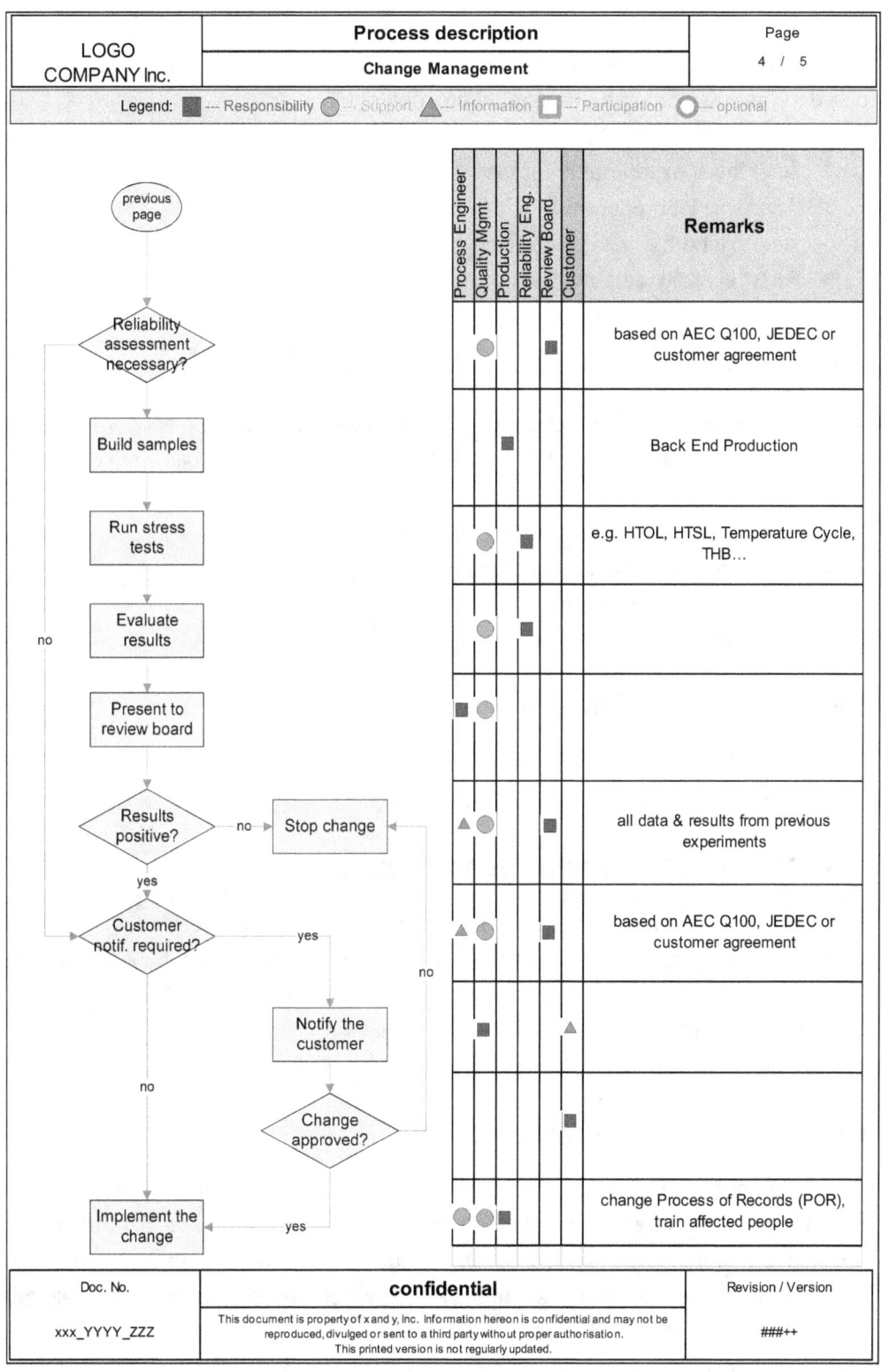

Fig. 8.16 Generic Process Front End Change Management – Product Part

After each step of the change all results are presented to the process change review board for assessment and decision on proceeding:

- After initial experiments (proposal)
- After split lot production
- After pilot run
- After reliability data availability

Depending on those results the change process may continue or stop.

After full production is approved by the process change review board and prior implementation in production all formal requirements have to be fulfilled, e.g.

- Adjust and release manufacturing documentation
- Adjust and release control plans
- Execute operator training
- Document the change and the respective results

### 8.5.5. Customer Notification

Customer notification may be required and is handled as described in the next chapter "Customer Quality Management".

## 8.6. Control of Measurement Equipment

Inspection, measurement and test equipment (here: "measurement equipment") used in manufacturing has to be subject to active management. "Mechanical, electrical, electronic, and physical measurements"[1] have to be performed with stable and accurately measuring equipment in order to meet standards or customer requirements and in order to ensure stable product quality.

### 8.6.1. Calibration Traceability

Calibration of measurement equipment is a regular task to ensure correctness of the measured values. Measurement standards used for calibration have to be traceable to national or international standards. If in-house standards are used traceability to national standards has to be ensured, e.g. "by certified employees with calibration training"[2].

---

[1] Intel (2014)
[2] Seiko Epson Corporation (2009)

Calibration of measurement equipment is required for instance to characterize drift, accuracy and linearity. Calibration results have to be documented and data retention time needs to be defined according to customer requirements or standards requirement. If official standards are not available consensus standards (between organization and customer) may be used. Calibrations are performed both by internal laboratories and by external calibration organizations, e.g. an external accredited laboratory or, if not possible otherwise, by the manufacturer.

The initial point of a calibration standard is the International Committee of Weights and Measures in Paris, France, which establishes adequate reference standards and coordinates comparisons of measurement standards on a national level. Hence the national standards represent the top of the gauge hierarchy within a country. They are the starting point for all lower level standards in this country. Often this task is assigned to federal authorities like

- NIST – National Institute of Standards and Technology, USA
- PTB – Physikalisch Technische Bundesanstalt, Germany
- NPL – National Physical Laboratory, Great Britain
- NMIP – National Metrology Institute of Japan (part of National Institute of Advanced Industrial Science and Technology (AIST)NIM – National institute of metrology, China

Typically subordinate organizations which have to be accredited according ISO/IEC 17025 execute the actual operational activities like calibrations. Eventually companies tend to use working secondary or tertiary standards for the daily business.

Fig. 8.17 shows as an example Japanese measurement standard traceability system, which is based on the Fujitsu company specific traceability system.

## 8.6.2. Calibration Interval

Measurement equipment as well as measurement standards are subject to regular calibration. The intervals chosen are, at the beginning, typically based on the manufacturer's recommendation separately for each piece of equipment.

If the measurement equipment turns out to be very stable, i.e. no drift and no change over time, then the intervals may be stretched. Since this is a substantial change, a risk assessment is required. Typically the intervals are found to be roughly in a range from 6 to 24 months. In addition to the equipment calibration, daily or weekly inspection cycles may be required, which have to be determined for each piece of equipment, too[1].

---

[1] Fujitsu Semiconductor Limited (2012)

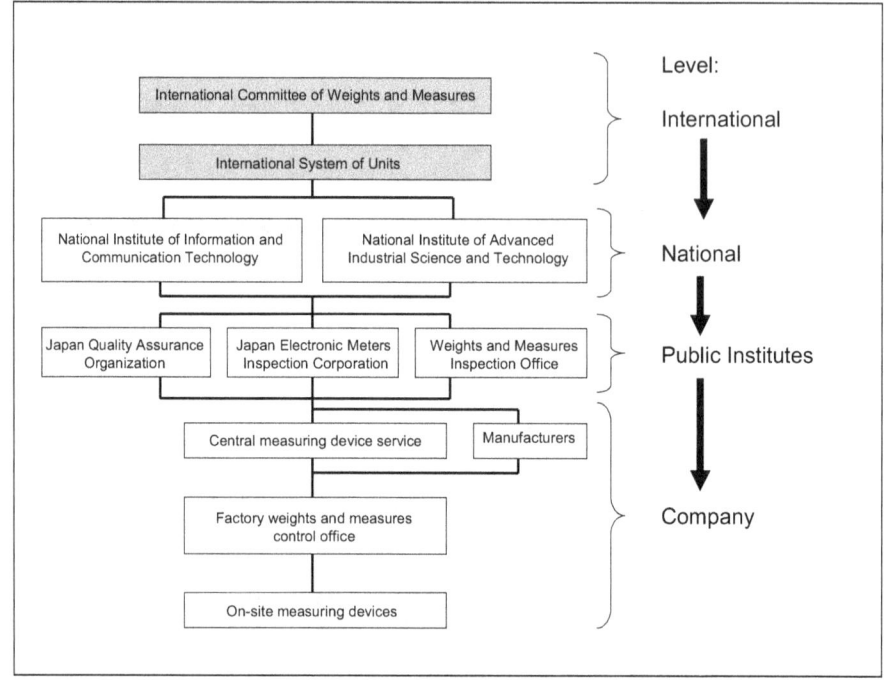

Fig. 8.17 Exemplary standards traceability[1]

### 8.6.3. Calibration Record

Calibration records have to contain[2]:

> - "Unique identification of unit tested
> - Environmental condition under which the unit is tested
> - Deviations, correction factors or limits of calibration, if applicable
> - Equipment owner/user department
> - Calibration interval
> - Last calibration date and next due date
> - Calibration procedures
> - Calibration results
> - Repair information, if applicable
> - Identity of person performing the calibration
> - Standards used to perform the calibration"

In addition to the actual calibration record, the measurement equipment itself may have a tag attached, which indicates the calibration status of the equipment[3], e.g.

---

[1] Fujitsu Semiconductor Limited (2012)
[2] Intel (2014)
[3] Seiko Epson Corporation (2009)

- Calibrated gauge (calibrator, date of calibration, next scheduled calibration date)
- Engineering use (gauge is not to be used for mass production)
- Calibration fail (gauge is not to be used due to calibration non-conformance)

The calibration data retention time is frequently results in legal implications and is related to compensation. Therefore, since the required retention time depends on different factors, a concrete number of years cannot be stated for all different possible cases.

### 8.6.4. Measurement System Analysis

In addition to calibration measurement system analysis may be required in addition. Please consult the chapter "capabilities" for details.

## 8.7. Traceability of the Product

Customers and international standards require traceability of the product, often on a component level. Hence, traceability chips on a wafer, the wafer itself and the lot level in Front End production and individual chips and/or modules in Back End production.
The purpose of traceability is to enable to recover the entire history of a component, e.g. in case of failure, from the incoming material level through the entire Front End and Back End production as well as test areas and distribution centres.
Fig. 8.18 illustrates the scope of traceability based on Seiko Epson.

A system for marking a manufacturing unit (batch, lot, chip, module etc.) has to be in place in order to enable proper identification.
Typical tracing information contains at least:

- Manufacturing year
- Manufacturing week
- Country of production / assembly
- Date of sealing the batch
- Product identification (e.g. part number)
- Identification number for a lot or batch of material
- Quantity
- Traceability code

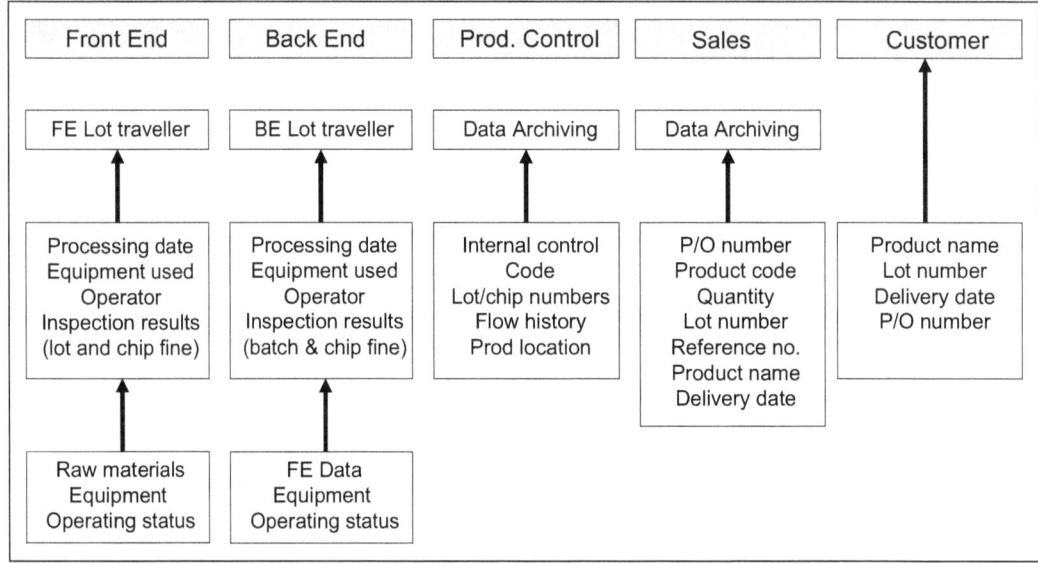

Fig. 8.18 Traceability scope on various levels of the supply chain[1]

Additional requirements regarding traceability may include:

- Minimum records retention (e.g. 7 years (INTEL))
- Completion time of each process step
- Completion time of inspection
- Test status & pass / fail quantities
- Custom marking of products as required by a customer

# 9. Customer Quality Management

> A quality management system should aim "to enhance customer satisfaction through the effective application of the system, including processes for continual improvement of the system and conformity to customer and applicable statutory and regulatory requirements"[2]

In other words the entire quality management system eventually serves the customer. This does not mean to neglect the company's own interests, but clearly makes aware that customer satisfaction is one of the priority one topics in quality management. In this

---

[1] Seiko Epson Corporation (2009)
[2] DIN EN ISO9001:2015

context customer satisfaction is the "customer's perception of the degree to which the customer's expectations have been fulfilled".[1] Generally a customer can be an internal or an external customer.

However, the following sections do not describe the general and comprehensive customer perception, but only specific customer related requirements of the standards and customer related processes.

## 9.1. Requirements of the Standards

In this section both important ISO9001 as well as ISO/TS16949 requirements regarding specific customer related requirements are summarized.

### 9.1.1. Customer Focus

ISO9001: The company, in particular the top management, has to assure that

- the customer's requirements are identified
- the capability for increasing customer satisfaction is identified and respective activities are started
- risk and opportunities with potential impact on product conformity are identified and respective activities are started
- the company focuses on a stable delivery of conforming products
- eventually increasing customer satisfaction is in focus

### 9.1.2. Determination of Product Requirements

ISO9001: The company has to define and to implement a process for the determination of product requirements. In addition the company has to assure to be capable to meet the previously defined requirements.

ISO/TS16949
The customer may define special characteristics, i.e. product parameters which are particularly important for the product quality.

### 9.1.3. Customer Communication

ISO9001: The company has to establish processes for customer communication regarding e.g.

- information about the product

---

[1] DIN EN ISO9001:2015

- contracts, orders, requests and respective changes
- how to receive the customer's opinion and customer complaints
- how to handle the customer's property
- handling of contingencies

ISO/TS16949: Language and format of the customer communication is defined by the customer. The organization has to assure compliance to this format.

### 9.1.4. Customer Property

ISO9001: The company has to take care of customer property. This includes for instance marking and protection of those items. In case of problems, e.g. misuse or damage, the customer needs to be informed.

### 9.1.5. Customer Satisfaction

ISO9001: The company has to gather information regarding customer's impression and opinion about the product and the organization in order to clarify whether or not the customer requirements were met. Methods for these activities have to be defined.

ISO/TS16949: Customer satisfaction shall be monitored on a regular base by using data based performance indicators at least for quality of delivered parts, product related problems at the customer or in the field, delivery performance and customer notification in the case of quality and/or delivery problems.

In addition to the above mentioned items ISO/TS16949 does contain several additional requirements which are not reflected in the ISO9001:

Customer representative
The customer representative, defined by the management, takes care of customer requirements within the company. This includes e.g. the definition of special characteristics, the definition of quality objectives and the definition of preventive and/or corrective actions.

Customer approved suppliers
If agreed upon in the contract, the company has to purchase products from customer approved suppliers only.

Customer information
In case the company delivered defective products to the customer, the customer has to be informed about it immediately.

Special release

A special release by the customer is required prior proceeding with production, if the product or the production process deviates from the process of records (= POR, i.e. released process).

In the following sections a small selection of customer related processes is depicted.

## 9.2. Customer & Product Requirements

### 9.2.1. Specification

Customer and product requirements have to be determined by the company. In the business-to-business case (b-to-b) frequently customers inform the supplier about their needs. This may or may not be in a structured way and includes technical discussions, common development projects as well as documented requirement lists.

Eventually, in order to define a specification, the customer has to provide information at least about

- required performance parameters
- operating conditions
- range of use

Often the requirements can be classified in a basic requirement (= a must requirement), a quality- and/or performance requirement (= a should requirement) and so called customer's delightment reqirements.

In addition to this technical information the company also needs information about

- the potential market volume and
- the customer's timeline
- statutory requirements
- standards requirements
- the market

Otherwise it is not possible to make a realistic judgement on project and product cost as well as manufacturability.

The information given by the customer and from other sources is then applied to create the engineering concept, which is later on elaborated and ends up in a more functional specification.

The functional specification should contain[1]

- detailed operational functions
- product set up
- material requirements
- technical solutions
- performance parameters
- geometrical dimensions
- module and assembly structure

## 9.2.2. Quality Function Deployment (QFD)

Quality Function Deployment is a quality management tool for a systematic and comprehensive translation of customer requirements into product requirements and product specifications.

Being an important pillar of the product development process QFD has shown to be used successfully in other fields, too[2], e.g.:

- service development
- ERP system selection
- Semiconductor system-on-a-chip product design planning

The method Quality Function Deployment was invented by Yoji Akao in the ship building industry of Japan in the nineteen sixties and was transferred to America and Europe in the nineteen eighties[3]. Today it is widely used in the automotive industry. It comprises four phases or steps starting with gathering the customer requirements and ending with a full blown production plan.

Those process steps are worked off step-by-step supported by a table called House of Quality[4] (fig. 9.1). The purpose of the house of quality is to transfer "what is needed" into "how do we achieve this" in each single phase of the QFD process.

The four phases of the QFD process consist of an initial preparation step and following three "translation" steps[1], namely

(1) Translate the customer's requirements
Customer needs are translated into product characteristics and parameters. It deals with the identification of the product attributes which contribute to the fulfilment of the

---

[1] Boutellier, R., and Biedermann, A., (2007)
[2] Raharjo, H. et al. (2011)
[3] Akao, Y. (1997)
[4] Hauser, J. R. (1993)

customer needs. Those needs, however, are often expressed in an ambiguous and unclear way and have to be prepared, e.g. by the use of workshops, interviews or market analysis activities.

(2) Plan the product
Product parameters identified in step 1 are now used to identify technical design parameters, i.e. parts parameters and tolerances. It deals with critical parameters and their target values and defines the parameters which have to go through a more detailed process planning process.

(3) Plan the process
Now the same way design parameters are used to determine the respective critical process parameters. The purpose of this step is to determine the target values of the process parameters under investigation and the determination of parameters for a more detailed production planning process.

(4) Plan production
Eventually in step 4 a detailed production planning process takes place. It covers the general concept, e.g. use of SPC, JIT etc., as well as more detailed tasks like preparation of the quality plan or of required standard operating procedures.

Being most suitable for step (1) a more detailed explanation of the practical execution of the house of quality is given using the identification of the customer's requirements (fig. 9.2).

(5) Customer Requirements
Up to roughly twenty customer requirements have to be collected and listed. Afterwards they may be grouped by topics and classified, e.g. fundamental (legal), key and enthusiastic requirements and weighted according the importance for the customer. Preferentially the weighting is provided directly by the customer. It this is not possible, the weighting should at least be determined using a systematic approach.

(6) Customer Perception
Now from a customer point of view the company's product is compared with the competitors' products. The basic question in this category is whether or not the own product is better or worse than the competitors' products for each customer requirement. Again it is useful to have the customer's direct judgement available. Otherwise an internal judgement becomes necessary.

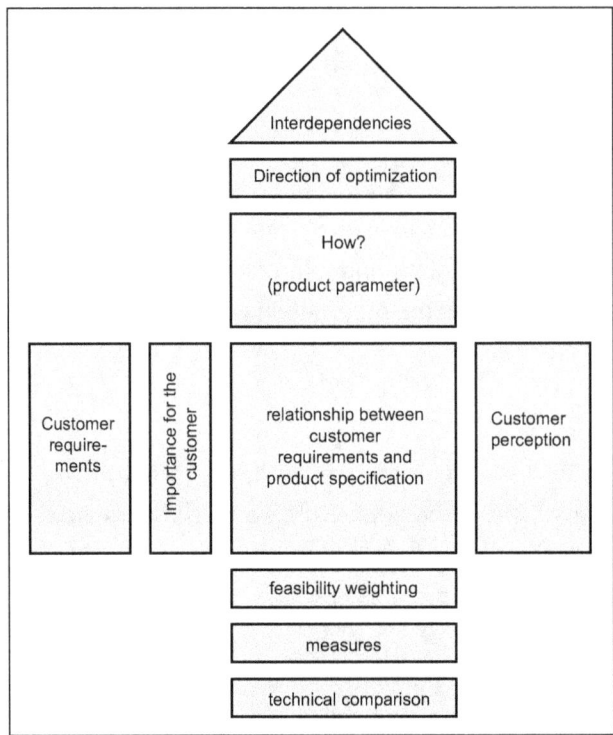

Fig. 9.1 House of Quality for step (1) "Identification of the customer's requirements"[1]

(7) Product Parameters
Product characteristics and parameters translated from customer requirements are used to describe how the customer requirements can be achieved. The most important parameters are documented here. Another purpose is to "quantify" the often somewhat diffuse customer requirements.

(8) Interdependencies
This category looks into the interdependencies of product parameters. For instance a safety relevant parameter, e.g. special material, may have a negative impact on the product weight.

(9) Direction of Improvement
The purpose of this step is to identify the direction of change of a product parameter in order to achieve an improvement.

(10) Relationship between customer requirements and product specification

---

[1] Linß, G., (2005)

The main task here is to determine direction and strength of the correlation between the customers' requirements and the product parameters.

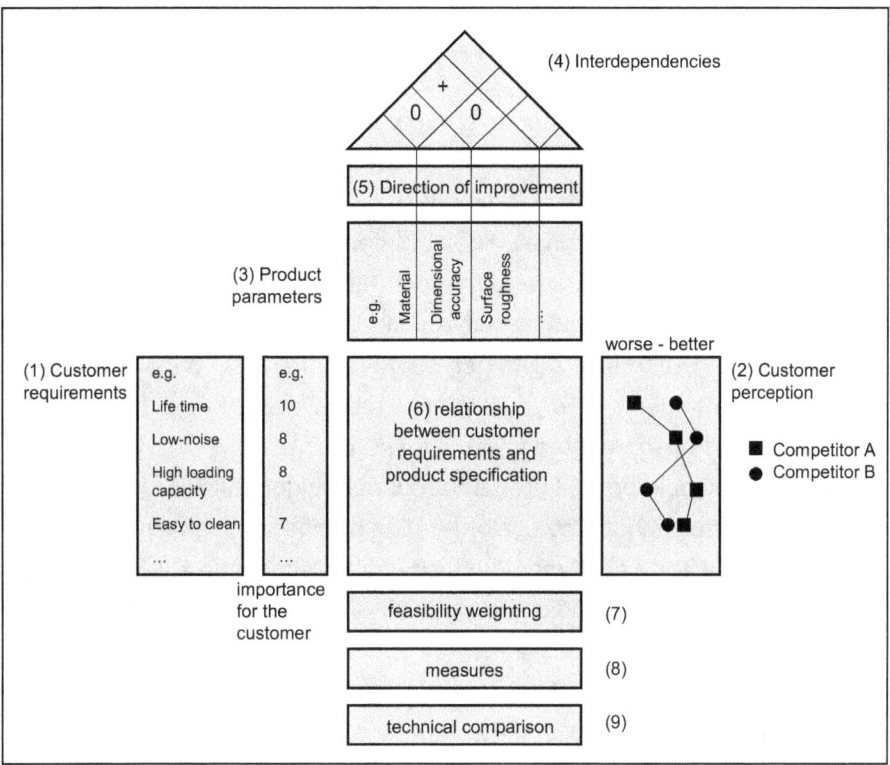

Fig. 9.2 House of Quality for step (1) "Identification of the customer's requirements"[1]

(11) Feasibility Estimation
The realization of improvements (see "direction of improvement") or an entirely new product development is associated with various degrees of complexity. The feasibility of the respective change has to be estimated.

(12) Measures
This category describes the measured value in the form of target values and criteria for measurement.

(13) Technical Comparison
The values of the product parameters of the own products are compared with the competitors' products. It's a "better – worse" like comparison.

---

[1] Linß, G., (2005)

Eventually the House of Quality is a comprehensive and systematic tool to document how customer requirements found their way into the functionality of a product.

### 9.2.3. Customer Survey

Monitoring of customer satisfaction may be executed e.g. by analyzing and reacting on customer complaints or the customer's supplier evaluation results. A more direct way to gather information about the customer's perception regarding product or company performance is to ask the customer directly, i.e. by performing a customer survey.

Due to the fact that the respective customer interface for sure has to deal with a larger number of suppliers and also due to the fact that a sound evaluation of the customer responses including triggering corrective actions might take several months it is not recommended to execute a customer survey more than once per year and per customer. With a larger period of time in between two surveys the willingness of a customer to provide a thorough answer is for sure higher.

The survey itself should be well prepared regarding right definition of the questionnaire and timely announcement to the customer. The number of questions should be limited. Otherwise eventually the customer interface might discontinue answering all questions. Questions with reference to the company itself or the product may be:

- How satisfied are you with our product in regards to e.g. functionality, reliability
- How satisfied are you with the company's delivery performances
- How satisfied are you with our product in regards to
    - Functionality
    - Reliability
    - Technology
    - Cost of ownership
    - Consumption
    - Service
- How satisfied are you with the company's customer orientation in regards to
    - Friendliness
    - Availability
    - Content of received information
    - Flexibility
    - Response time
    - Reliability
- From a customer's point of view: what are the company's main strengths and weaknesses?
- How satisfied are you with the company's overall performance?
- etc.

The questionnaire should offer the customer a grading system as well as a comment field. An example for an equipment maker's project related questionnaire is given in tables 9.1 and 9.2.

It is mandatory to define and execute corrective actions based on the customer's response. Communication of those activities, in particular a successful execution, to the customer serves to increase customer satisfaction. However, an actual improvement of the gap reported by the customer, supports a long term customer satisfaction and his confidence towards the supplier.

## 9.3. Planned Changes

> **PCN**
>
> **Process Change Notification**

Frequently information to the customer is required if changes in production or along the supply chain are planned or if the company decides to change their product portfolio. In order to ensure that this information arrives at the customer early enough to give him the opportunity to react in time and to ensure that this information arrives at the right person a formalized process of communication should be agreed upon between the company and the customer. In other words the customer needs to be notified about planned process and / or production process changes.

> **Major Change**
> "a change that may affect the form, fit, or function of the product or adversely affect the quality or reliability of the product"
>
> **Minor Change**
> "a change that does not affect the form, fit, function, quality, or reliability of the product" [1]

---

[1] JEDEC46

|  | No | Topic | Judgement* | | | | Comment |
|---|---|---|---|---|---|---|---|
|  |  |  | 1 | 2 | 3 | 4 |  |
| Project Management | 1 | Was the name of the responsible project manager (PM) clearly communicated to you at the beginning of the project? | | | | | |
| | 2 | Was the project manager introduced to all involved members of your team? | | | | | |
| | 3 | How satisfied were you with the communication skills and competence of the PM? | | | | | |
| | 4 | Have you always been sufficiently informed about the actual project status by the PM? | | | | | |
| | 5 | Were requests / complaints processed promptly? | | | | | |
| | 6 | Were requests / complaints resolved to your satisfaction? | | | | | |
| | 7 | Did you receive sufficient feedback regarding progress while your request / complaint was in process? | | | | | |
| Project Preparation | 8 | Was your staff well prepared for equipment delivery by our PM? | | | | | |
| | 9 | How satisfied were you with the Hook-up documentation? | | | | | |
| | 10 | How satisfied were you the the facility specification? | | | | | |
| Equipment | 11 | Was the equipment delivered on time? | | | | | |
| | 12 | Did you get exactly what you ordered? | | | | | |
| | 13 | Was the equipment delivered complete and without damage? | | | | | |

Table 9.1 Project related customer questionnaire – part I

| Category | No | Topic | Judgement* | | | | Comment |
|---|---|---|---|---|---|---|---|
| | | | 1 | 2 | 3 | 4 | |
| Installation & Final Acceptance | 16 | Was the final acceptance test carried out by the agreed time? | | | | | |
| | 17 | How satisfied were you with the results of the Final Acceptance Tests? | | | | | |
| | 18 | How satisfied were you with our staff regarding competence, flexibility and result orientation? | | | | | |
| Documentation & Training | 19 | How satisfied are you with the operation manual? | | | | | |
| | 20 | How satisfied are you with the offered training program? | | | | | |
| | 21 | How satisfied are you with the training already carried out? | | | | | |
| | 22 | How satisfied are you with the competence of our trainer? | | | | | |
| | 23 | How satisfied are you with the quality of our training material? | | | | | |
| General | 24 | What went particularly well? | | | | | |
| | 25 | What went particularly bad? | | | | | |

*1 = bad performance
4 = excellent performance

Table 9.2 Project related customer questionnaire – part II

This process has to include the customer's response, e.g. to show awareness of the company's information or to provide objections to or release of changes. On the other hand, this process does not cover the technical part of the change, i.e. it does not describe risk assessment and introduction of the change itself.

The need to inform the customer about planned changes may be based on JEDEC46, on individual contracts or quality assurance agreements or other documented and agreed upon customer requirements, e.g. PPAP (see Section "PPAP"). Typically a change has to be classified into a minor or a major change and the customer has to be notified in case of major changes in any case and has to be notified in case of minor changes only if required by the customer.

Major change lists may be commonly agreed lists and, e.g. in case of material suppliers, may include items like

- Change of a production site
- New production sites
- Change of a crystal pulling process (in case of silicon)
- Change of an epitaxial process

Towards the customer of the microelectronics industry major changes may include items like

- Change of chip size or density of devices
- Change of fab or assembly production site
- Change of layer thickness (metallization, passivation, …)
- Change of material (e.g. gate oxide, wafer substrate, lead frame, die attach, mold compound, …)
- Change of testing equipment or location

The starting point for this process is the need for a change, for instance due to a change of material or equipment. The technical change process has to be started in order to generate all required information like yield and reliability data. All this data is collected and put together to a predefined Customer Information Package. Once the data is available and information to the customer is required the customer is notified ahead of time, e.g. 90 or 180 days prior planned introduction of the change in production.

After receiving of all data the customer evaluates the planned change and provides a response to the supplier within a predefined period of time.

The PCN form and the Customer Information Package should at least contain the following information:

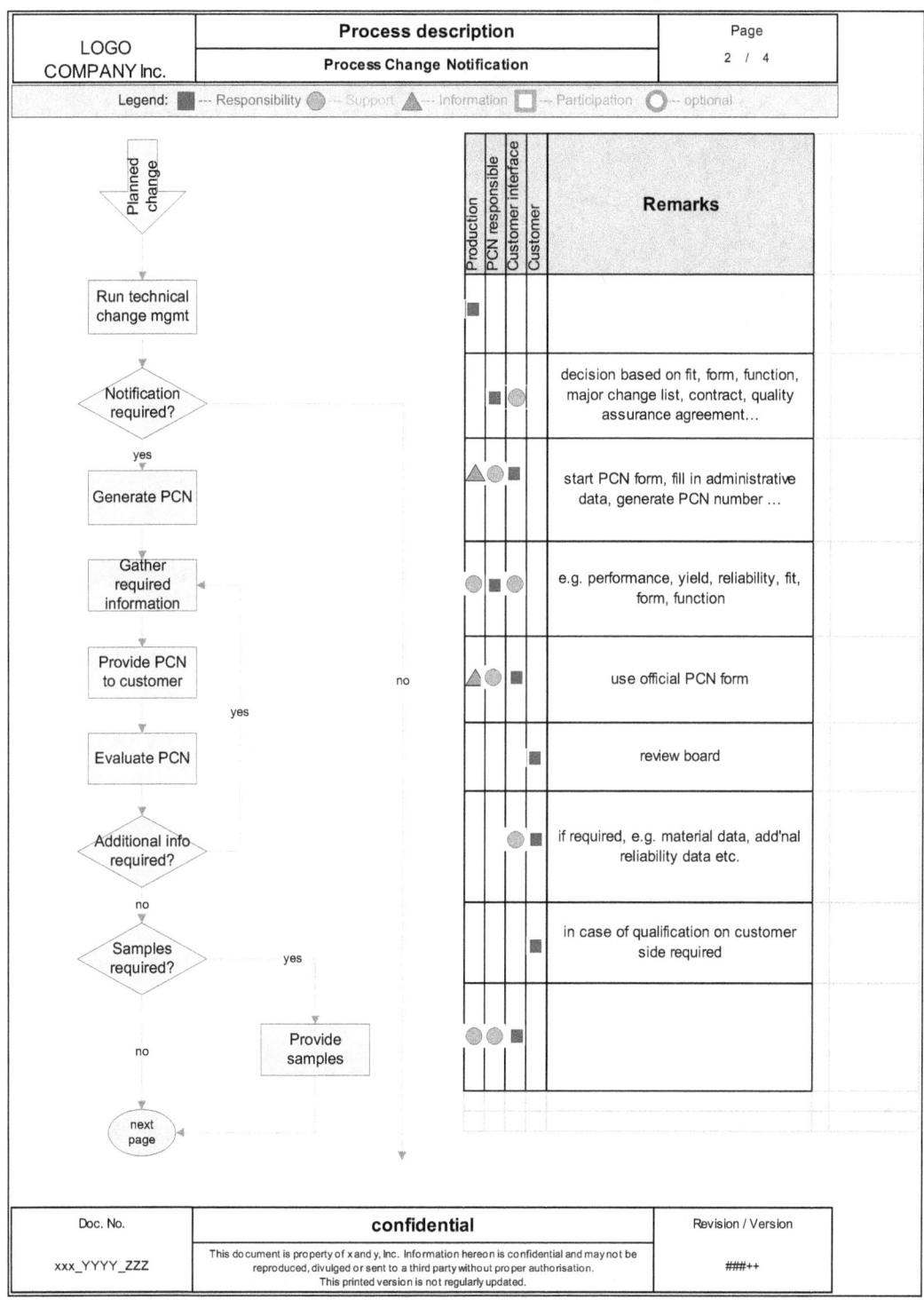

Fig. 9.3 Generic PCN process – part I

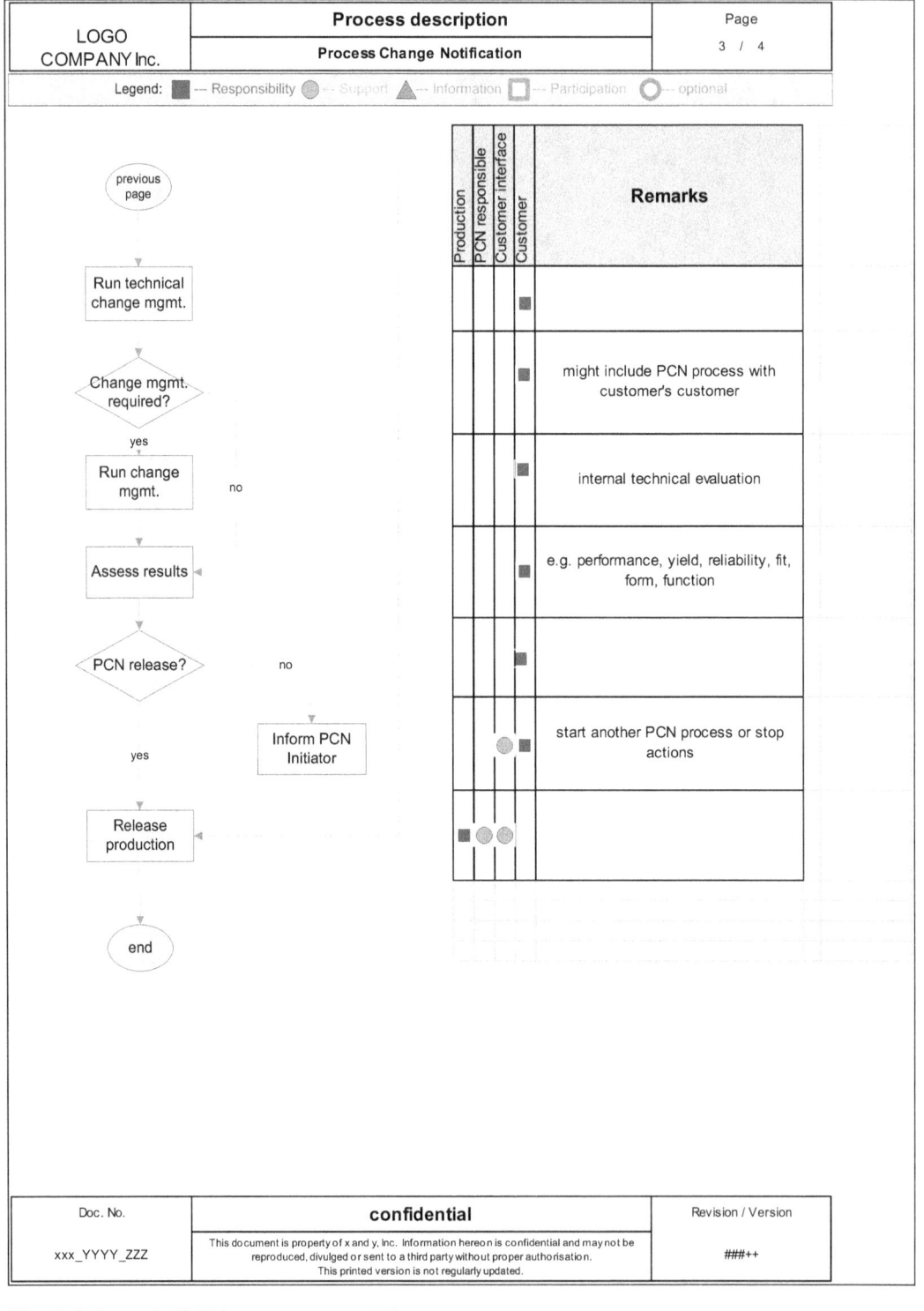

Fig. 9.4 Generic PCN process – part II

- PCN-nr.
- Date of issue
- Required response date
- Kind of change with description
- Reason for change
- Time schedule (qualification & sampling schedule, planned release date)
- Qualification results (as soon as available)
- List of products which are affected by the planned change

In addition to the PCN process, other product or product quality related processed have to be defined and introduced. For instance a PTN process (product termination process) should be implemented which ensures the customers are informed about the termination of a product early enough. Depending on the contractual situation a PTN process may result in a 6 month advance notice (or more) or even preproduction covering several years of supply. Preproduction stock may be the company's or the customer's responsibility.

Highly complex production environment might also require a so called "Early Warning process", which ensures that the customer is informed about a potential quality or delivery issue affecting his production. Again this allows the customer to react early enough, hence to keep his own production running without interruptions. However, it's in the nature of things the Early Warning Process needs to be well controlled. Otherwise the customer might loose confidence in the supplier's competences.

## 9.4. Complaint Management

### 9.4.1. General Process

The objective of the complaint management process is to ensure that customer complaints are handled effectively within a company. This includes for instance to identify and to address the root causes of complaints, i.e. of product malfunction, in order to learn how to optimize the product as well as the production or logistics processes. The purpose of this approach is to reduce the failure rate of the products (e.g. dpm rate or FIT rate). A well established way of complaint management also helps to provide root cause analysis results and corrective actions to the customer in a reasonable or contractually required turnaround time, eventually to increase customer satisfaction.

In the course of this section we will focus on technical issues, i.e. product malfunction or product performance issues etc. Of course logistical issues, e.g. delays of delivery, also have to be managed by the company.

The general process flow of complaints consists of three major steps:

(1) receive & register
The complaint is received by the company and receipt is confirmed to the customer. The customer-supplier path of communication has to be defined ahead of time and could e.g. include a complaint hotline, an email address or an interface of ERP (Enterprise Resource Planning) systems. Internally the complaint triggers many actions like technical analysis and risk analysis.

(2) analyze & correct
Failing samples from the customer have to undergo a technical assessment. This includes confirmation of the fail by means of production type test, simulation or lab test. Analysis of the fail root cause needs to be performed and based on the results corrective actions are derived and started (in case of a new type of fail).

(3) report & closure (see also 9.4.2.)
An 8D-report is created including root cause analysis results, containment actions and corrective actions and is sent to the customer for approval (if required). Depending on the severity of the deviation also direct customer communication may be required. It is recommended to assign this task to experienced and trained employees.

In addition to the handling of the actual deviating product accompanying actions have to be initiated by the company. These include for instance a comprehensive risk assessment – this kind of problem could also occur at other customers – or repair or replacement of the product. Systematic quality issues can be identified by the use of a complaint register with classification by root cause and severity or impact.
Professional management of customer complaints is important for the long term economic success of a company. Therefore the company's management should be involved in this topic. At least this should be ensured by a formalized and regular reporting system towards the management.

## 9.4.2. 8D: Eight Disciplines

Frequently the 8D approach (eight disciplines) is required by the customer to manage, solve and report all relevant findings regarding a failure at the customer. The methodology as well as the 8D report consists of the following eight steps[1]:

D1: Set up of the cross functional 8D team

---

[1] Neagoe, L. N., Klein, V. M. (2010)

8D team leader as well as team members with relevant knowledge are selected and have to have enough resources (time!) to contribute to the 8D team. For major topics it is beneficial to have access to a sponsor who can provide management support if needed.

D2: Description of the problem
A detailed and comprehensive description of the problem is mandatory. Without it, it is very difficult to solve complex problems. Hence the following minimum information has to be included:

- Product number, name and description
- Production date of material concerned
- Customer
- Customer's failure description
- Early data and physical analysis results verifying the failure reported by the customer
- A detailed description: when, where, how often etc. the problem appears and when, where, how often etc. the problem does not appear. This includes the information where the problem is NOT found. This aspect is often neglected with a focus only where the problem is found, leaving "uncharted" territory where the problem may have occurred but was not noticed yet.

D3: Implementation of containment actions including containment assessment
Prior to definition and implementation of containment actions, a risk analysis has to be performed. Based on the result of the risk assessment containment actions are defined in order to reduce the customer's risk soon and significantly.
Those actions could be

- Stop of shipment to the customer
- Quarantine of affected manufacturing equipment
- Quarantine affected material
- Start return of material process (initiate the return of the delivered material)
- 100% sorting instead of sampling
- Additional screening activities
- Stop of production from a particular tool or facility
- Tightening of screening limits
- …

D4: Root Cause Analysis

A fast and constructive root cause analysis is crucial for a successful 8D problem solving. Various methods and tools are in place supporting the identification of the root cause (see next section). Frequently there is an immediate root cause for the specific problem and a more systematic, general root cause which could lead to the same problem again. It is desirable to identify and to eliminate both.

D5: Definition of permanent corrective actions
Corrective actions eliminating the immediate root cause of the current problem are defined and tested for effectiveness, e.g. by running production with and without the corrective action implemented. In addition, it is necessary to ensure that the corrective action planned does not have any detrimental side effects somewhere else.

D6: Implementation of permanent corrective actions
After proving effectiveness of the corrective actions those have to be permanently implemented. This includes for instance

- Change of processes and/or products
- Change of respective documentation (e.g. process flow)
- Training of all relevant employees

D7: Preventive Actions
The problem solved may appear very specific, but often similar products or applications might have the potential to experience the same or a similar problem. Hence it is useful to think proactively where and when this problem could occur again and to define and implement countermeasures ahead of time. The elimination of the systematic root cause is one option to execute preventive actions.

D8: Information to the team
Finally the 8D team is rewarded, the 8D report is documented and distributed to all interested parties and the 8D project is closed.

## 9.4.3. Root Cause Analysis[1,2]

As long as the actual root cause of a problem is not identified correctly corrective actions may or may not be successful regarding prevention of the problem. Typically root cause analysis (RCA) starts with brainstorming activities alone or in a team, more or less intense data analysis and does often include a variety of electrical test and physical

---
[1] Rooney, J. J., Heuvel, L. N. V. (2004)
[2] Dogget, M. (2009)

failure analysis activities. Understanding a root cause means that not only known what event lead to the problem and how it came up, but also looks into why this happened. The following section introduces a guideline of root cause analysis:

### 9.4.3.1. Fishbone / Ishikawa Diagram

The cause and effects diagram, often called fishbone diagram or Ishikawa diagram, is a very common method to illustrate and determine the main impact factors and sub (secondary) impact factors for a particular event.

Frequently, the "main" starting branches are "material", "machine", "method" and "man" (4M). Other branches like "management", "measurement", "environment" and others may be added as needed. Those main branches are also triggered by various impact factors which then are branched off from the main branch.

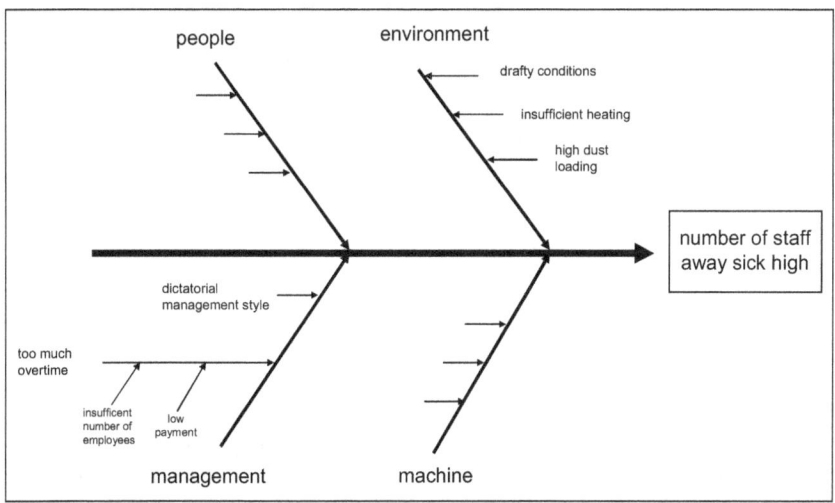

Fig. 9.5 Fishbone diagram

### 9.4.3.2. 5-Why

This method simple requires the team to ask five times why this or that happened which in the end leads to the primary root cause. The method is not limited to 5 Why's but could end after four Why's or even need more than five.

Example:
A chip failed in the field and initial analysis showed the root cause to be an open metal line:
   1. Why
   Why did the metal line fail after two years?
   Answer: The metal line suffered from electromigration.

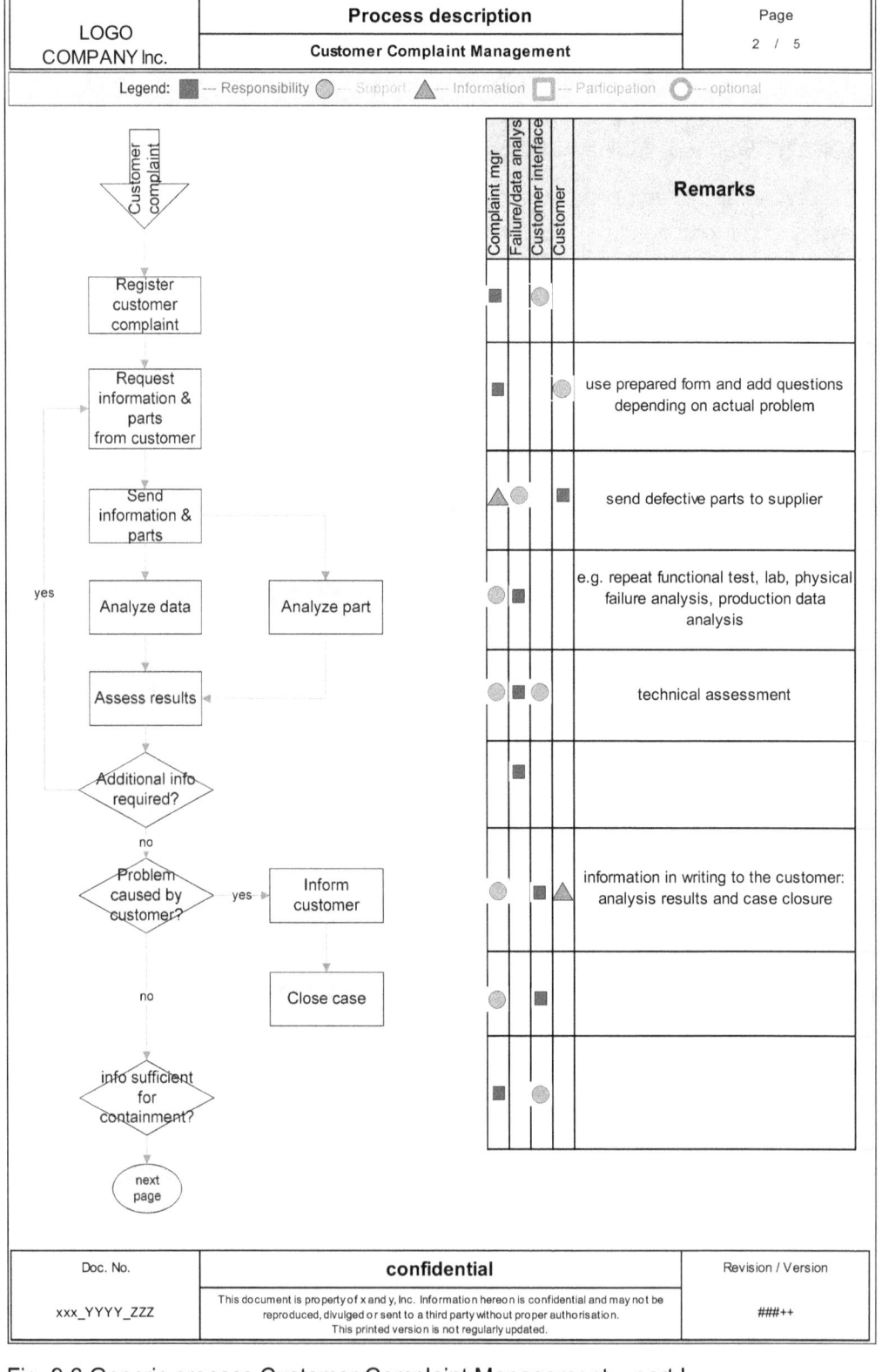

Fig. 9.6 Generic process Customer Complaint Management – part I

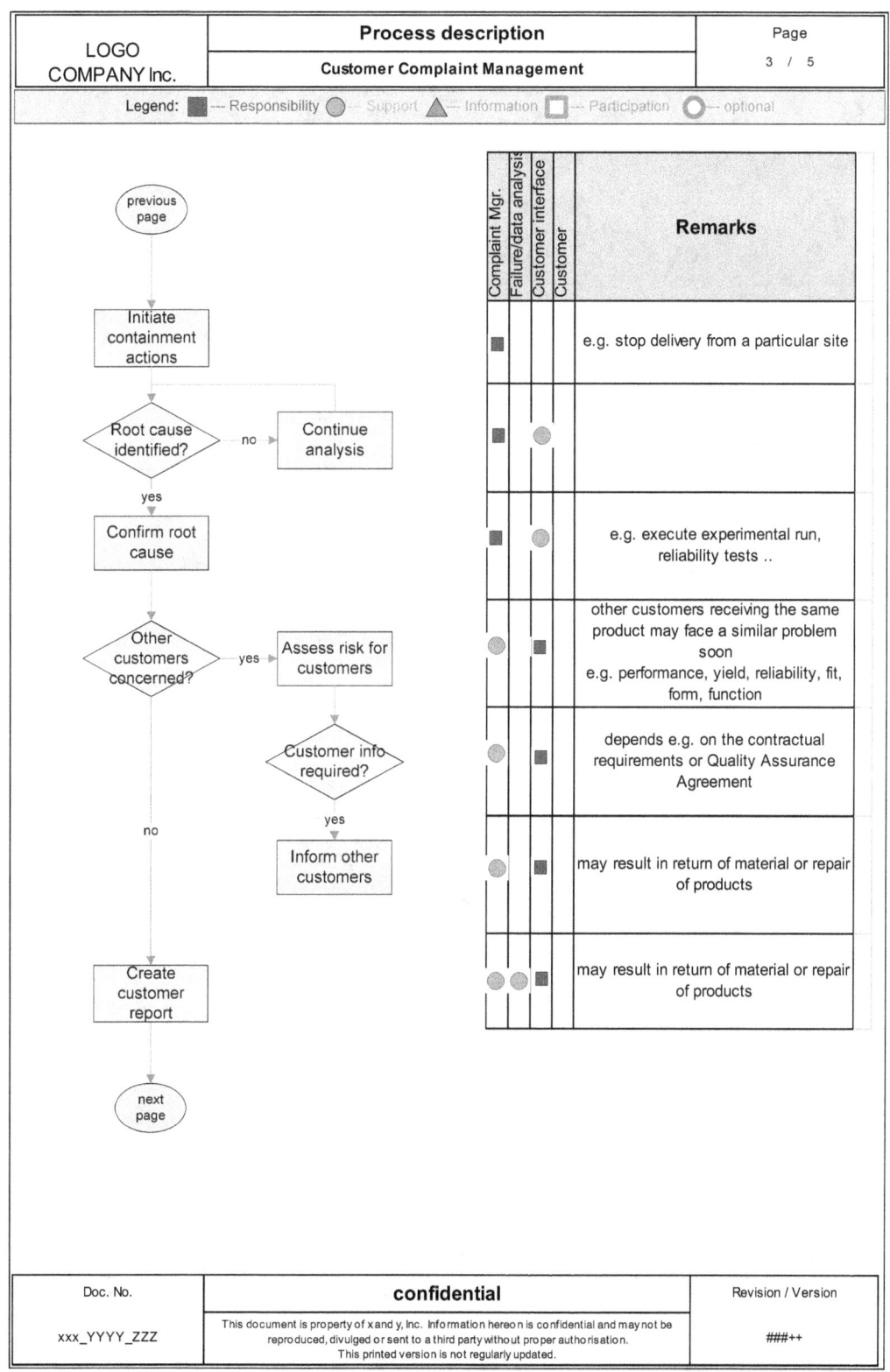

Fig. 9.7 Generic process Customer Complaint Management – part II

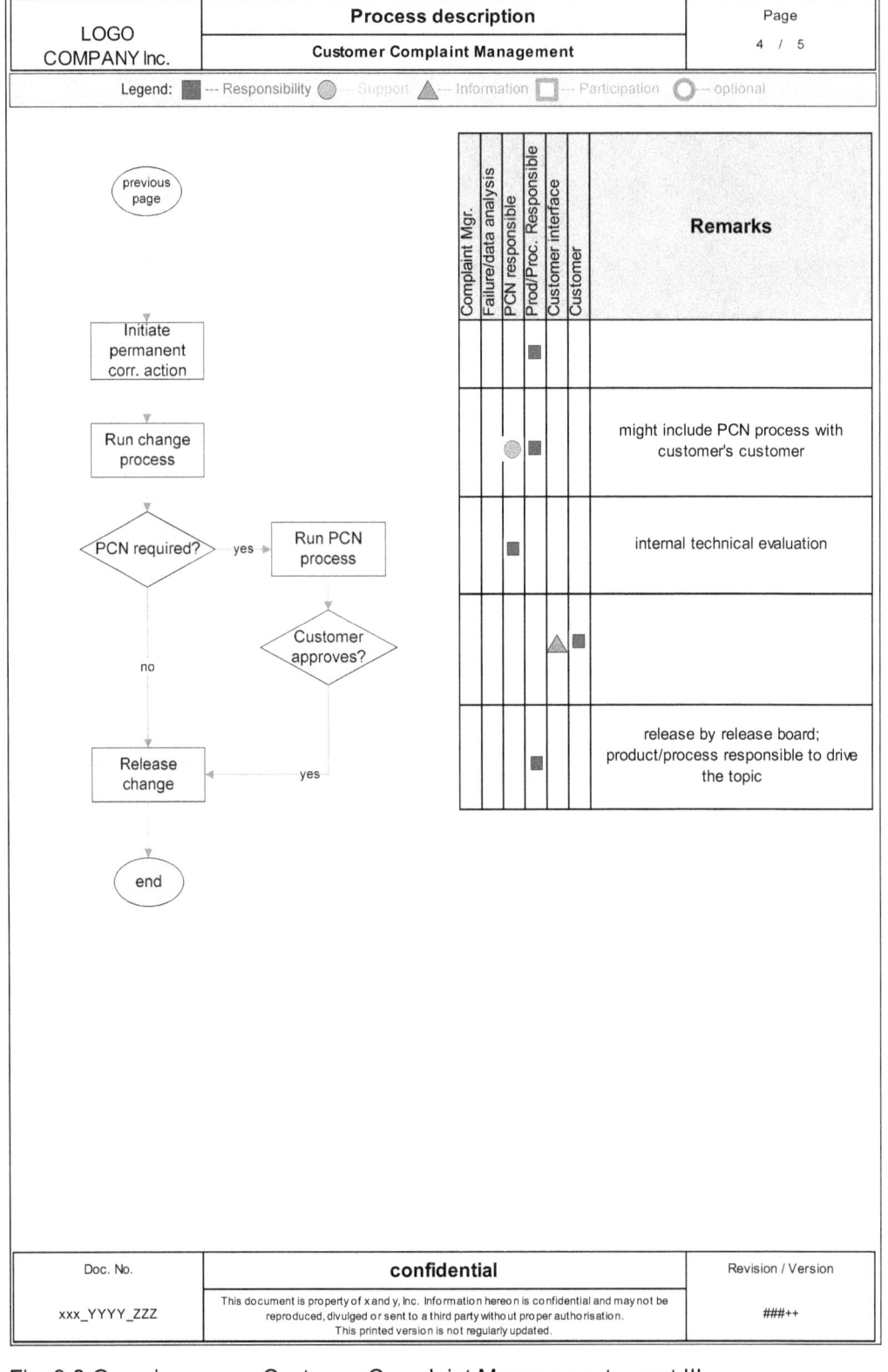

Fig. 9.8 Generic process Customer Complaint Management – part III

2. Why
Why did the metal line suffer from electromigration
Answer: the Cu content in the aluminum line is too low

3. Why
Why is the Cu content in the aluminum line too low
Answer: the sputter target Cu content was lower than typical, but still within the specification limits

4. Why
Why was the target Cu content lower
Answer: the target of a cheaper second source was used with stronger target to target Cu content variations (but always within specification)

5. Why
Why was this not considered when introducing the new target supplier
Answer: the process window for this parameter was not known, it was assumed that the specification is tight enough, but actually the first supplier's superior performance ensured stable process results.

## 9.5. Product Recall

The worst case for a company is to be obliged to call products back from the customer due to performance, reliability or even safety issues. In this chapter product recall also includes product repair at the customer site.

The main reasons for a recall are safety issues with the product or major quality and/or reliability issues. However, the interpretation of legal regulations is explicitly not subject of this chapter. It is recommended to seek legal advice in all where there is a potential violation of legal requirements. Here we only provide an overview on how to handle the recall topic from a Quality Management point of view.

The recall process typically comprises at least the following process steps:

- Monitoring of the market for potential quality issues
- Receiving information
- Assessing Risk
- Planning of the corrective action strategy
- Informing the customer
- Executing corrective actions (e.g. recall)

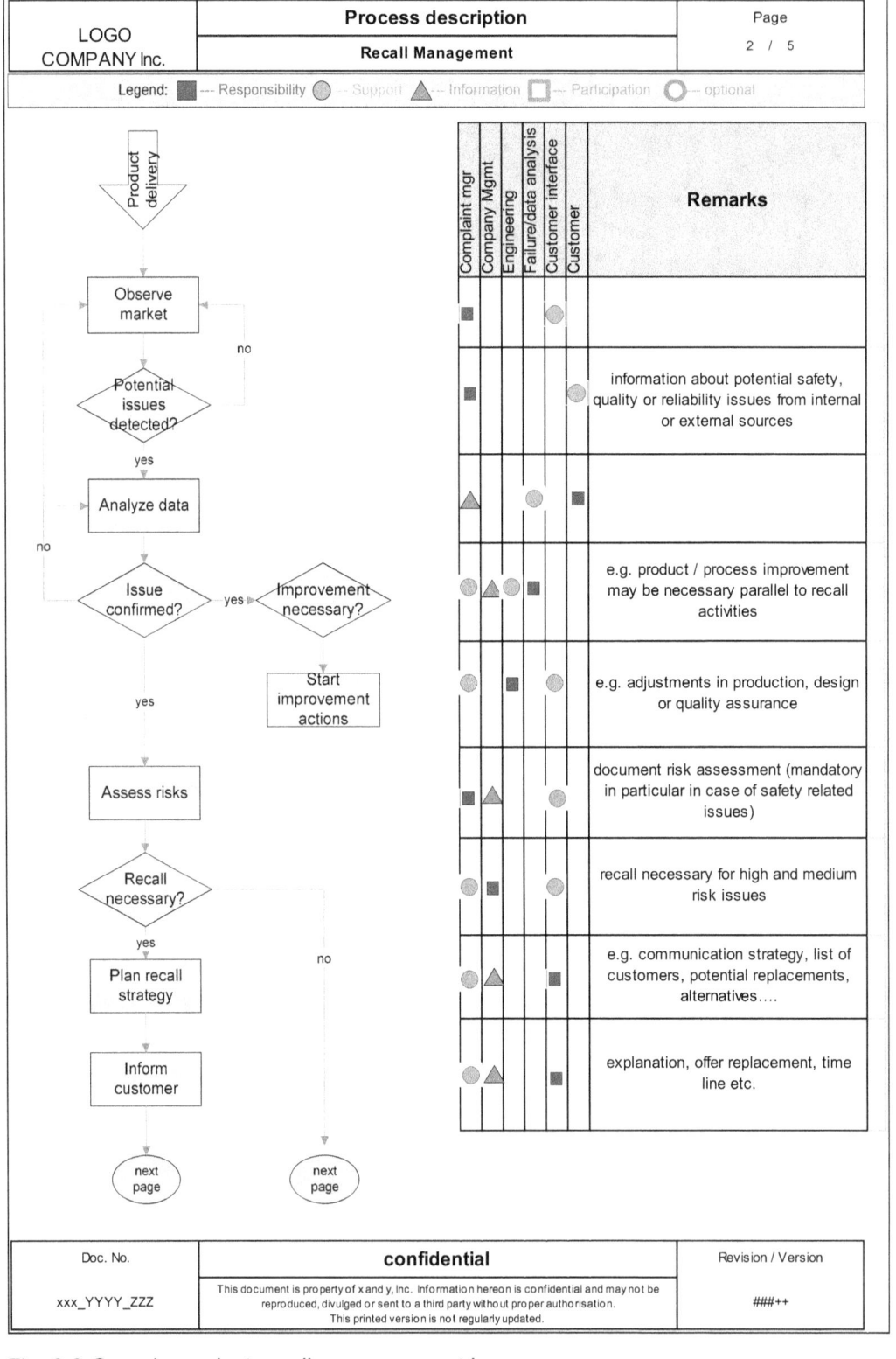

Fig. 9.9 Generic product recall process – part I

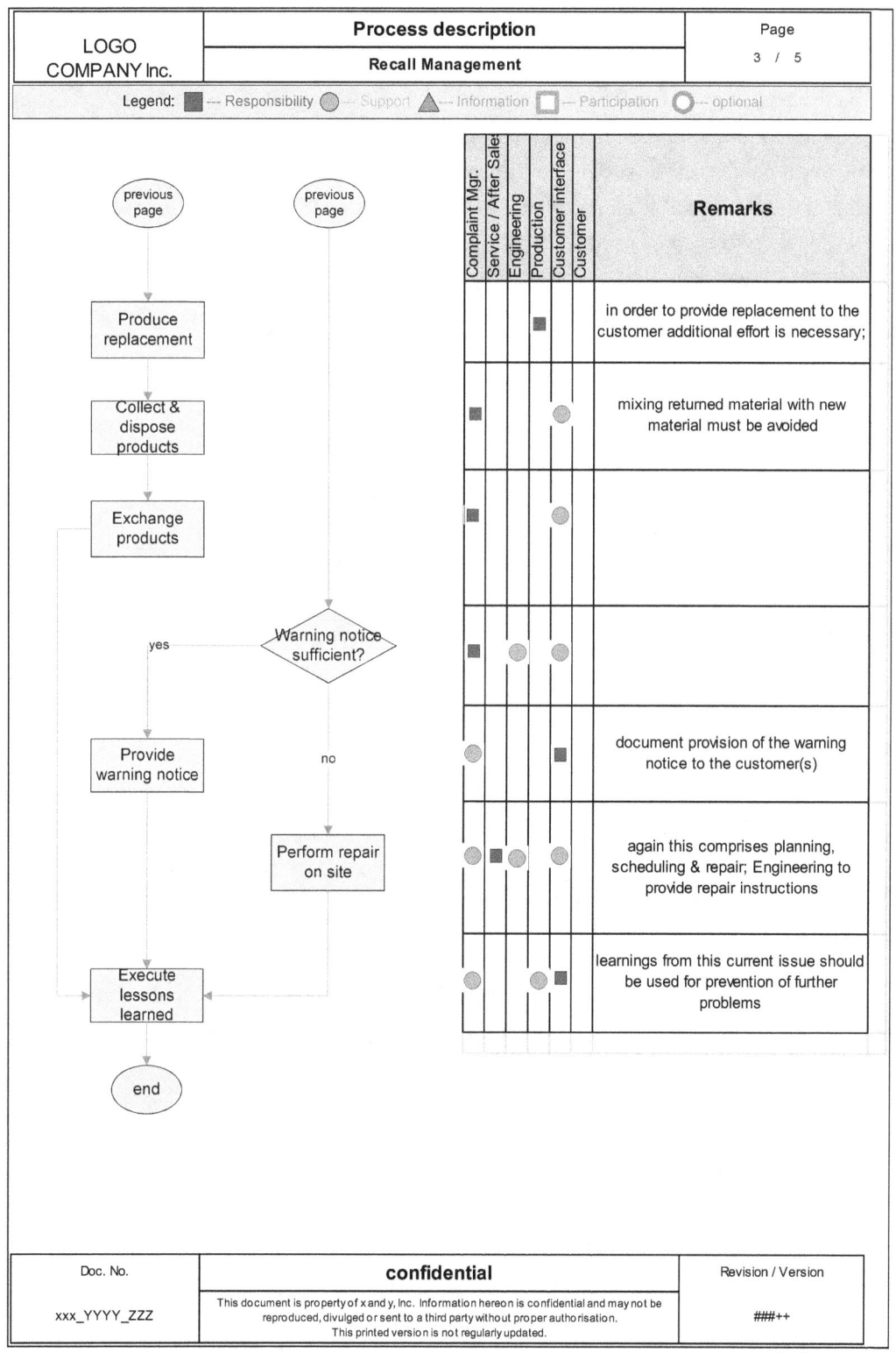

Fig. 9.10 Generic product recall process – part II

Again, safety related recalls may be managed in a different way due to legal requirements. For instance this may include early information of the respective authorities.

The risk assessment is a crucial part of the recall process. It has to be performed as soon as possible. Depending on the severity of the issue corrective actions range from an immediate recall of all products ("serious or high risk") over a partial recall ("medium risk") down to corrective actions at the partner site[1,2] or even to issue related warning notices ("low risk").

Independent of a potential product recall corrective actions may be necessary anyhow.

Those include e.g.

- Changes of the product or the production process
- Reinforcement of quality assurance activities
- Provision of warning notices

Most likely, due to the fact that a recall may have a major detrimental effect on the company's economic situation, the decision whether or not to perform a recall requires a top management decision.

## 9.6. PPAP

The Production Part Approval Process[3] (PPAP) according to ISO/TS16949 is an integral part of quality assurance of supplied parts in the automotive industry. The focus of PPAP is to ensure that high volume production parts consistently meet the clients' requirements regarding quality and reliability.

> All internal and external production sites are required to apply PPAP for production parts and spare parts.

Prior starting volume production of new or modified parts the manufacturer has to receive approval by the customer, i.e. a signed off PPAP package. The products used for PPAP have to be manufactured in a representative production run, e.g. 300 subsequently produced parts.

---

[1] WEKA (2012)
[2] PROSAFE (2011)
[3] Bureau Veritas (2008)

## 9.6.1. PPAP Requirements

PPAP requirements as well as additional customer requirements have to be met by the company. This also includes requirements of the authorities, e.g. safety relevant requirements.

If requirements cannot be fulfilled or are not applicable to a particular product, this has to be discussed with the customer. Corrective actions may be required.

In this section the required elements are listed:

1. Design Records

A printed copy of development records, e.g. drawings, CAD data, has to be available. References to other documents are possible and make those documents development documents. Development records contain the part material composition together with a confirmation of compliance to customer specific requirements. More concrete, data sheets with information like block diagram, functional block description, PIN (electrical connectors) configuration, application circuits, product characteristics etc. may be part of this section.

2. Authorized Engineering Change Documents

This is a detailed description of approved changes which are not yet part of the design documents.

3. Customer Engineering Approval

Technical approval by the customer.

4. Design FMEA

Design FMEA to be performed by the company. Similar parts or materials may be covered by one design FMEA.

5. Process Flow Diagram

A clear description of the production process flow is necessary.

6. Process FMEA

Process FMEA to be performed by the company. Similar processes may be covered by one process FMEA if the similarity has been assessed by the company.

7. Control Plan

The control plan provides more details on how the manufacturing process quality performance is monitored. Similar parts may be covered by one control plan if the similarity has been assessed by the company.

8. Measurement System Analysis Study

New and modified gauges have to undergo a measurement system analysis (GR&R).

9. Dimensional Results

Verification of dimensions has to be executed. Results have to meet requirements and results have to be documented.

10. Records of Material / Performance Test Results

Material and performance tests are required according to design records and/or control plan. Test results have to be available and have to contain e.g. date of test, number of tested parts, actual test result etc.

11. Initial Process Studies

Critical characteristics (special or key control characteristics KCCs) have to be identified. They have to meet acceptance criteria (to be discussed and agreed upon with the customer).

Typically initial process studies are based on process capability analysis resulting in $c_p$ and/or $c_{pk}$ data. The following values have to be achieved:

| | |
|---|---|
| $c_p$, $c_{pk}$ > 1,67 | the process meets acceptance criteria |
| 1,33 ≤ $c_p$, $c_{pk}$ ≤ 1,67 | the process may be acceptable; discussion with the customer required |
| $c_p$, $c_{pk}$ < 1,33 | the process does not meet acceptance criteria; discussion with the customer is required |

Those limits are defined for two sided normal distributions. Limits for deviating distributions have to be discussed with the customer. For processes not meeting acceptance criteria a corrective action plan is required. Non-complying processes might need a 100% control.

12. Qualified Laboratory Documentation

Laboratories have to meet customer requirements, e.g. qualification or accreditation. A copy of the laboratory certifications have to be available.

13. Appearance Approval Report

The customer has to approve the appearance of a product (if applicable).

14. Sample Production Parts

As defined by the customer a sample is taken, e.g. from the initial production run. This sample has to be available. Documentation of this activity is part of the PPAP documentation.

15. Master sample

A master sample has to be kept and be available. This sample needs to be signed off by the customer.

16. Checking aids

The customer is authorized to request part specific gauges for submission. Those gauges have to pass the measurement system analysis requirements.

17. Customer specific requirements

Customer specific requirements have to be fulfilled and documented.

18. Part Submission Warrant (PSW)

When meeting all PPAP requirements, the company has to fill in the PSW form. It confirms compliance of test results and documentation as required.

In addition to those general requirements the customer is authorized to come up with additional customer specific requirements. General Motors Company[1], for instance, is asking

- for submission of two sample parts (for level 2 and 3)
- for submission of a pre-launch control plan
- the supplier to obtain GM approval for various items (e.g. if not all items are completed)
- etc.

---

[1] General Motors Company (2015)

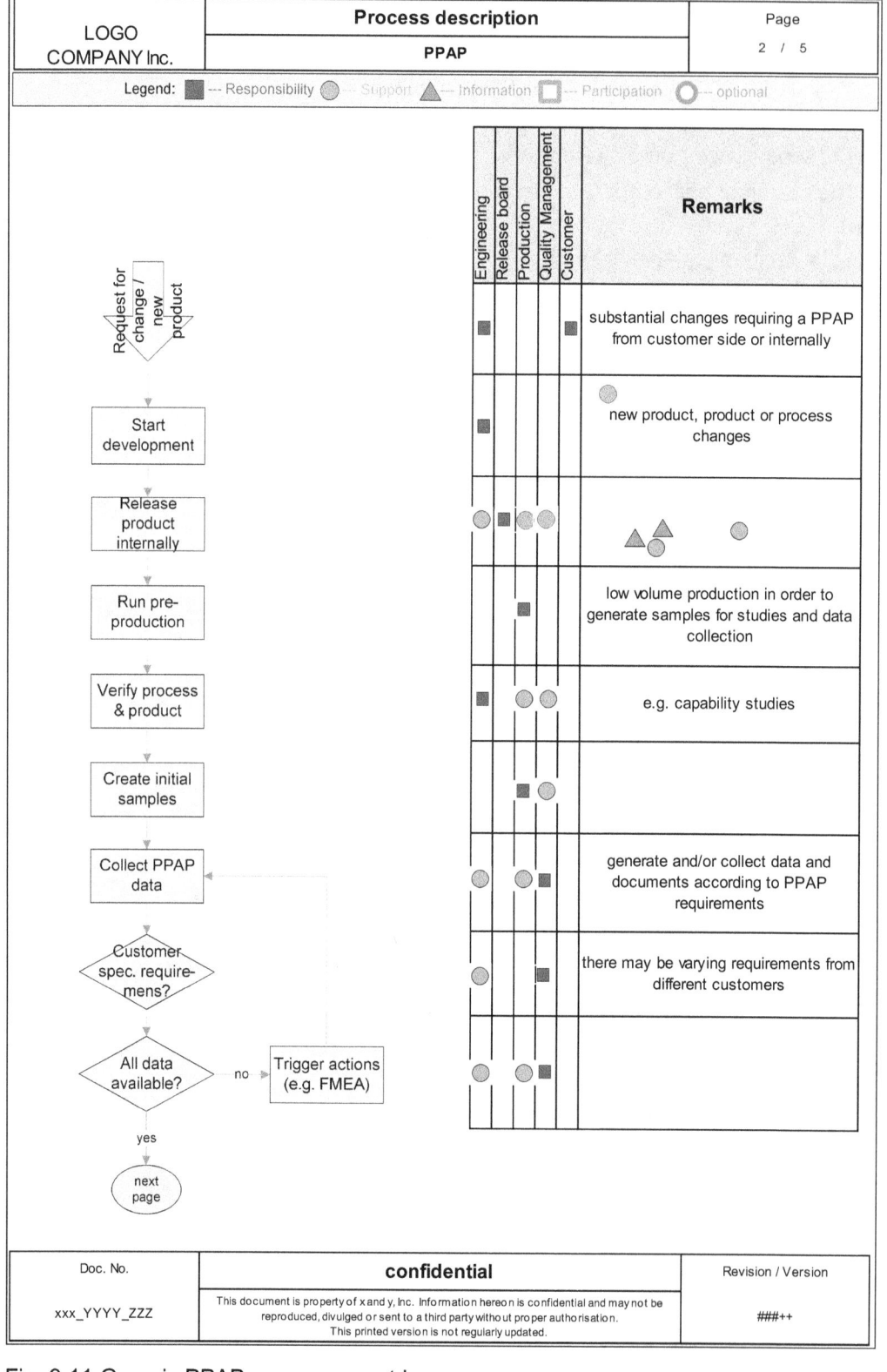

Fig. 9.11 Generic PPAP process – part I

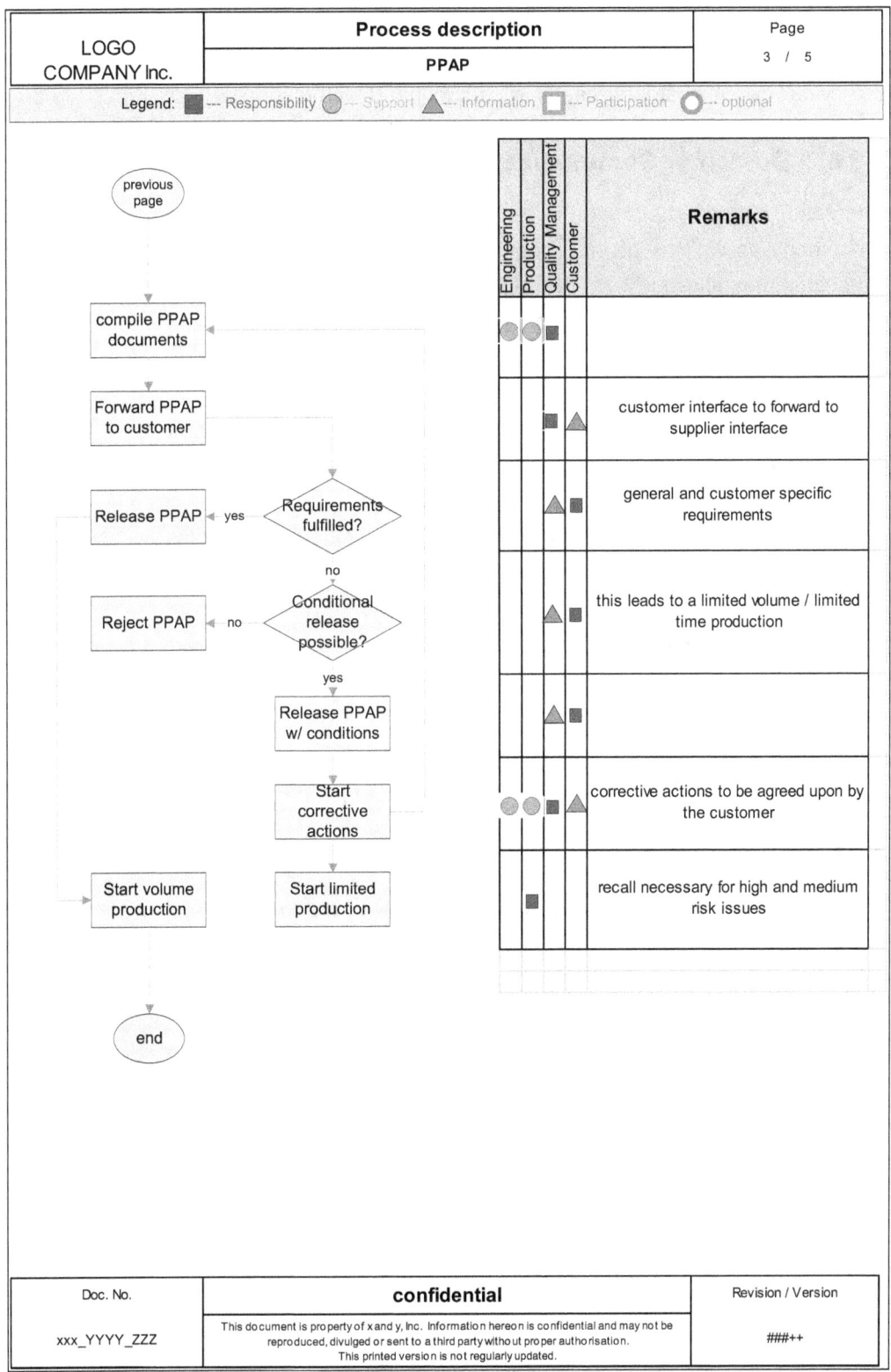

Fig. 9.12 Generic PPAP process – part II

Other customers, e.g. ZF, a manufacturer of gearboxes and electronic car systems, have special requirements regarding electronic components[1]. The mission profile provided by the customer needs to be confirmed by the supplier.

### 9.6.2. Customer Communication

In case of planned changes to the product or the manufacturing process (within the company as well as at the supplier) etc., the company has to notify the customer ("customer notification").

Reasons for customer notifications are e.g.[2]:

- Change of material
- Use of different, modified or improved manufacturing equipment
- Manufacturing using equipment transferred from a different site
- Change of suppliers of parts and materials
- Use of equipment which was not in use for at least 12 months
- …

Typically a PPAP documentation package is required for a new part, a new product or a change as listed above. The need for PPAP submission to the customer is defined using various levels of evidence and ranges from level 1, confirmation only, up to level 5, confirmation with master sample and a full set of data.

Eventually the customer approves or temporary approves or rejects the PPAP.

### 9.6.3. PPAP Process

In order to consistently provide compliant PPAP document packages to the customers it is worthwhile to standardize all activities in the form of a documented process description.

---

[1] ZF (2011)
[2] Bureau Veritas (2008)

# 10. Sub dpm Quality

## 10.1. Introduction

Frequently the phrase "Zero Defect" is used when referring to a very low failure rate in the defects per million (dpm) range. Primarily the concept of "Zero Defect" was brought up by Philip B. Crosby in the 1950s for the low defect production of Pershing cruise missiles when working for Martin Company, Orlando Div[1].

However, due to the fact, that lowering the defect rate typically requires additional efforts somewhere else (fig. 10.1, left), hence at one point in time, except for safety issues, it makes no sense to further reduce it, because it becomes too expensive and the customers might not be willing to accept this additional cost.

> FOR PHYSICAL REASONS ZERO DEFECT IS NOT POSSIBLE, BUT TO STRIVE FOR IT IS A MUST

On the other hand it is noteworthy, that the literature also suggests achievability of 100% good products at a non-infinite cost per good unit level[2] (fig. 10.1 right).

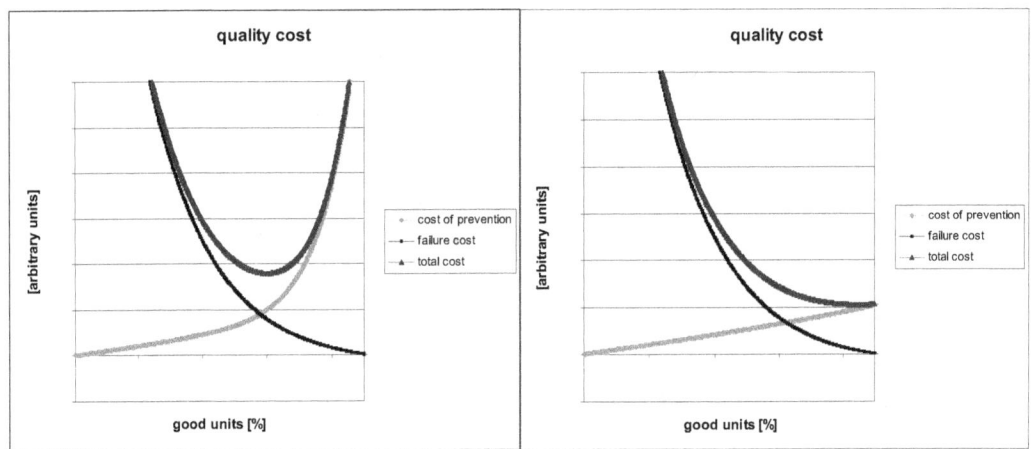

Fig. 10.1 Different views of total quality cost

Common sense and ultimately the second law of thermodynamics however suggest that except for very special circumstances and limited time this is unrealistic.

Due to continuous improvement activities of all kind the general trend towards lower defect rates is expected to continue on as an exponential curve, which was shown for

---

[1] Zollondz, H.-D. (2011)
[2] Schiffauerova. A., Thomson, V. (2006)

defects in IC metallization in the ppm range, accident induced absenteeism in "hours lost per million", defective vendor lots in the percentage range etc. However, the speed of improvement, i.e. the time it takes to reduce defects by 50%, differs from months to years for those examples[1].

Today, failure rates in the industry have already reached (or are asked to reach by the customers) single digit dpm rates or even sub-1-dpm-rates the phrase "Sub dpm Quality" is used in this context.

## 10.2. Automotive Semiconductor Customer Expectation

The automotive semiconductor segment is chosen as an example for low defect requirements for two reasons. On the one hand the automotive application is obviously relevant for safety and on the other hand the number of semiconductor components in a car is increasing leading to higher and higher quality requirements from the customer perspective.

> "Semiconductor companies focusing on the automotive industry as a key market will need to have a laser-sharp focus on quality from product design through to Production"[2]

As mentioned earlier the automotive semiconductor annual growth rate is currently in the 9% range. This growth is driven by the following contributors[3],[4]:

➢ Increasing number of vehicles

As people in developing countries become more affluent the demand for motor vehicles is growing especially in China and India.

➢ Increasing electronic content within a vehicle

Public regulations regarding safety and environment protection require more semiconductor devices, e.g. electronic stability control, blind spot detection, tire pressure monitoring, start-stop system, advanced motor control systems

---

[1] Schneiderman, A. M. (1988)
[2] PwC (2013)
[3] PwC (2013)
[4] Germany Trade & Invest (2015)

> Increasing number of application enabled by semiconductor components
Hybrid and electric cars have significantly higher semiconductor content than traditional cars. E-mobility, driver assistance systems etc. are asked for by the customers and are supported by electronic components. Others like traffic management systems or fully electronic dashboards and so on will further increase the number of semiconductors in motor vehicles.

Hence, more and more semiconductor devices in safety or non-safety applications are built into a motor vehicle. From the customer's point of view, however, the overall quality performance must not change or even get better.

> "Major goal is a constant or even decreasing total system failure rate. Therefore, as the number of electronics is increasing, the single component failure rate has to decrease dramatically."[1]

Assuming 100 PCBs in a car with an average of 100 semiconductor components on a single PCB a defect rate of 1 dpm (per single semiconductor component) would result in 10000 dpm, i.e. 1%, for the semiconductor based parts of the car.

Shortly, the customer expectation is or soon will be to go below the 1 dpm failure rate for semiconductor components. The Continental company, for instance, clearly states: "Target is a zero-defect production"[2].

## 10.3. Sub dpm Quality Measures in the Semiconductor Industry

In 2006 the Automotive Electronics Council Component Technical Committee (AEC) proposed a draft version for a Zero Defects guideline for automotive semiconductors[3]. The methods and tools proposed are not mandatory but are rather meant to be a "tool box" to achieve low defect rates during the entire product lifetime.

In addition to the use of relevant quality tools, often specific for each phase in the products life time, and overall quality related programs like continuous improvement as

---

[1] Glueck, J. et al. (2006)
[2] Continental (2015)
[3] AEC Q004 (2006)

proposed in AEC Q004, defect reduction etc., a sub dpm quality program has to be embedded into the company strategy (fig. 10.2), too.

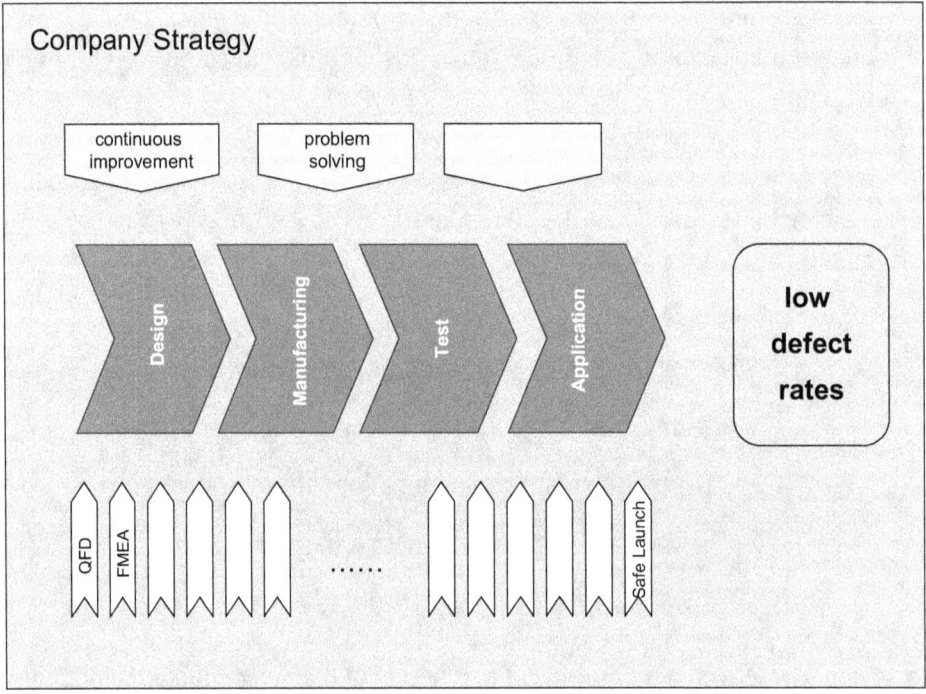

Fig. 10.2 Sub dpm quality program embedded in the company strategy

Full commitment of the management and company target setting including quality as a major contributor for competitiveness are necessary to make this approach successful. Altera, for instance, claims to have reached less than 1 dpm regarding customer return rate for their automotive products. The identified six key elements to make their top management driven Zero-defect strategy successful[1]:

➢ Design for Test and Manufacturing
Test coverage larger than 99% is the key to sort out bad parts

➢ Defect reduction on wafer level
Continuous defect reduction using FPGA test vehicles

➢ Yield improvement in assembly

---
[1] Altera (2), (2015)

Tight manufacturing control and long-term strategic relationship with suppliers supports assembly yield

> Detection of outliers

Identification of outliers which escape the production tests applied relies on methods like Statistical Yield Limit (SYL), Part Average Testing (PAT) and High Voltage Stress Test (HVST)

> Tight supplier management

Lot quality and yield are constantly monitored and respective data is communicated to the suppliers.

> Continuous improvement

Continuous improvement is monitored using key performance indicators. Next generation products benefit from learning from resolved current quality issues.

Overall a Quality Office is implemented which "monitors Key Performance Indicators (KPI) and manages ongoing continuous improvement efforts"[1]. The Quality Office meets on a regular base to keep the focus on quality and its chair reports directly to the CEO of the company.

In addition to these more general and global recipes for a successful defect reduction program, the application of quality tools to "nitty gritty" process problems is equally important. Therefore, in the following sections we focus on various quality tools and methods used to achieve low defect rates. Methods detailed in other chapters of this book are explained only shortly and referenced to the respective section.

## 10.3.1. Design Phase

As a typical engineering task the design phase is the first actual step of the product life time. Prior to the design phase the market and customer requirements have to be collected and assessed and later on translated into the system specification. Right after completion of the circuit design activities the circuit design has to be translated into the layout, i.e. the particular components like transistors, capacitors etc., have to be represented mask layer by mask layer by physical structures.

---

[1] Altera (2), (2015)

> IC design summarizes microelectronic engineering activities starting from system specification through architectural, functional and logic design and ending in actual circuit design.

In this section all those additional steps prior actual start of manufacturing are considered.

### 10.3.1.1. Quality Function Deployment (see 9.2.2. QFD)

The purpose of QFD is to translate customer requirements into concrete product characteristics. Quality Function Deployment contributes to sub dpm quality because it helps to avoid design errors which are often due to insufficient knowledge about the customer requirements. Hence this method is used in the very early phase of product definition and primarily increases customer satisfaction by identifying and meeting the real customer expectations.

In this process, an important objective is to identify the engineering solution with the lowest development and production risk, and it makes sure that the "to be developed product" is competitive. Another purpose is to shorten the development time through selection of materials and processes which are known to be robust and feasible.

### 10.3.1.2. Design FMEA (see chapter "FMEA")

As described earlier the purpose of a Failure Mode and Effects Analysis (FMEA) is to identify potential failures and their effects both on the product and on the customer. For this purpose root causes are identified, too, and severity, occurrence and detection are assessed and scored.

In the early phase the FMEA tool is a used as a design FMEA which is trying to identify and eliminate potential design related failures.

### 10.3.1.3. Redundancy

The purpose of redundancy in a system is obviously to increase the mean time to fail (MTTF), which is important to achieve high quality performance and in particular to achieve the safety standards required. The use of redundancy however results in two main disadvantages

> Additional area required for redundant structures reducing the number of chips on a wafer which directly reduces productivity and increases cost
> Additional support for the redundancy required, e.g. additional or more complex software and control of more functions on a chip

> Description of Redundancy
> "A parallel system of duplicate cells or components that can replace faulty ones seamlessly during the final test or actual use of a part" [1]

As a consequence this approach only makes sense for safety application, products with mandatory long life times when cost per die size is still found to be in a reasonable range.

Being very important for the safety of motor vehicles the following section focuses on automotive electronics. In the early days of automotive electronics redundancy used to be made of mechanical back up systems or mechanical sub-systems. However this turned out to be prohibitively expensive, heavy and big in size which is why it became clear that the redundancy itself has to be based on electronics, too. Today there are different options available to realize redundant structures, e.g. complete system on a single chip (SoC), complete systems in a single package (SiP), multi chip solution or multi package solution.

Redundancy on silicon[2,3]

Even though single chip solutions suffer from the fact that they are sensitive to common mode failures, it is a typical design practice. Common mode failures are failures hitting the structure and the redundant structure at the same time, e.g. because they are very close to each other on the wafer. Different redundancy levels are in place:

> Device Level Redundancy

Single devices, e.g. heavily loaded VIAs are duplicated and placed on the wafer in order to keep up the functionality of the IC once the primary component fails.

> Component Level Redundancy

---

[1] AEC Q004 (2006), p. 9
[2] Shivakumar, P. et al. (2012)
[3] Xiang, Y. et al. (2013)

A component consists of several devices and the component's reliability depends on each single device's reliability. Identical components may be placed on the chip in order to enhance the lifetime of the system.

➤ Systems Level Redundancy

Several components can be grouped to a system. An identical system or a system with a different structure, but the same functionality, may be placed in addition to the primary system. The latter case reduces the risk of failure for structure related or structure sensitive defects.

Redundancy in a package (system in a package SiP)

Two or more separate ICs with the same functionality and the same physical principle are placed in one package as redundant systems. Infineon's GMR-Based Dual Die Angle Sensor[1] (giant magneto resistance) represents a typical example in automotive applications (two GMR sensors in one package).

The same functionality with different physical principle would be to combine a GMR based sensor with an AMR (anisotropic magneto resistance) or TMR (tunnel magneto resistance) based sensor in one package. This way principle based problems would be avoided in addition.

Multi processor redundancy[2,3]

Beyond the scope of the component manufacturer, i.e. at the system manufacturer side, further reduction of the risk of failure can be achieved by the use of more than one IC. Multi processor architectures are less prone to common mode failures. There are different alternative options to realize redundant structures with more than one processor, for instance:

➤ Lock-step Dual Processor Architecture

So called master CPU and checker CPU work fully synchronized. A third circuit constantly compares master's and checker's results. Discrepancies are interpreted as failures. In addition to the redundant structure also the comparing device requires additional area on the wafer and additional software. Due to the fact, that both CPUs process in an identical way this architecture, in addition, is not sensitive to software design issues, i.e. software design issues remain undetected.

➤ Loosely-Synchronized Dual Processor Architecture

---

[1] Infineon (2015)
[2] Baleani, M., et al. (2003)
[3] Gillen, C. (2014)

Two different CPUs work in parallel, but independent from each other. Both have access to an individual memory subsystem. After completion of a task the output is exchanged and both CPUs check for consistency. Inconsistencies are interpreted as failures and this triggers a self test of the CPUs. In case only one CPU detects a fault, the other one becomes the only one to be used.

> Triple Modular Redundant Architecture

As the name implies in this case three different CPUs work fully synchronized (lock-step). A fourth circuit constantly compares the CPUs' outputs and interprets an output as correct, if at least two outputs are identical. Triple modular systems may also use loosely-synchronized architectures.

Although being used in the aviation industry triple redundancy is not cost competitive enough in the automotive industry.

Many other processor solutions are available with even more CPUs, which can be found in the relevant microelectronics design literature[1,2].

### 10.3.1.5. Built in Self Test (BIST)

BIST is typically used for complex products. The IC design contains additional circuits and the product requires additional programming, but this enables the system to test itself initially as well as regularly during its life time. Different variations of Built in Self Tests are in place for logic and memory devices, i.e. LBIST and MBIST, even on one product[3]. Figure 10.3 shows a general model of a BIST structure.

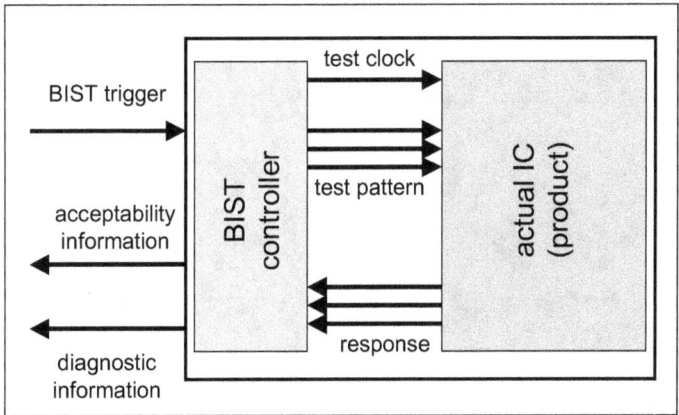

Fig. 10.3 BIST – general model[4]

---

[1] Weis, S. et al. (2014)
[2] Ren, Y. et al. (2013)
[3] Turner, A., McMenamin, D. (2015)
[4] Ahmad, A. (2005)

The purpose of the initial Built in Self Test is obviously to detect process or design related failures in particular in case an external test is not possible. In addition to failure detection another benefit of BIST in this phase is a potentially significant reduction of expensive test time. Since test time is a significant cost factor, it may actually be overall more cost efficient to invest in additional real estate on the wafer and software for self test.

> "The practice of designing the circuitry such that inputting a logic solution will allow the part to test itself." [1]

In addition to the initial test the regular or constant use of the BIST during product life time allows to detect latent defects induced malfunctions. Due to additional efforts in design and programming, however, so far "LBIST found its use mainly in safety-critical (automotive, medical, military), mission-critical (deep-space, aviation) and high-availability (telecom) applications"[2]. Quick diagnosis of failing, but self tested ICs even reduces the availability of systems by reducing the mean time to repair (MTTR).

### 10.3.1.6. Design for Testability

Increasing complexity of integrated circuits makes it more difficult to access all nodes and components of the IC directly through the external pins. One consequence of this fact is that those cannot be tested for functionality, either. Hence, the test coverage, i.e. the percentage number of functions and nodes of an IC which can be tested, is decreasing.

> "The practice of designing the circuitry such that as many nodes as possible can be tested in a reasonable amount of time." [3]

The purpose of "Design for Testability" is therefore to add additional circuits and the respective test patterns in order to keep or even increase test coverage, i.e. both observability and controllability, for instance also on embedded circuits without direct

---
[1] AEC Q004 (2006), p. 10
[2] Li, N. et al. (2015)
[3] AEC Q004 (2006), p. 11

access. DFT is applied to different applications like chip diagnostics and chip failure localization or for parametric testing.

Various approaches and techniques are available of which only a few are mentioned:

- Use of scan chains
- Use of Automatic Test Pattern Generation (ATPG) tools
- Use of Built-in Logic Block Observer
- Partitioning for test (physical and electrical partitioning)
- Normal and test modes

In addition design for test also includes rules which consider various boundary conditions in order to adapt and optimize design and specification of product parameters etc.:

- Specification of product parameters and respective tolerances based on given production process capabilities
- Use of test interfaces based on available test equipment, preferably standard equipment
- Provide access to suitable test points and chip area for those test points
- Determination of engineering tolerances using Geometric Dimensioning and Tolerancing (GD&T)

### 10.3.1.7. Design for Manufacturability (DfM)

The semiconductor manufacturing process suffers from technological constraints like limitations of the lithography resolution or simple defect density. The purpose of DfM is to minimize the effect of those constraints by changes of the design.

> "The practice of designing the circuitry so that the part can be more easily manufactured via larger design margins." [1]

For an effective DfM approach several quite obvious principles of systematic design should be implemented within an organization[2], for instance:

- Use of standardized components

---

[1] AEC Q004 (2006), p. 13
[2] Da Silva, M. G. et al. (2002)

The manufacturing process can be optimized for a limited number of standardized components which reduces the risk for failures, e.g. introduced by weaknesses of customized components (customized components are only manufactured in small numbers; therefore there is no opportunity to detect and eliminate defects which occur relatively infrequently which only manifest themselves if a larger number of components has been manufactured)

➢ Use of modular designs

Modules with standard interfaces which have shown to work well should be used in other complex systems, too

➢ Use of standard flows

Production flows which proved to be working well in the product as well as for manufacturing should be used for other products as well (again: "standardization")

➢ Design for manufacturing standards

In general materials, processes, equipment, geometry etc. etc. should be standardized as much as possible. This allows manufacturers and designers to optimize process and design to those given boundary conditions.

Further miniaturization of circuit geometry challenging the lithography process also leads to several design techniques supporting the lithography process, e.g.:

➢ Optical Proximity Correction[1]

Optical diffraction improves the resolution of advanced lithography processes significantly. One way out of this restriction is to alter the shape of the masks using additional serifs at the corners of a feature or by removing inside corners. This change results in an improved process result (fig. 10.4). Optical Proximity Correction "is the deliberate and proactive distortion of photo mask shapes to compensate for systematic and stable patterning inaccuracies"[2].

In the very beginning of OPC corrections were made manually to the layouts. Meanwhile, with a very high number of devices on a chip, the OPC process has been automated using specialized software. OPC is a rough analogue to equalization in audio engineering. In both cases deficiencies in the respective transfer functions are compensated.

➢ Alternating Phase Shift Masks (altPSM) and off axis illumination (OAI)[3]

---

[1] Liebmann, L. W. et al. (2001)
[2] Liebmann, L. W. et al. (2001)
[3] Liebmann, L. W. (2003)

The principle of phase shift masks is to change the phase of the light of two adjacent features on the photo mask by 180°. This can be achieved by reducing the light path on one feature while leaving the other one untouched. On the wafer right in the spot between the two features destructive interference suppresses the light intensity resulting in an increased resolution. Designer and in particular the layouter have to adjust the circuit layout accordingly.

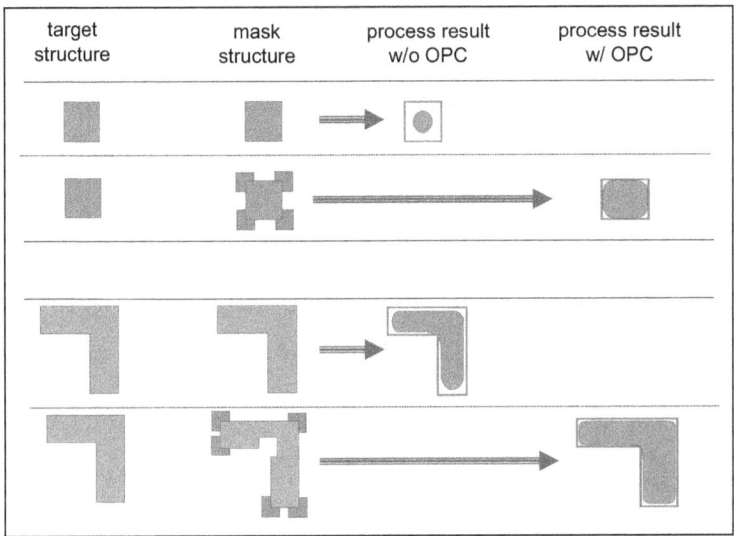

Fig. 10.4 Lithography process result with and without OPC

A similar process result can be achieved by Off Axis Illumination (OAI). In this case the illumination of the regular (non-PSM) mask occurs with an appropriate illumination angle.

Fig. 10.5 Illustration of sub resolution assist features

- Sub resolution assist features (SRAF)[1,2]

The Off axis illumination process is optimized for a particular pitch (= spatial frequency of the pattern to be printed). Resolution enhancement is reduced as soon as features with a different pitch are used. In this case additional dummy structures are placed right next to the intended structure. The dummy structures are too small in size to be effective on the photo resist but lead to supporting diffraction phenomena with them which stabilize the actual target structure.

- Multiple Patterning techniques[3,4,5]

More recently double, triple and quadruple patterning techniques came up which require additional efforts already during the design phase. The basic idea is to split the target structure in a way that the actual exposure is not done in one step, but in two or more steps. This way each exposure step uses a mask with somewhat relaxed structures. The complete exposure combines those single steps to the target exposure (fig. 10.6 and 10.7). Despite major challenges to overcome overlay issues this kind of technology is being used or at least investigated down to the 7nm node.

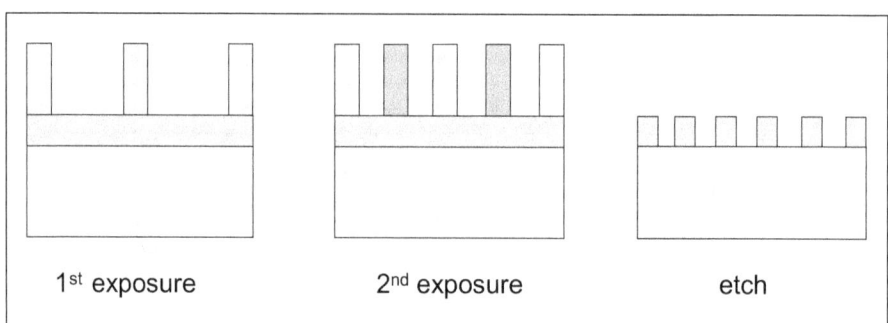

Fig. 10.6 Cross sectional illustration of double exposure

**10.3.1.8. Design for Reliability**

Design for reliability requires sound knowledge of failures and failure mechanisms and methods to determine those. Keeping these facts in mind, the design of new devices or circuits targets for less demanding design rules where ever possible. Of course, this must not decrease the circuit's performance.

---

[1] Liebmann, L. W. (2003)
[2] Lin, C. W. (2007)
[3] Pan, D. Z. et al. (2013)
[4] Zimmerman, P. (2009)
[5] Cho, M. (2008)

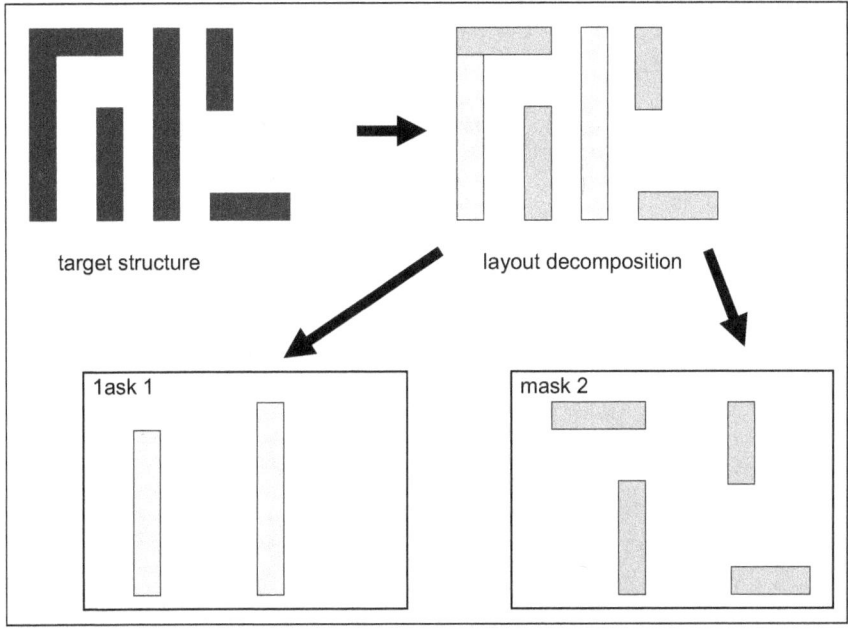

Fig. 10.7 Mask decomposition for double exposure

Substantial investigations are necessary to identify the most relevant design variables and their impact on product reliability. As an example analysis of packaging reliability of Insulated Gate Bipolar Transistors (IGBT's) are quoted here[1]. For several parts in the package, i.e. solder joints, wire bonds and bus bars, failure criteria and potential design variables were determined which might have an impact on the reliability of the product when applying the given mission profile.
With this physics of failure life time model for the chose variables the impact on reliability was determined:

- Solder joints: solder thickness increase leads to longer life time
- Wire bond: for instance, increasing wire diameter and decreasing loop height lead to more strain, hence to decreasing life times
- Bus bar: damage increases with increasing Cu thickness

From those individual life time predictions the life time of the system may be estimated and the relevant design variable can be changed accordingly. Wang et al. proposed the following procedure to predict the life time of power electronic systems (PEM):

---

[1] Lu, H. et al. (2009)

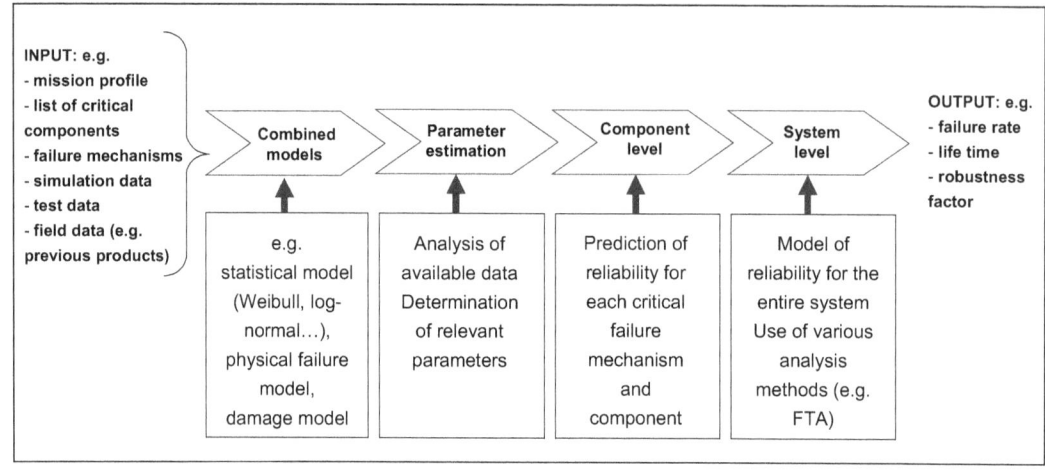

Fig. 10.8 Process proposal for life time prediction method for PEM[1]

### 10.3.1.9. Simulation

Simulation of components and systems is a matter of course during the design phase and later on for failure analysis during the production phase.

> "Recreating the functioning of the component through computer modeling using established engineering and physics-based relationships to functionality, construction and reliability" [2]

Variations in the design, process fluctuations and other parameters like supply voltage and currents may be input to simulation experiments. Outputs of these experiments are expected to identify process, design or operating sensitivities.

### 10.3.1.10. Design for Analysis

Failure analysis consists of several steps including electrical measurement as well as a variety of chemical and physical techniques (e.g. Scanning Electron Microscopy, X-ray, X-ray fluorescence, FIB, SIMS etc.).
Putting crucial for functionality elements of the circuit more easily accessible for later analysis during the design phase helps to significantly reduce analysis time, i.e. time to find the root causes for potential failures.

---

[1] Wang, H. et al. (2012)
[2] AEC Q004 (2006), p. 16

> "The practice of designing the circuitry such that failure analysis can be performed as efficiently as possible for elimination of no defect found" [1]

### 10.3.1.11. Characterization

> "The process of collecting and analyzing data in order to understand the attributes, behavior and limitations of a process, product design and the package" [2]

Characterization means determination of the device's functionality and parametric performance depending on process influence or electrical / environmental input. This way, having e.g. a mission profile in mind, the limits of the device performance can be identified.

## 10.3.2. Manufacturing

### 10.3.2.1. Process and Product FMEA

As described earlier the purpose of a Failure Mode and Effects Analysis (FMEA) is to identify potential failures and their effects both on the product and on the customer. For this purpose root causes are identified, too, and severity, occurrence and detection are assessed and scored.

Based on an existing design FMEA the product and process FMEA intends to identify product or process related weaknesses and potential failures and to address those proactively, e.g. by implementing additional process controls in the production process or other suitable prevention.

### 10.3.2.2. Statistical Analysis of Variance

During the semiconductor manufacturing process a large amount of data is accumulated like process data (layer thickness, doping concentration, line width ...), equipment data (see Advanced Process Control etc.), product data (parametric

---

[1] AEC Q004 (2006), p. 12
[2] AEC Q004 (2006), p. 17

performance, functional test data) and eventually also reliability data inhouse (e.g. WLR) or even product / process qualification data.

The intention of statistical analysis of variance is "determining the variables in a process that most influences the output characteristics of a given product depending on the manufacturing parameters"[1].

### 10.3.2.3. Control Plan

The control plan is a typical quality tool to keep the production process under control. It defines the way the process steps are arranged as a process flow and contains all process control activities required to determine the process performance.

"The control plan shall

- list the controls used for the manufacturing process control,
- include methods for monitoring of control exercised over special characteristics defined by both the customer and the organization,
- include the customer-required information, if any, and
- initiate the specified reaction plan when the process becomes unstable or not statistically capable.

Control plans shall be reviewed and updated when any change occurs affecting product, manufacturing process, measurement, logistics, supply sources or FMEA"[2].

> "A plan to control the product/process characteristics and the associated process variables to ensure capability (around the identified target or nominal) and stability of the product over time" [3]

It is a controlled document which has to contain data like control plan number, date of document issue and revision, part / process information (number, name, description) and organization's name. More detailed information would refer to the relevant customer or the product status (engineering, prototype, production…)

### 10.3.2.4. Statistical Process Control

---

[1] AEC Q004 (2006), p. 20
[2] International Automotive Task Force (2014)
[3] AEC Q004 (2006), p. 21

Statistical Process Control is one of the key elements to keep a production process stable and to avoid failures in a proactive manner. This topic is covered in detail in the respective section of this book.

### 10.3.2.5. Wafer Level Failure Mechanism Monitoring

In between the actual integrated circuits the scribe line (tens to hundreds of μm in width) is required for later separation of single chips. During the production process this area can be used for test structures for

- inline monitoring (layer thickness, line width, doping profiles)
- electrical monitoring (electrical parameters like sheet resistance or transistor parameters)
- reliability monitoring (Wafer Level Reliability)

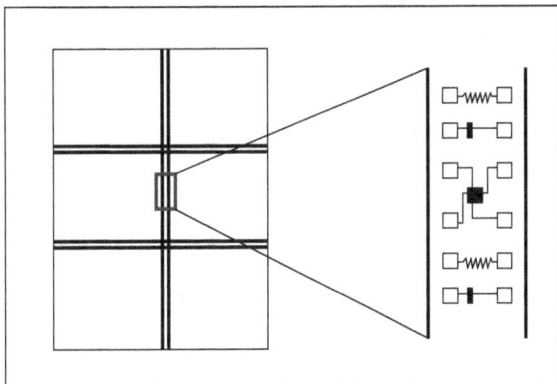

Fig. 10.9 Scribe line to be used for wafer level monitoring

## 10.3.3. Test

Testing and, in particular, handling of test data including the conclusions and actions generated upon those is a very wide field. In this chapter only a few methods are shown in order to provide the general idea.

### 10.3.3.1. Good Die and Bad Bin in a Bad Cluster[1]

In the semiconductor process defects typically tend to form clusters, i.e. they are typically not statistically distributed across the whole surface of a wafer. Hence, the probability of a good die in a bad cluster to have latent defects is fairly high. Latent defects result in a failure later on at the customer application and are therefore even more critical for quality.

---

[1] Moreno-Lizaranzu, M. J., Cuesta, F. (2013)

In order to reduce the risk of delivering potentially bad parts to the customer, those good dies in a bad cluster are considered as defective chips. The threshold for this action may depend on the number of bad chips surrounding the good chip.

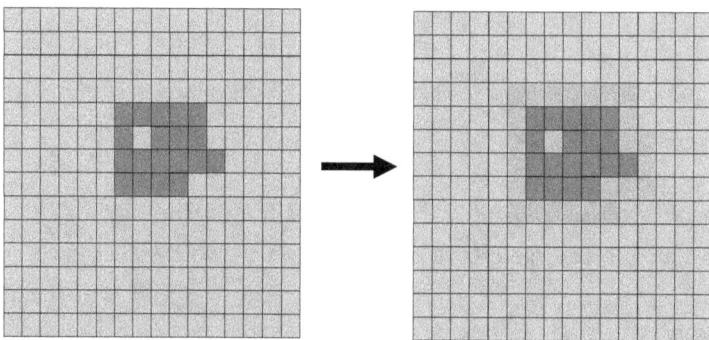

Fig. 10.10 Good Die and Bad Bin in a Bad Cluster (red: originally bad dies, orange: in addition considered bad)

The "Bad Bin in a Bad Cluster" is basically the analogy to the "Good Die in a Bad Cluster". The underlying assumption here is that it is assumed that a cluster of defects does not have a hard boundary. The probability that good dies neighbouring the defective cluster might suffer from latent defects is high. The consequence of this assumption is to consider nearest neighbours of a defect cluster as bad chips as well.

### 10.3.3.2. Part average testing[1,2,3]

The purpose of part average testing is the identification and removal of outliers from the population. Variations of part average testing have been developed:

➢ Static Part Average Testing

The die undergoes parametric test and is considered fail as soon as one parameter is out of the defined constant specification limits.

➢ Dynamic Part Average Testing

The specification limits for the parameter test are calculated for each wafer separately according the following equations:
with

    σ    : standard deviation for the

$$UpperLimit = \mu + 6\sigma$$

$$LowerLimit = \mu - 6\sigma$$

---

[1] Moreno-Lizaranzu, M. J., Cuesta, F. (2013)
[2] Nahar, A. et al. (2009)
[3] AEC Q001 (2011)

particular parameter over the entire wafer

μ   : respective mean value

A die is considered a fail if the chip value is outside those dynamic specification limits.

➤ Automotive Electronics Council Dynamic Part Average Testing

Additional exclusion of outliers and taking into account skewness of the distribution results in a more robust way of limit calculation:

$$UpperLimit = Median + 6(p_{99} - Median)*0.43$$

$$LowerLimit = Median - 6(Median - p_1)*0.43$$

with   $p_1$   : first percentile
   $p_{99}$   : 99th percentile

➤ Nearest Neighbour Residual (NNR).

The nearest neighbour residual is the chip parameter value minus the average of the neighbouring chip parameter values. In order to be insensitive to outliers the median of the neighbours could be used instead of the average. This residual is then subject to dynamic part average testing. NNR is used when there is a gradient of the parameter value over the wafer.

Knowing that these outlier tests lead to both type I and type II errors the methodology is continuously under development. Further approaches like "Robust Dynamic Part Average Testing" or "Location Averaging" are explained in more details in the literature referenced.

### 10.3.3.3. Statistical Bin Yield Analysis[1]

From experience it is well known that low yielding lots or wafers tend to have lower quality than wafers/lots in the usual yield range. Statistical bin yield analysis helps to identify lots and wafers with an untypical low yield. Alternatively also untypical high bin failure rates can be used for analysis.

The statistical bin yield analysis follows the following steps ("lot" stands for wafer, FE lot or assembly lot):

(1) process and test at least six lots and collect all yield relevant data

---

[1] AEC Q002 (2012)

(2) determine the statistical distribution of yield and selected fail bins

> "A system for analyzing and controlling manufacturing variations through measurements of critical test parameters/bins with the goal of ensuring final product quality" [1]

(3) if the distribution is somewhat close to the normal distribution calculate mean standard deviation of the yield and selected fail bins

(4) from there calculate statistical yield limits (SYL) and statistical bin limits (SBL)

$$SYL_1 = Mean - 3\sigma$$
$$SBL_1 = Mean + 3\sigma$$
$$SYL_2 = Mean - 4\sigma$$
$$SBL_2 = Mean + 4\sigma$$

Lots below $SYL_1$ or above $SBL_1$ require engineering attention and judgement. Lots below $SYL_2$ or above $SBL_2$ have to be put under quarantine and require a risk assessment. In an early production phase preliminary limits may be used, for non-normal distributions statistical methods may be applied to calculate the limits[2].

### 10.3.3.4. Screens

Various tests have the potential to identify defective parts:

➤ $I_{DDQ}$ leakage test[3,4]

$I_{DDQ}$ is the supply current ($I_{DD}$) in the quiescent status. The consumption of electric current of defect free CMOS integrated circuits in a steady state is very low. Defects however, e.g. shorts in the gate oxide or in between metal lines, create a conducting

---

[1] AEC Q004 (2006), p. 28
[2] AEC Q002 (2012)
[3] Rajsuman, R. (2000)
[4] Chang, C. L. (2013)

path between power supply and ground, which leads to comparably high leakage currents.

> "Testing of every manufactured part for functionality or parametric conformity to the device specification" [1]

With decreasing feature size the $I_{DDQ}$ values and process induced variations increase which makes it more difficult to distinguish between defective and defect free circuits which requires this approach to be improved continuously. Since this test is very simple and fast, it is good practice to do this test first and skip all further tests for a fail die to save costly tester time.

➢ High voltage stress test (HVST)

Supply voltages ($V_{DD}$) higher than application supply voltages are applied to the integrated circuit. Circuit testing is done prior and after voltage stress.

➢ Very low voltage test (VLVT)[2]

The use of very low voltages (e.g. $2 - 2.5 \times V_t$) reveal defects which do not lead to any malfunction as long as the regular supply voltage is applied.

## 10.3.4. Application

### 10.3.4.1. Part Qualification & Robust Validation

As explained in more detail in the Development & Qualification Section the objective of product / process qualification and robust validation is to ensure product reliability and to identify the margins available.

### 10.3.4.2. Part Derating

Another method to suppress defects showing up at the customer application is to scale down ("derate") the conditions the part is exposed to as compared to the designed conditions, if this is warranted by the real conditions for use. This includes environmental as well as operational conditions (temperature, humidity, voltage, current …). This way the operating margin opens up leading to less stress and/or lifetime related failures.

---

[1] AEC Q004 (2006), p. 30
[2] Chang, J. T.-Y., McCluskey, E. J. (1996)

### 10.3.4.3. System Engineering

> "Alignment of the system design with the user application through co-engineering activities between supplier and user" [1]

The integrated circuit is typically part of a major system inhouse or at the customer. Interactions of the IC with other parts of this system are necessary for the functionality of the system as a whole. Weaknesses of the IC or other parts on the system, in particular if those weaknesses amplify each other, lead to system failure. Tuning and synchronizing all part on the system decreases the failure rate of the system.

---

[1] AEC Q004 (2006), p. 50

# III. Business Excellence

# 11. Process Management

## 11.1 Introduction

Over the last years the active management of processes has become very important for companies to improve their efficiency and their cost situation and at the same time the process stability and failure rates, eventually to increase competitiveness.

It is now generally understood that department oriented companies are less successful than process oriented companies[1]. Hence many companies started to describe their processes in detail and implemented programs to improve their processes regularly and align the hierarchical organization to the process organization (process landscape). The main drivers for a company wide process management are:

- to increase transparency within the company
- to clarify responsibilities and authorities
- to improve internal procedures
- to minimize cost and risk
- to identify improvement potentials

### 11.1.1. Basic Definitions

**Process**

Processes are typically defined as an activity or a sequence of activities that transforms inputs into outputs. Processes can also be split up into a sequence of smaller sub-processes (fig. 11.1).

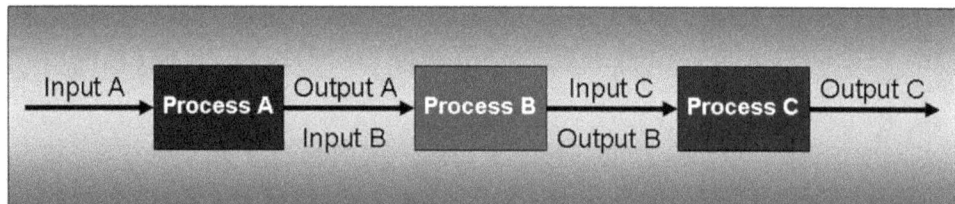

Fig. 11.1 Process consisting of a sequence of sub-processes

The process input includes also required resources, e.g. manpower or data. The output is often considered the result of the process.

**Business process**

Many authors do not make a difference between processes and business processes. However, business processes are often understood as processes within a company targeting for a value increase for the customer.

---
[1] Aguilar-Saven (2004)

In this book not only technical processes like product development or production, but also non-technical processes, e.g. so called "overhead" processes like sales or procurement are meant to be part of a company business process landscape.

> "A **business process** is the combination of a set of activities within an enterprise with a structure describing their logical order and dependence whose objective is to produce a desired result." [1]

In many cases effectiveness as well as efficiency of processes are subject to monitoring. The effectiveness of the process describes the potential of the process to produce the desired results, also called quality of the output. On the other hand, efficiency of the process describes process results in relation to the resources used. This is also called performance of the process.

Business processes are not specific to a certain level of detailing.

## Business Process Management

The management of processes in a company means

- to identify processes and process interactions
- to structure and describe processes
- to form and shape processes towards a target achievement
- to control and to apply processes
- to evaluate and to improve processes

These activities are frequently supported by the use of key performance indicators for each process and a company wide process target setting, which is needed to align the process targets with the overall company targets or the company strategy, respectively. For the above mentioned activities continuous planning, organizing and improving of processes is required.

The above mentioned set of activities requires the right organizational set up within the company and includes definition and execution of improvement measures. Structured business processes help people to deal with the increasing complexity of today's economical and technical processes.[2]

---

[1] Aguilar-Saven (2004)
[2] Wolf (2008)

> "**Business Process Management** (BPM) includes methods, techniques, and tools to support the design, enactment, management, and analysis of operational business processes."[1]

The industry distinguishes various kinds of business processes which are listed here:

- Management process — Process for strategy development and business planning and control
- Operating process — Process, which executes strategy and business plan
- Core process — Operating process which describes the value chain from supplier to customer (development, production, logistics)
- Support process — Process which supports other processes in particular aspects, e.g. IT, safety, …
- Key process — Process which is particularly important for the business success. Core processes are typically regarded key processes.

**Process owner**

The process owner has the overall responsibility for the process as well as the respective sub-processes. He is responsible for review and release of the process and also has the duty to remove obstacles (e.g. lack of resources).

His responsibilities also include:

- introduction of the process in the organization
- process target definition
- training of participants
- implementation of performance indicators
- process monitoring and coordination
- identification of improvement potentials
- documentation of the process

---

[1] Van der Aalst et al. (2003)

In reality some or all of the above mentioned tasks are delegated to the process leader or are executed together. Occasionally the process owner and the process leader is the same person.

For the execution of the above mentioned tasks the process owner has to receive the respective budget and authority by the management.

**Process leader (or process responsible)**

The process leader is responsible for the process execution. He directs the process using pre-defined performance indicators. It is also part of his responsibility to coach and coordinate the process team.

The process leader reports to the process owner and supports the process owner's process improvement activities.

**Process team member**

The process team member supports the process leader executing the process. That includes e.g. to perform the actual required activity as well as to document results or deviations. In addition the process team member supports process improvement activities.

> The **process owner** releases the process.
>
> The **process leader** executes the process.
>
> The **process team member** supports the process.

## 11.1.2. Motivation for Process Management

The motivation for process management is – theoretically – simple and clear: a company needs to have transparent processes, people like to know what they are responsible for, processes are controlled and can be improved when they are known and have defined performance indicators and so forth.

However, reality in many companies is different. Processes

- are frequently not documented
- are documented, but the documentation not up do date
- are not known
- do not have defined targets
- are not monitored and controlled using KPIs
- are not systematically optimized
- ...

In many companies there is an urgent and significant need to improve processes to reduce cost, to increase productivity and to align process targets the with overall company strategy. The reason for this is that in many cases the benefit of process management cannot directly be measured in a cost reduction of x and y Euros, but will only materialize over time. Since process management needs a small amount of extra resources, process management is often deficient or non-existent.

Process oriented companies name the benefit of process management as follows[1]:

- Significant improvement in transparency within the company
- Clearly identified opportunities for quality improvement, e.g. by reduction of the process variation
- Significant and sustained improvement of efficiency
- Clearly defined responsibilities and reduction of non-value add activities
- Higher productivity and clearly defined interfaces
- Adherence to delivery dates
- Higher customer satisfaction
- Higher employee motivation

In addition to those more qualitative benefits there are also quantitative figures available. The 2011 Gartner "Hype Cycle for Business Process Management" shows:

- "Lincoln Trust, a leading U.S. provider of trust and custodial services — including self-directed individual retirement accounts and open-architecture 401(k) plans — realized a first-year ROI of 120%, reduced cycle time on key processes of 50% to 70%, and a 90% decrease in customer complaints.

- The Carphone Warehouse, a major European technology retail organization, increased sales by 120%, increased customer satisfaction by 25% and achieved 1,100% ROI.

- The New York State Department of Taxation and Finance realized a 60% reduction in cycle time, a $100 million increase in tax revenue and $1.2 billion in savings by preventing fraudulent refunds during a two-year period. "[2]

---

[1] Kohlbacher, M., (2009)
[2] Dixon, J., Jones, T., (2011)

Other quantitative benefits of business process management, summarized by J. Rudden support the statement that process driven companies do have a significant cost and performance advantage:

- "Eliminate Manual Data Entry Reduction in time to add a new employee record into the HR system from 9 hours to 10 minutes.

- Reduce Process Cycle Time Reduction in compensation processing timing for 12,000 sales reps from 33 days down to 7.

- Reduce Manual Analysis/Routing Elimination of 80% of the manual work previously required to route invoice exceptions to the appropriate resolution teams.

- Evolve process from saving 5% of distressed shipments to saving 70% - yielding $2M per quarter in saved revenues.

- Make better decisions better review process results in $3M saved in billing dispute write-offs that would formerly have just been processed because the process was poorly controlled.

- Consistent Execution Customer satisfaction improvement to 92% based on proactive tasks that help ensure the home loan process executes better and faster"[1]

- etc.

## 11.1.3. Modelling Techniques & Process Description

In general there are many different ways to model a process. A summary was provided by Aguilar-Saven[2]. A variety of software tools is on the market supporting the modelling of company processes. Here we will focus on the illustration of a process. The description of a process has to provide the reader with sufficient information to perform the task. Typically the way those processes are described in a company depends on the process (detail) level.

The so called process landscape of a company consists of a few more strategic oriented processes on the highest level, followed, depending on the size of the company, by

---

[1] Rudden, J., (2007)
[2] Aguilar-Saven, R. S., (2004)

various mainly strategic oriented processes lower levels, down to the operative processes and eventually ending at so called standard operating procedures (fig. 11.2). The process landscape is a graphical illustration of the top level processes on one page, i.e. it is not a detailed description, but a general overview. An example is shown in fig. 11.3. In most cases this level is described using simple arrows and represents the basic structure of the company process set up. Depending on the size of the company there might be a process level underneath the general management which is still using an arrow based illustration. The example (fig. 11.3) shows mainly top level processes. However the process "product realization" also includes three sub-processes.

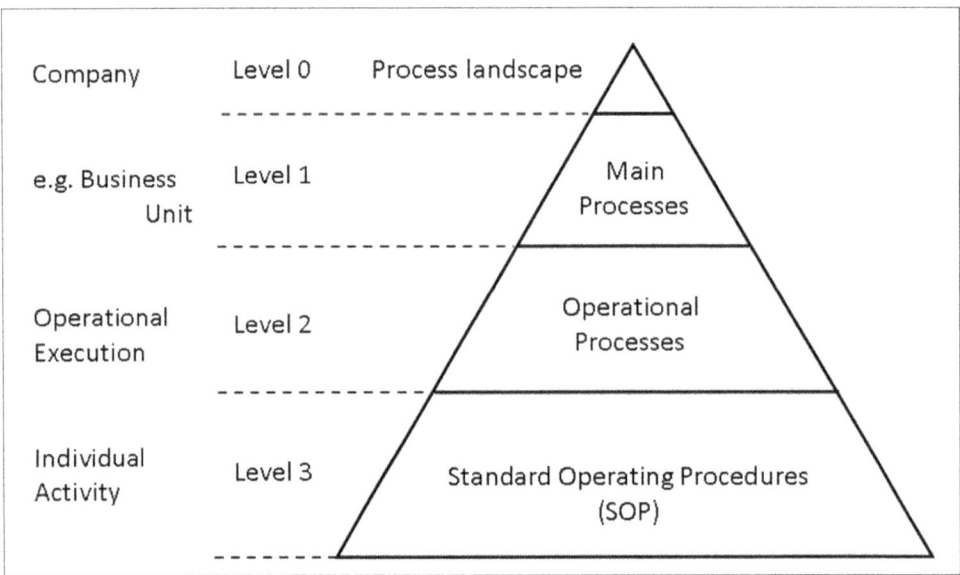

Fig. 11.2 Hierarchy of company business processes

The next step further down to the actual working level is the process flow based step wise description of activities. Often top-down flow charts are used for this purpose. Alternatively also process swim lane diagrams with the process flow from the left to the right are used (see example fig. 11.4).

The lowest level, e.g. a standard operating procedure (SOP), is frequently using flowing text with detailed description of activities to be performed. Process descriptions typically involve participants from more than one department, whereas SOPs often focus on activities within one department.

However, in reality, process descriptions as well as SOPs are used on different levels of the process hierarchy. In addition they are often overlapping, i.e. an SOP might also contain process flows and process descriptions might also contain detailed explanations of activities.

Flow charts based process descriptions usually have right next to the graphical illustration showing responsibilities, support functions, information receiver etc. for each step (fig. 11.5).

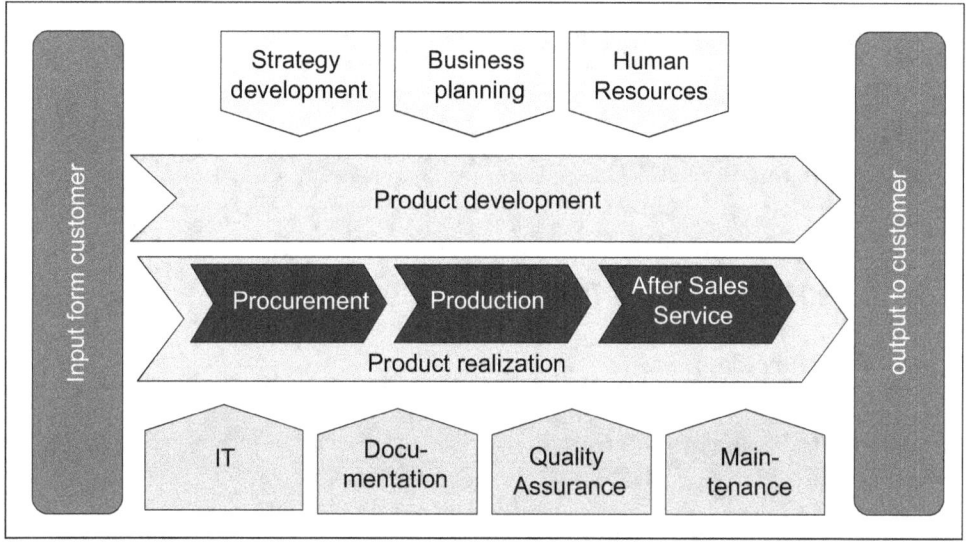

Fig. 11.3 Generic company process landscape

Independent from the kind of modelling technique being used the following items need to be clearly defined:

- input and output of the process
- detailed activities in each process step
- way of information exchange
- process owner
- process leader
- process team members
- performance indicators
- monitoring and/or review frequency

Process descriptions in a company are binding for the participating employees, i.e. they have to follow the described process flow in very much the same way as law has to be observed by individuals or corporations. Therefore a process description has to be a controlled document (fig. 11.6).

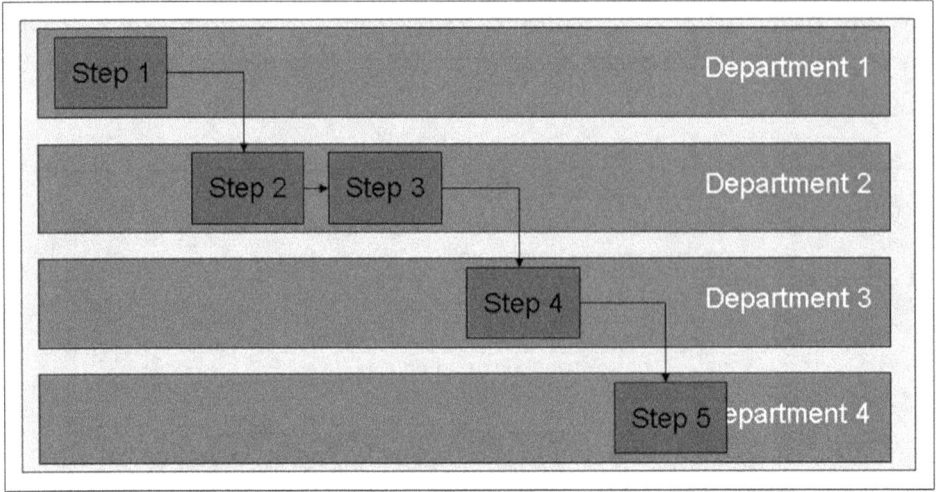

Fig. 11.4 Process swim lane

This includes for instance

- originator name, signature & date
- reviewer name, signature & date
- approver name, signature & date
- if required: also a formal approver name, signature & date
- revision number
- revision history
- document name & number
- defined storage place

## 11.2 Process Management

Process management consists of several steps:

(1) Identification of required processes
(2) Documentation of processes
(3) Definition of performance indicators
(4) Process Target setting
(5) Continuous process improvement

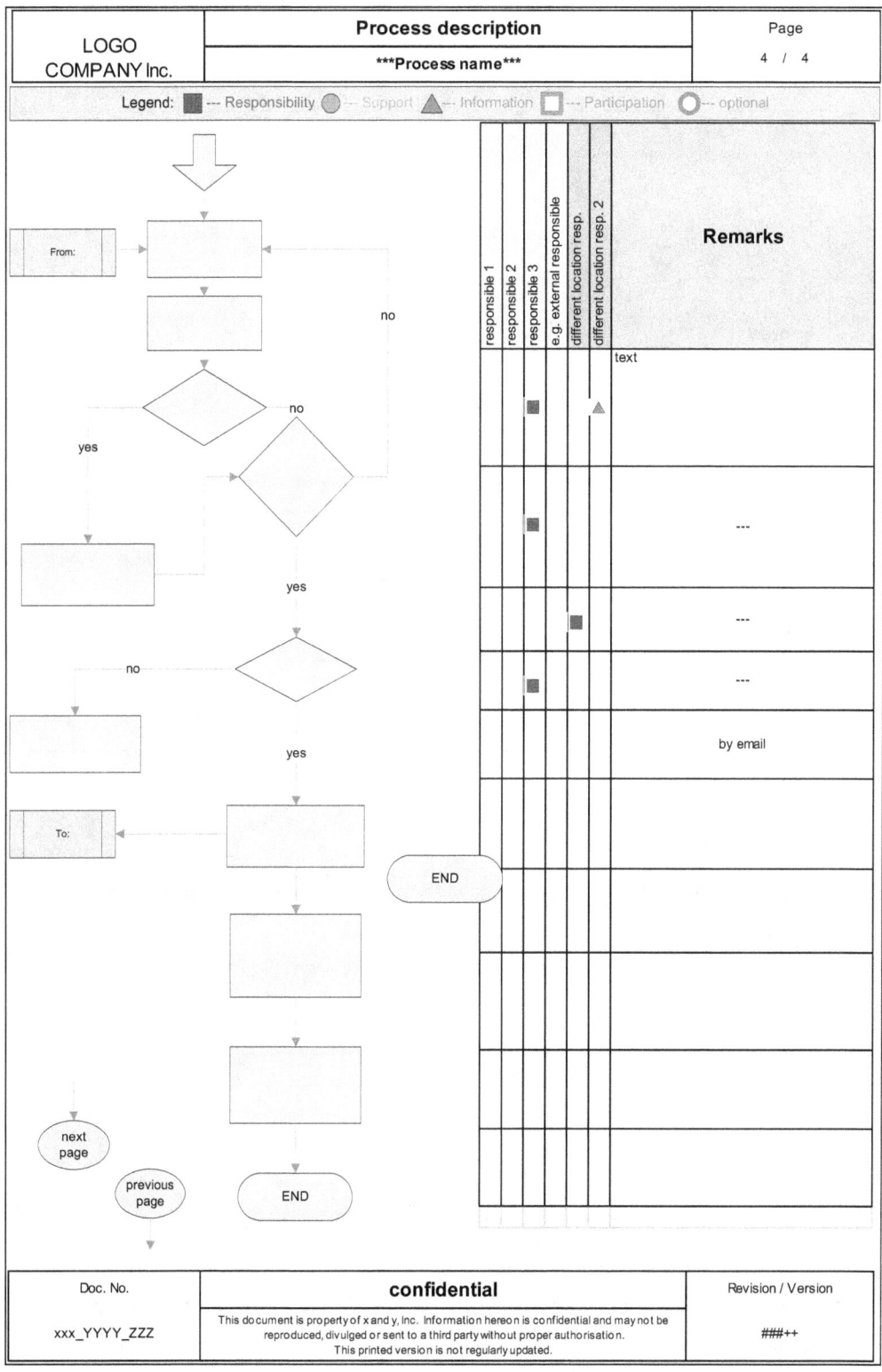

Fig. 11.5 Example for process flow description

| LOGO<br>COMPANY Inc. | Process description | Page |
|---|---|---|
| | ***Process name*** | 1 / 4 |

Legend: ■ --- Responsibility  ● --- Support  ▲ --- Information  ☐ --- Participation  ○ --- optional

# Process Name

xy GmbH

**Approval**

| Originator | Reviewer | Approval | Formal Approval |
|---|---|---|---|
| Name, Department | Name, Department | Name, Department | J. Wittmann, QM |
| Date and Signature | Date and Signature | Date and Signature | Date and Signature |

**Document Information**

| Originator: | |
|---|---|
| Original file name: | only basic-file-name |
| Created on: | 20xx-yy-zz |

**Legend of Amendments**

| Revision | Version | Date | Name | Description of Amendments |
|---|---|---|---|---|
| | | | | |
| | | | | |
| | | | | |
| | | | | |
| | | | | |

| Doc. No. | confidential | Revision / Version |
|---|---|---|
| xxx_YYYY_ZZZ | This document is property of x and y, Inc. Information hereon is confidential and may not be reproduced, divulged or sent to a third party without proper authorisation. This printed version is not regularly updated. | ###++ |

Fig. 11.6 Example for the cover page of a controlled document. It is also possible, but not recommended, to place the revision history at the end of the document to have a "clean" title page

## 11.2.1. Process IS-Analysis & Documentation

The starting point of process management activities in an existing company is the identification of existing processes or process-like activities. It is useful to illustrate the result of the process identification graphically. Subsequently small groups of process participants need to discuss their daily activities. Under the lead of a moderator the following items have to be clarified:

- which activities are part of the process
- or what sequence of activities, respectively
- what are inputs & outputs
- what is the way of information flow
- what interfaces (people / departments) need to be considered
- what kind of documents and data is needed and/or generated

During this initial discussion phase in many cases it turns out that different people are doing things differently. It makes sense to use this discussion to align on one single way the process should be executed. Since this is the first time the process is visualized, weaknesses of the process or double work might become visible. Hence the resulting process is not necessarily the way it was done so far, but could also be an optimized process.

This way of process execution is to be documented in the defined way (as described above) and now there is also the opportunity to define forms and checklists required for the proper execution of the process.

## 11.2.2. Performance Indicators

According to DIN EN ISO9001:2015 a process description requires the definition of performance indicators, which make the performance of the process transparent, i.e. help to measure the process results. Typically there are two perspectives to look at process performance:

(1) Efficiency – "doing things right"
Indicator to judge the efficiency of the process, e.g. turn-around-time, cost per activity etc.

(2) Effectiveness – "doing the right things"
This indicator provides quantitative information about how good the actual process results meet the target, e.g. internal or external customer requirements. This could be an actual failure rate, yield or customer complaint frequency.

In addition for the definition of performance indicators additional different points of view should be considered, too, like early or late indicator or indicators. It is also beneficial to focus more on steering indicators, i.e. indicators which can be influenced by the process owner directly rather than counting disruptions etc.

Table 11.1 provides a few examples of performance indicators for various typical processes:

| Process | Indicator for Effectiveness | Indicator for Efficiency |
|---|---|---|
| Production | (1) Number of failure free products<br>(2) Rework rate<br>(3) Yield<br>(4) Process capability ($c_p$, $c_{pk}$)<br>(5) Defect density | (1) Cycle time<br>(2) Flow factor<br>(3) Cost / product<br>(4) Equipment uptime<br>(5) Equipment utilization |
| Customer Management | (1) Number of customer complaints | (1) Cost per complaint<br>(2) Time to complaint resolution |
| Development | (1) Customer rating for innovative ranking of the company<br>(2) Product portfolio as defined by the company<br>(3) Energy consumption of a new product | (1) Delay to defined milestones<br>(2) Time to market<br>(3) Development cost |

Table 11.1 Examples for process performance indicators

However, depending on the kind of process and process targets, process performance indicators could be either effectiveness or efficiency indicators. In table 11.1, for instance, equipment uptime is considered an efficiency indicator for the production process. If we defined an equipment maintenance process equipment uptime would be an indicator for effectiveness, because the process target is clearly to improve uptime.

## 11.2.3 Target Setting

Within companies the purpose of processes is to support the overall company strategy or the overall company targets, respectively[1]. Hence process targets have to be aligned with other personal or department targets.

---

[1] Kavakli, V. (1999)

Processes are defined to have inputs, activities and outputs. In order to benefit from a process the target output has to be defined by the process owner based on customer requirements.

Based on the previously described process hierarchy (see fig. 11.2), process targets are derived from the respective higher level process target (fig. 11.7). This way the company management makes sure that activities within the company are focussed on the company targets and are aligned to each other to lead to an overall optimum for the company. If the target setting is done in an uncoordinated way, local optima will result, which are always inferior to overall optima.

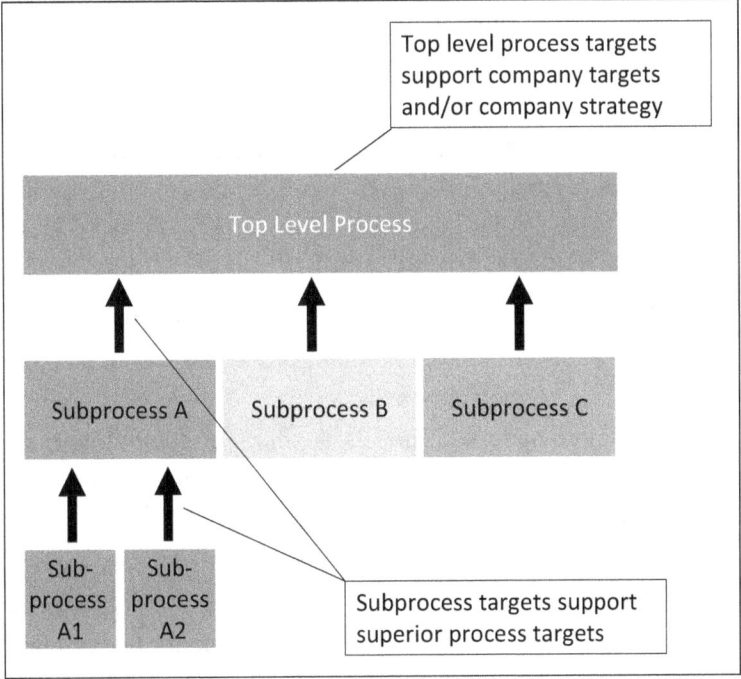

Fig. 11.7 Process target hierarchy

In addition to a clear target orientation also a clear definition for each target is mandatory. Hence it is useful to apply the following, here slightly adjusted, "SMART" rule[1,2], i.e. targets have to be phrased in a way that they are useful and recognizable etc.:

S   Specific    Targets have to be phrased explicitly and unambiguously

M   Measurable  Targets have to be measurable

---

[1] Storch, M. (2009)
[2] Locke, E. A., Latham, G. P (2002)

A   Accepted   Targets have to be accepted by the receiver of the targets, which includes that targets are actively influenceable by the receiver

R   Realistic   Targets have to be realistic and accomplishable

T   Timed   It has to be clearly defined when the target fulfilment has to be achieved

## 11.2.4. Process Improvement

There are basically two methods for process improvement: either an evolutionary approach is used, i.e. many small improvements eventually have a significant positive impact on the process, or the existing process is entirely replaced by a new optimized process. The first method is typically called process optimization, the second method is the so-called process re-engineering. Both are described in the following chapters.

### 11.2.4.1. Continuous Process Optimization

The more evolutionary approach requires the contribution of the participants of the process. Their integration in the improvement process increases openness to accept changes to the previous sequence of activities.

The way to identify improvement potentials depends on the nature of the process. On the one hand for processes which are continuously repeated, e.g. in production, various feedback loops help to identify suboptimum conditions. Those can be uncovered by regular improvement meetings with brainstorming on the process as wells as in form of a review of performance indicators. The time frame for improvements based on feedback loops ranges from minutes in case of an operator feedback over several months in case of feedback from internal or external audits or other information sources related to processes with long cycle times to up to a year in case of customer feedback. An alternative way to identify improvement potentials is to analyze causes of process failure and /or inferior process performance and implement ways to prevent process failure/deterioration.

On the other hand for processes, which are more or less "one-time-events", e.g. development of a particular product, but are generically repeating themselves, tools like "best practice sharing" or "lessons learnt" are used to improve the process.

Approach:
Potential process improvements are also identified by checking if one of the following measures can be executed:

➢ **Elimination of process steps**

This requires the identification of value adding vs. non-value adding process steps. Value adding steps are appreciated by the customer, i.e. the customer is willing to pay for it, e.g. production, service.

Obviously no value adding are activities like rework of products, handling of waste or rejected material or repair of equipment. Hence the main focus should be to reduce these activities by removing the respective root cause.

➢ **Optimization of single process steps**

The use of new materials, a new technology or just by applying a new approach single process steps can be improved. Optimization could be acceleration of the process as well as reduction of failures etc. Many minor improvements are then adding up to an overall process improvement.

➢ **Combination of several process steps**

In some cases several process steps can be combined into one step. This approach reduces the number of interfaces in the process and reduces the chance to exchange insufficient or wrong information. In addition, it will, as a rule, result in a reduced cycle time for the combined process.

➢ **Change of the sequence of process steps**

Here a check is required if a change of the process sequence leads to better information flow, speeding up of the process or reducing failure rate.

➢ **Run process steps in parallel**

Generally sequential activities take more time than parallel activities. If one step can be run without inputs from the previous step, those process steps can also be run in parallel. However, if the sequential flow of activities results from limited resources, then additional commitment of resources have to be weighed up against the benefit of the process optimization.

➢ **Automate a process step**

Manual handling of activities typically generates more failures than automated handling. In particular in a production environment automating a process might be possible using more advanced production equipment.

➢ **Add process steps**

The main purpose of a process is to achieve the defined targets. Besides process optimization like speeding up the process or reducing its failure rate, additional process steps might be required to ensure the defined process output.

### 11.2.4.2. Re-Engineering

"Business Process Reengineering (BPR) concerns the fundamental rethinking and radical redesign of a business process to obtain dramatic and sustained improvements in quality, cost, service, lead time, flexibility and innovation"[1]. The concept of BPR was first introduced by Hammer in 1990[2].

In contrast to the above described continuous evolutionary approach business process re-engineering stands for replacing the existing process by a completely new process. Typically the new process was not developed by the existing process team but by the management or by management mandated external consultants. This approach might provoke open or hidden resistance by the process team. Therefore it is important to convince them of the benefits of the new process or at least to integrate the process introduction phase into the project.

The idea behind is to fundamentally challenge the existing way of doing things in the company by asking[3]:

- Why are we doing what we are doing ("doing the right things")
- Why are we doing things they way we are doing them ("doing things right")
- What are the processes which are relevant to company success
- Which of those need to be replaced

Once the management is convinced that only an entirely new process improves the company performance, a new process is set up without considering existing procedures or organizations.

The fundamental concept of process reengineering includes[4]:

- The use of modern information technology to achieve a significant improvement of performance
- No constraints in redefining a business process; organizational structures and management systems need to be changed as well to adapt to the new processes and process landscape

---

[1] Gunasekaran, A., Kobu, B. (2002)
[2] Ahadi, H.R. (2004)
[3] Freidinger, R. (2003)
[4] Ahadi, H.R. (2004)

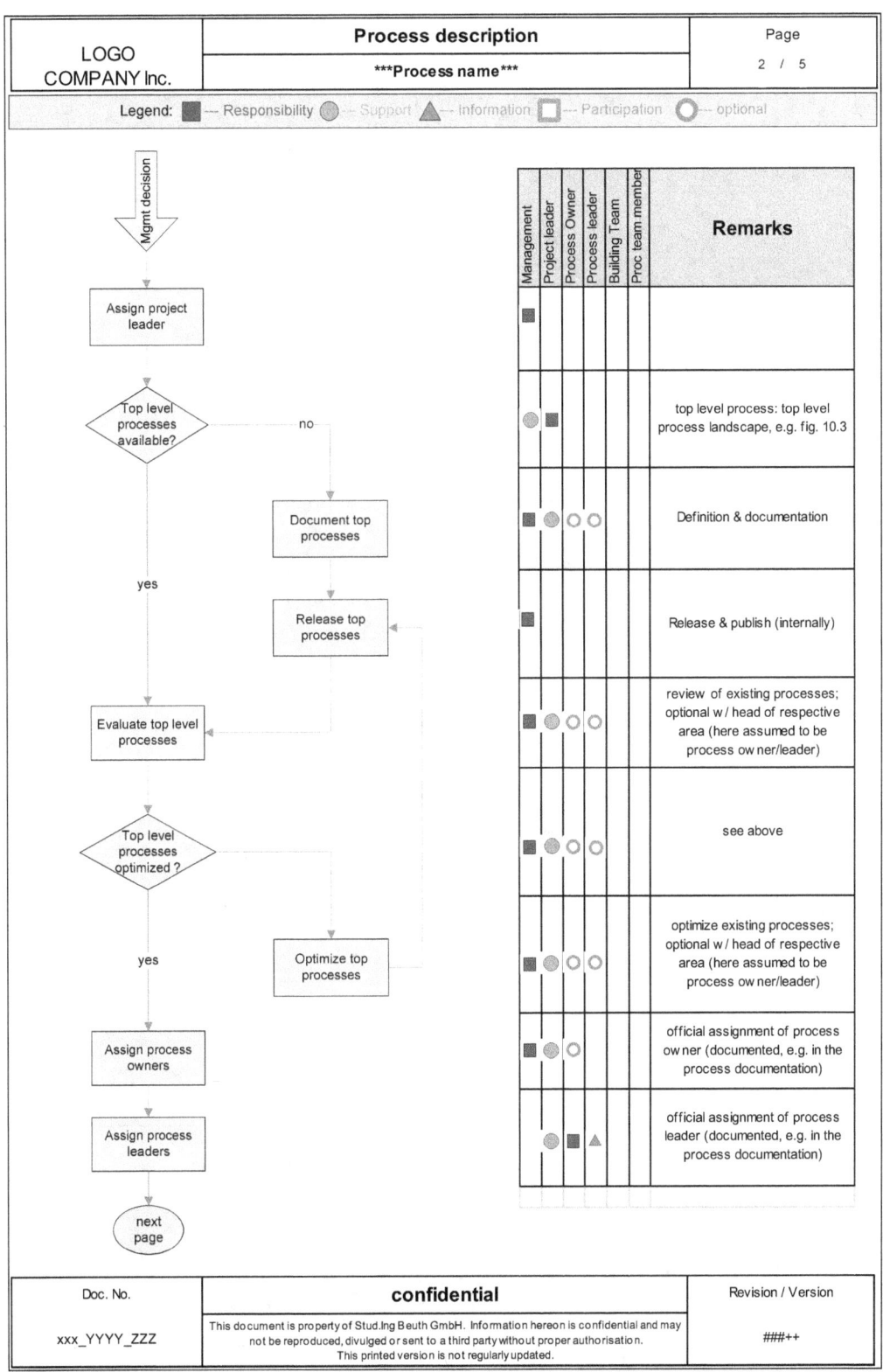

Fig. 11.8 Generic process management rollout process: starting phase

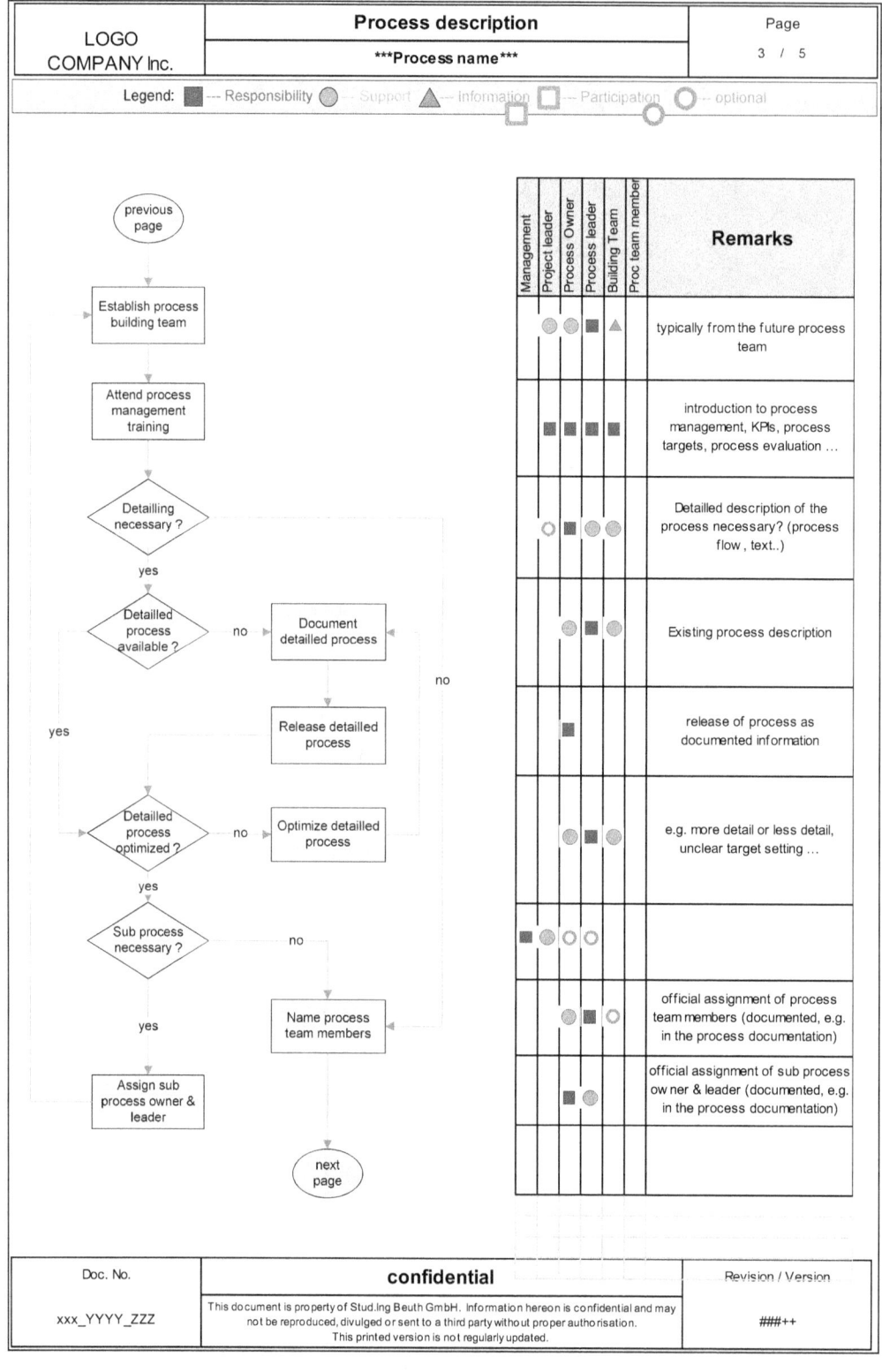

Fig. 11.9 Generic process management rollout process: detailing phase

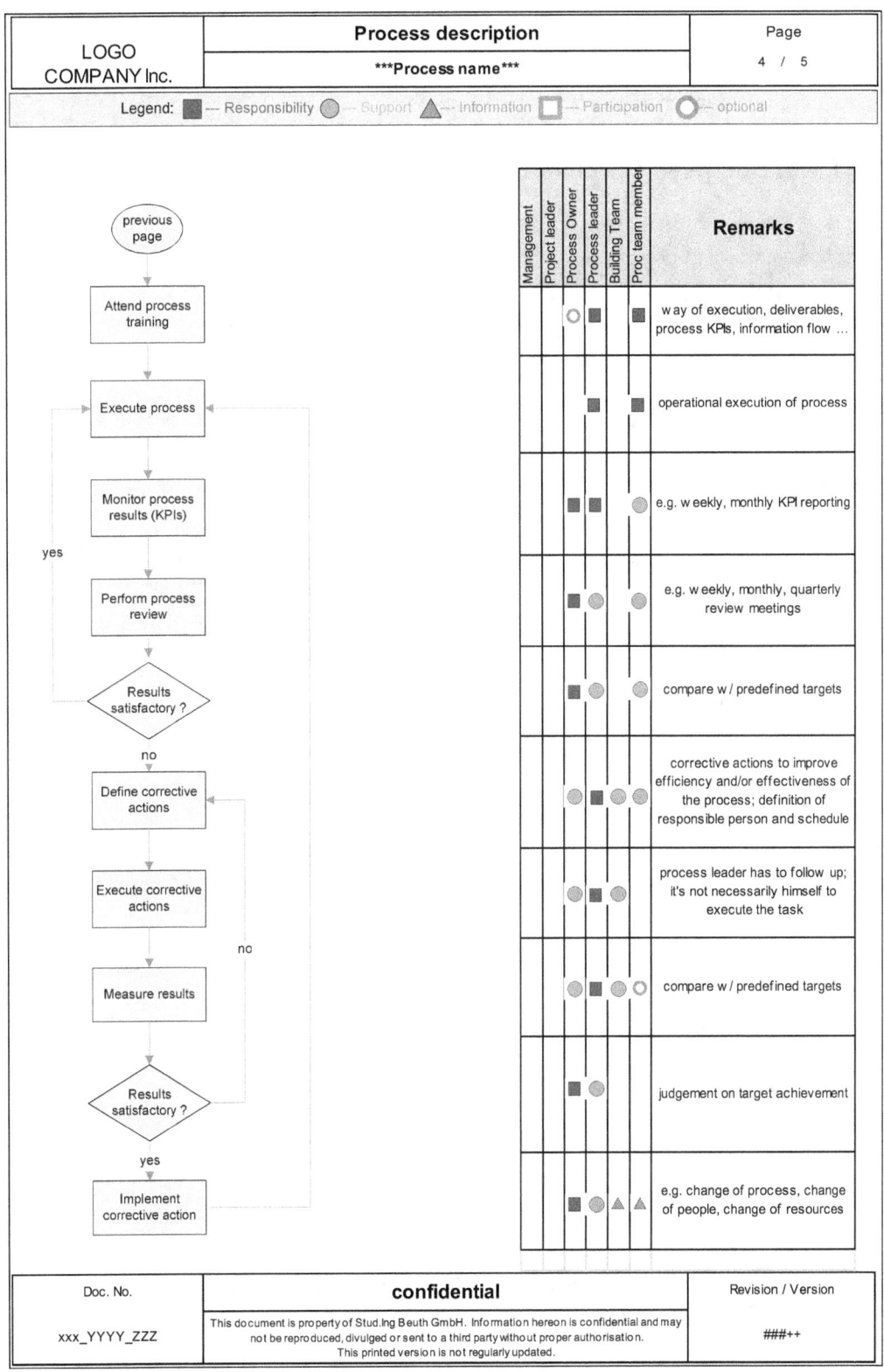

Fig. 11.10 Generic process management rollout process: execution & improvement

> Significant changes in performance (e.g. output, speed, quality) require significant changes of the processes.

### 11.2.6. Business Process

In this section an example for a business process in process management is given for demonstration. It describes a "process management rollout" process (fig. 11.8, 11.9, 11.10) at the very beginning of process management in a company.

## 11.3. Process Assessment

In addition to quantitatively monitoring and controlling the process output and process target achievement, e.g. using defined key performance indicators, process performance can also be increased using a more qualitative process evaluation approach, in which the generic process quality is assessed.

This approach is based on the assessment of e.g. particular features of the process description or the way process improvement is formally established. Often the maturity of a process is described using various levels of maturity.

### 11.3.1. Maturity Models

Qualitative assessment models using maturity levels are e.g.[1]

> CMMI (Capability Maturity Model Integration)
> SPICE (Software Process Improvement and Capability dEtermination)
> PEM-Model (Process and Enterprise Maturity)-Model

In addition, the quality of a process can also be assessed by the RADAR scoring algorithm of the EFQM models, which will be explained in detail in chapter "Improvement, Excellence and Self Assessment".

CMMI

Driven by the US ministry of defence the CMMI model was developed by SEI (Software Engineering Institute of the Carnegie Melon University). It provides reference models in order to compare the company processes against them. The main target of CMMI is to optimize and to stabilize processes. Those are grouped in various process areas which are assessed based on criteria which themselves are assigned to five maturity levels and four capability levels:

---

[1] Däubner, T. (2013)

| Level 1 | Initial | Process unpredictable, poorly controlled, and reactive |
| Level 2 | Managed | Process characterized for projects and is often reactive |
| Level 3 | Defined | Process characterized for the organization and is proactive |
| Level 4 | Quantitatively managed | Process measured and controlled |
| Level 5 | Optimized | Focus on process improvement |

Table 11.2 CMMI maturity levels (cited from Dayan, 2006)[1]

| Level 0 | Incomplete | The process is not or only partially performed. Goals are not achieved. |
| Level 1 | Performed | The process results in required output. Goals are achieved. |
| Level 2 | Managed | The process is planned and executed in line with overall targets and policy. Output is achieved with right set of skills and resources. The process is monitored and controlled |
| Level 3 | Defined | In addition to level 2 the process is described in detail, is pro-actively controlled and monitored. From the company standard process project (or organizational unit) specific processes are customized. |

Table 11.3 CMMI capability levels[2]

The CMMI assessment is performed in the "Standard CMMI Appraisal Method for Process Improvement". The method is rather lengthy, and it will become obvious that it resembles the other techniques for assessing process maturity.

### SPICE[3]

SPICE stands for Software Process Improvement and Capability dEtermination. It is an international standard for company process assessments (ISO/IEC 15504). SPICE provides reference models the assessor can compare the company processes with. It's a two-dimensional approach to assess company processes using process dimension and capability dimension.

---

[1] Dayan, R. (2006)
[2] CMMI (2010)
[3] Steinmann, C. (2000)

The SPICE model provides various process categories and respective processes (p1, ..., pn) and various capability levels (cl1, ..., cl5) with nine process capability attributes.

For each process all nine process attributes are rated based on percentage achievement numbers[1]:

- N: 0 ... 15 %    not achieved
- P: 16 ... 50 %   partially achieved
- L: 51 ... 85 %   largely achieved
- F: 86 ... 100 %  fully achieved

Each process receives a process capability rating.

The process dimension covers the following five categories:

- Customer – Supplier
- Engineering
- Support
- Management
- Organization

For each process category several basic and component processes are available, e.g. for engineering:

- Development
    - System Requirements Analysis & Design
    - Software Requirements Analysis
    - Software Design
    - Software Construction
    - Software Integration
    - Software Testing
    - System Integration & Testing
- System & Software Maintenance

| Level 0 | Incomplete | Process performance & results are not complete, process is chaotic |
|---|---|---|

---

[1] Grambow, G. et al (2012)

| Level 1 | Performed<br>**Process Performance** | Processes are executed based on intuition; input and output are available |
|---|---|---|
| Level 2 | Managed<br>**Performance Management**<br>**Work Product Management** | Responsibilities are clear. The process is actively managed. |
| Level 3 | Established<br>**Process Definition**<br>**Process Deployment** | Predefined company processes are customized for specific areas (e.g. projects, organizational units). Resources are managed, too. |
| Level 4 | Predictable<br>**Process Measurement**<br>**Process Control** | Process performance and process results are monitored and controlled using a defined metrics |
| Level 5 | Optimizing<br>**Process Innovation**<br>**Continuous Optimization** | Continuous improvement is executed using quantitative measures. |

Table 11.4 SPICE capability levels and process attributes (bold)[1]

Eventually, over time, for these processes a varying number of reference practices is available for comparison with company processes.

PEM[2]

PEM stands for Process and Enterprise Maturity-Model and was developed by a consulting company (Hammer and Company). Its purpose is measurement and optimization of business processes.

The PEM model is based on two aspects, the process perspective (table 11.5) and the company perspective (table 11.6).

| Design | The way the process is to be executed |
|---|---|
| Performer | People executing the process including skills, knowledge |
| Owner | Person with overall responsibility for the process and its performance |
| Infrastructure | Process support, e.g. information, management system |
| Metrics | Indicators etc. being used to measure the process performance |

---

[1] Steinmann, C. (2000)
[2] Hammer, M. (2007)

Table 11.5 Process perspective in the PEM model[1]

Each process dimension has two or three sub-dimensions, e.g. purpose, context, and documentation for "Design". The maturity of a process is determined by rating all the sub-dimensions depending on the strength of fulfilment of various statements P1 to P4 for each respective sub-dimension statement.

The fulfilment rate is visualized using a colour code. If a statement is at least 80% true the colour is green, if the statement is to some extent true (20 ... 80%) the colour is yellow and if the statement is less than 20% correct, the colour is going to be red.

| Leadership | Senior management support for business processes |
|---|---|
| Culture | Company values like customer satisfaction, team, personal accountability, openness for change |
| Expertise | For process redesign skills and methodology is required |
| Governance | Established ways for changes in the company and management of complex projects |

Table 11.6 Company perspective of the PEM model[2]

## 11.3.2. PPA: Pragmatic Process Assessment[3]

As described in the previous section there are several more or less operational models available for assessing the maturity and in this way the quality of a business process. Although the models are standardized many companies tend to apply a simplified way of business process assessment. The reason behind is the fact, that in many companies the process management culture is not at a level which makes the use of the above mentioned models necessary or it is assessed that the benefits of the application of those models are not commensurate with the resources needed for the assessment. In the latter case attempts are made to implement a leaner process assessment.

Therefore in this section a simple and pragmatic approach is introduced, which was developed and tested in companies of various branches. It can be executed in companies of various sizes, does not need deep understanding of process maturity models and does yield fast, concrete and usable results. Like PEM this assessment method is not based on any reference models or reference processes.

Process Control Loop

---

[1] Hammer, M. (2007)
[2] Hammer, M. (2007)
[3] Wittmann, J. (2014)

The guiding process for the formalized procedure is summarized in the process control loop (fig. 11.11).

The process control loop can be executed once or continuously in order to improve single processes, or it could result in a general change of the company process strategy.

Planning

First of all the scope, i.e. the "process unit", of the process assessment has to be defined, e.g. entire company, business unit, production unit. This is mainly driven by the management of the concerned unit. Within this process unit processes have to be selected for the assessment. It makes sense to cover the entire portfolio of processes, which means various process levels, various sub units, management processes, support processes and, of course, core processes.

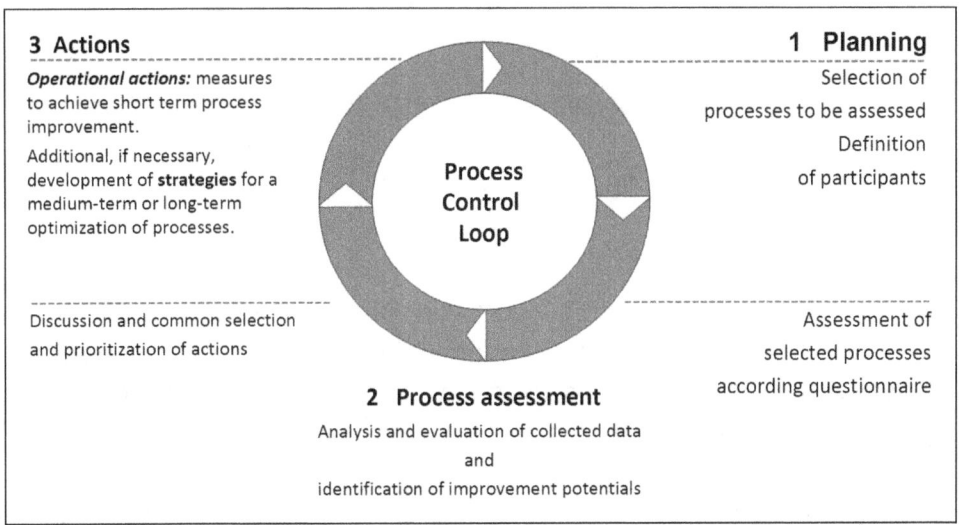

Fig. 11.11 Process Control loop

Secondly the extent of the assessment has to be defined. For an initial assessment, a baseline study, three to four days with an average of three processes per day has shown to be sufficient. Detailed planning of a particular process assessment session, however, depends on the size and the complexity of the process. For analysis and evaluation roughly the same period of time has to be expected.

The third step is to define the participants. Small teams of three to five team members plus the moderator are sufficient. The process leader should be part of the team.

The compilation of the concrete agenda is part of the planning phase, too. Below a generic agenda is given:

Day 0    Common selection of processes                Management & Moderator

Day 1    09:00 – 11:30    Process 1
            - short presentation of the participants    all
            - short presentation of the method    Moderator
            - short presentation of the respective    e.g. Processleader
              process
            - assessment of the process    Moderator & all
         12:30 – 15:00    Process 2    ditto
         15:30 – 18:00    Process 3    ditto

Day 2 ..x   ditto

Following days:
         Analysis    Moderator
Final day (results):
         - presentation of results    Moderator
         - discussion of results & measures    Management &
                                               Moderator
         - definition of further proceeding    Management

Process Assessment

After the selection of the process the actual process assessment is performed by the interviewer. The interviewer leads the discussion in form of a group interview. The interview is supported by questionnaires and a detailed description of different levels of a percentage wise target fulfilment.

Level 1 questions are supposed to confirm whether or not a workflow is implemented and people know it and comply with it. In level 2 the goal is to find out, if the process is described, i.e. if all aspects of a complete process description have been considered. On the next level the intention is to find out how the process is monitored and controlled and how the process targets are aligned with company targets. Finally, on the highest level the interviewer intends to check what measures or even formalized processes are executed in order to improve the process.

Maturity Levels

Based on the various levels of questions as described in the previous section the process maturity is defined in fig. 11.16. A maturity level is achieved as soon as the average rating on this particular level is at least 90% and in addition if the inferior levels also achieved at least 90%.

## Process Assessment

| Company / site | | | | | Date of assessment | |
|---|---|---|---|---|---|---|
| Process | | | | | ISO 9001 certified since: | |
| Process owner: | | | | | ISO/TS16949 certified since: | |
| Assessment team: | | | | | | |
| Questionnaire | \multicolumn{4}{c}{RATING [%]} | Result | |
| | 0% | 33% | 67% | 100% | [%] | Comments (Strengths / Weaknesses / Improvement potential / evidence) |
| Level 1: The process is implemented | | | | | 75 | |
| 1.1 The process structure is known | | x | | | 33 | |
| 1.2 The process flow is constant | | | x | | 67 | |
| 1.3 Where necessary the process is supported by ~~required~~ docum~~ents~~ ch~~ecklists~~ | | | | | 100 | |

| | Level 1: The process is implemented | 0% | 33% | 67% | 100% |
|---|---|---|---|---|---|
| 1.1 | The process structure is known | no evidence | the participants know their parts | the workflow responsible knows the entire process | all participants know the entire process |
| 1.2 | The process flow is constant | not visible | varying workflows visible | constant, if performed by the same person | constant w/ different persons |
| 1.3 | Where necessary the process is supported by required documents (e.g. forms, checklists) | required documents / records are not defined | required documents / records are partially defined | evidence for the use of documents / records is available | extensive evidence for the use of required documents / records is |
| | | the process is performed in chaotic way and ~~bea~~rs riks without ~~considerati~~on | a process description helps to reduce cost and risk | a process description helps to reduce cost | a process description is not considered necessary |

Fig. 11.12 Level 1 questionnaire examples with description of rating

Except for improvement the awareness of personnel or of the management the maturity level itself is not an additional benefit in terms of process performance or economic benefits. More important from the continuous improvement point of view is the identification of gaps in the description or the execution of the process and the identification of potential improvements.

## Process Assessment

| Company / site | | | | | Date of assessment | |
|---|---|---|---|---|---|---|
| Process | | | | | ISO 9001 certified since: | |
| Process owner: | | | | | ISO/TS 16949 certified since: | |
| Assessment team: | | | | | | |
| Questionnaire | RATING [%] | | | Result | | |
| | 0% | 33% | 67% | 100% | [%] | Comments (Strengths / Weaknesses / Improvement potential / evidence) |
| Level 2: The process is described | | | | | 76 | |
| 2.1 The process is sufficiently described. It's logical and understandable. | | | x | | 67 | |
| 2.2 The process leader is assigned | | | | x | 100 | |
| 2.3 The process is reviewed and released by the assigned process owner | | | | x | 100 | |

| | Level 2: The process is described | 0% | 33% | 67% | 100% |
|---|---|---|---|---|---|
| 2.1 | The process is sufficiently described. It's logical and understandable. | the process description allows the correct execution of the process | there are crucial gaps in quantity, quality or detailing | there are limitations in quantity, quality or detailing | quantity, quality and detailing are ideal |
| 2.2 | The process leader is assigned. | the process leader is not named | the process leader is orally named | the process leader is named in writing | the process leader is named in writing; this information is published |
| 2.3 | The process is reviewed and released by the assigned process owner | no evidence for review and release | parts of review and release are missing | review and release are insufficient (e.g. no four-eyes-principle) | review and release formally and content wise correct |
| | | no evidence | partially named | at least the responsibles for the particular process steps are completely | Responsibles and other participants (inf. legwork) con... |

Fig. 11.13 Level 2 questionnaire examples with description of rating

## Process Assessment

| Company / site | | | | | Date of assessment | |
|---|---|---|---|---|---|---|
| **Process** | | | | | ISO 9001 certified since: | |
| Process owner: | | | | | ISO/TS 16949 certified since: | |
| Assessment team: | | | | | | |

| | Questionnaire | RATING [%] | | | | Result | |
|---|---|---|---|---|---|---|---|
| | | 0% | 33% | 67% | 100% | [%] | Comments (Strengths / Weaknesses / Improvement potential / evidence) |
| | Level 3: The process is controlled | | | | | 85 | |
| 3.1 | process targets are defined and described | | x | | | 33 | |
| 3.2 | process targest are know by process participants | | | | x | 100 | |
| 3.3 | process ... SM... | | | | | | |

| | Level 3: The process is controlled | 0% | 33% | 67% | 100% |
|---|---|---|---|---|---|
| 3.1 | process targets are defined and described | not visible | at least one target documented in writing | one target or several perspectives documented in wirting (e.g. longterm, medium term, short term, finance, personell, environment) | extensively documented in writing |
| 3.2 | process targest are known by process participants | no evidence, not known | process targets are partially known (1/3) | process targets are known by most of the participants (2/3) | process targets are known |
| 3.3 | process targets are phrased SMART (specific, measurable, ... timed) | no evidence | at least one of the criteria / target met | at least 3 criteria per target met | all criteria for all targets met |
| | | no evidence | some evidence, missing alignment visible | distinct evidence, alignment visible | extencive evidence, clear alignment of process and compa... |

Fig. 11.14 Level 3 questionnaire examples with description of rating

## Process Assessment

| Company / site | | | | | Date of assessment | |
|---|---|---|---|---|---|---|
| Process | | | | | ISO 9001 certified since: | |
| Process owner: | | | | | ISO/TS16949 certified since: | |
| Assessment team: | | | | | | |

| | Questionnaire | RATING [%] | | | Result | |
|---|---|---|---|---|---|---|
| | | 0% | 33% | 67% | 100% | [%] | Comments (Strengths / Weaknesses / Improvement potential / evidence) |
| | Level 4: The process is optimized | | | | | 90 | |
| 4.1 | process targets are regularly reviewed and updated | | | | x | 100 | |
| 4.2 | defined performance indicators are regularly reviewed and optimized | | | x | | 67 | |
| 4.3 | the process is regularly reviewed f... effi... | | | | | | |

| | Level 4: The process is optimized | 0% | 33% | 67% | 100% |
|---|---|---|---|---|---|
| 4.1 | process targets are regularly reviewed and updated | no evidence | some evidence or at least once executed | evidence mostly available or more than one time executed | extensive evidence, yearly review |
| 4.2 | defined performance indicators are regularly reviewed and optimized | no evidence | some evidence or at least once executed | evidence mostly available or more than one time executed | extensive evidence, yearly review |
| 4.3 | the process is regularly reviewed for completeness, efficiency, effectiveness ... by audits, reviews, meas... | no evidence | some evidence or at least once executed | evidence mostly available or more than one time executed | extensive evidence, yearly review |
| | | the review ...ss is not | the review process is described | the review process is described and applied to this process | The review process is controlled and applied to... |

Fig. 11.15 Level 4 questionnaire examples with description of rating

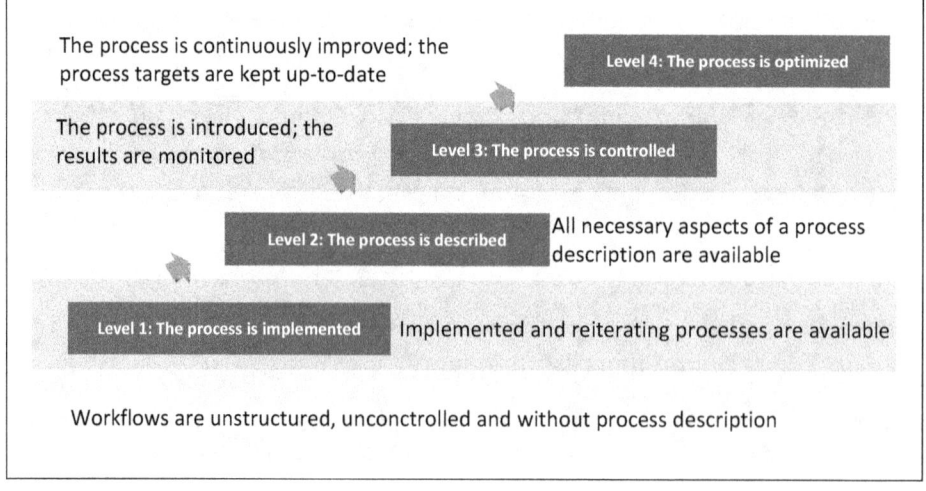

Fig. 11.16 Process maturity levels

Assessment Results

The analysis of the collected data has to be performed thoroughly in order to gain the most benefit from the process improvement. Hence it makes sense to plan for sufficient time for the analysis work.

The identification and documentation of gaps found during the interview is the starting point of a deeper analysis. In this phase gaps like deviations in the workflow, double work, unnecessary process descriptions, missing process descriptions, missing training, missing performance indicators etc. are described and summarized.

Looking at various perspectives is part of the analysis, e.g.

- Comparison of process levels
- Comparison of different areas
- Evaluation of processes by maturity level
- Etc.

The purpose of the detailed analysis is not only to find single process gaps, but, and this is more important, to identify company wide gaps in the process management system.

For instance, the pareto diagram of achieved average percentage per question is a strong indication of a general behaviour or "culture" in the company (see fig. 11.17).

The direct comparison of the percentage distributions of different process maturity levels provides a good representation of the company process maturity, too (see fig. 11.18).

Corrective Actions

Corrective actions are either operational or strategic. Operational corrective actions are typically short term measures to improve particular processes, e.g. process training, definition of responsibilities or process deliverables. On the other hand, strategic corrective actions are mainly long term activities like introduction of a company wide process management or an entirely new set up of the company target setting.

Both are useful. The more effective way of process improvement, of course in the long run, is the strategic approach, because in this approach all operative improvements will be executed sooner or later anyway.

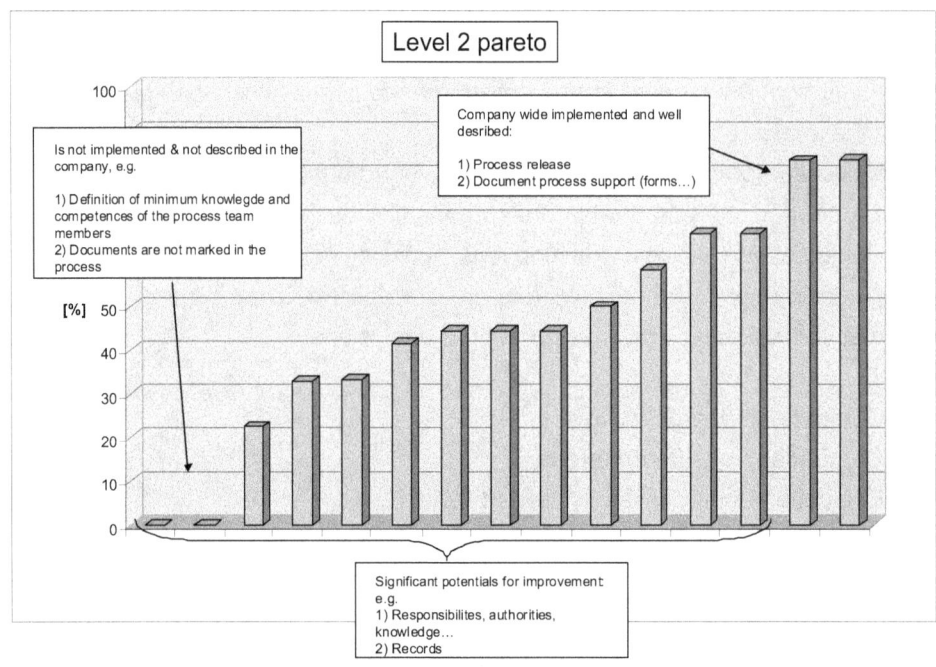

Fig. 11.17 Average percentage pareto over level 2 questions

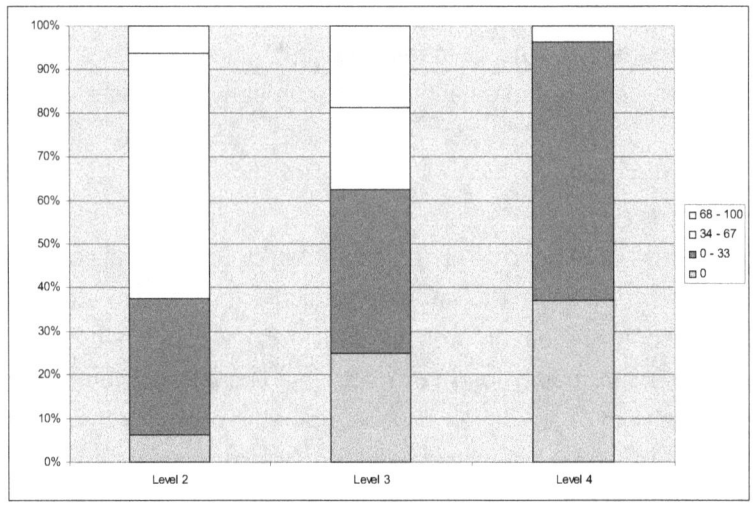

Fig. 11.18 Percentage distribution of different process maturity levels

### 11.3.3. Business Process: PPA

The following pages show a generic business process.

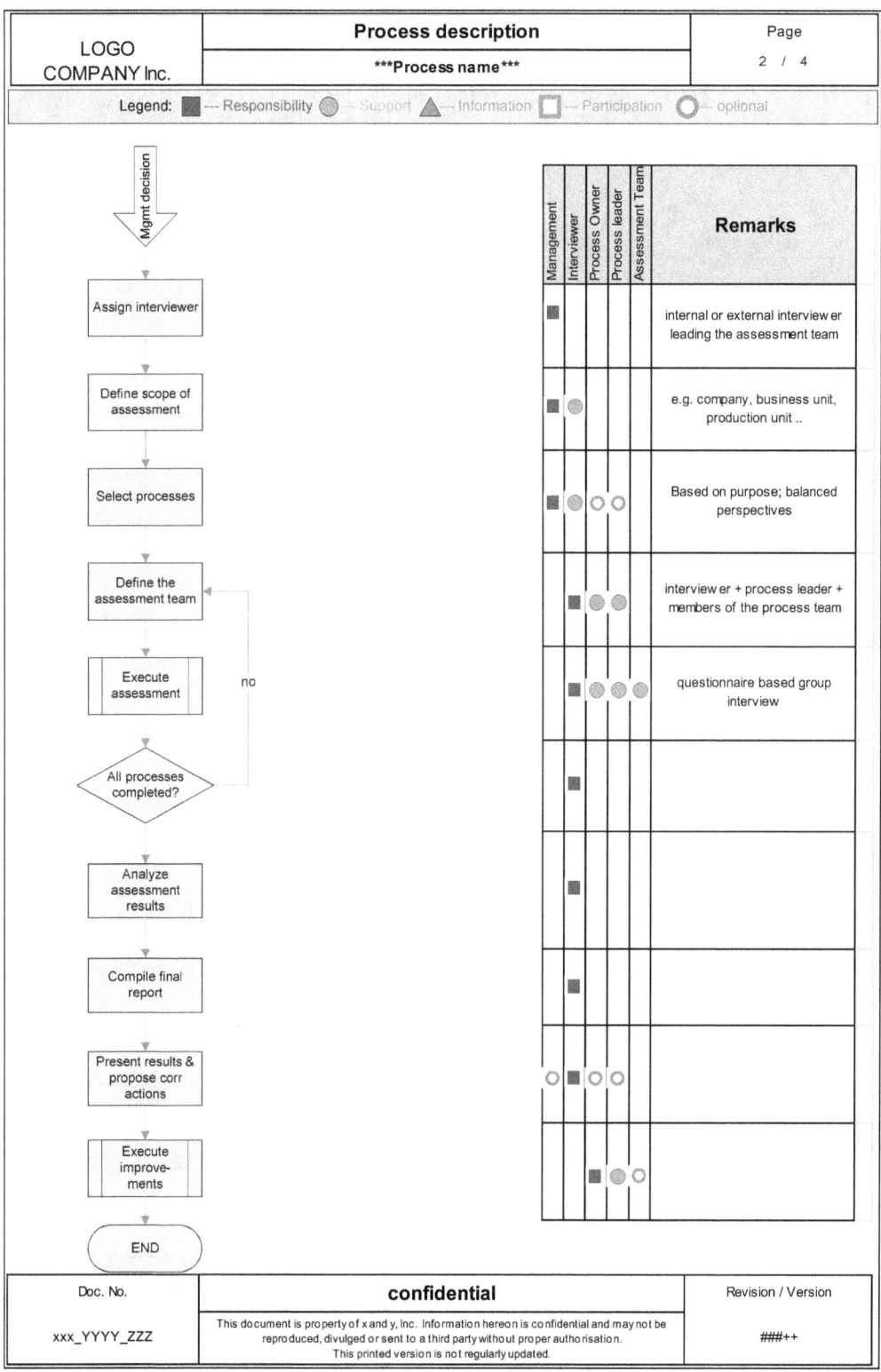

Fig. 11.19 Generic main process for a process assessment

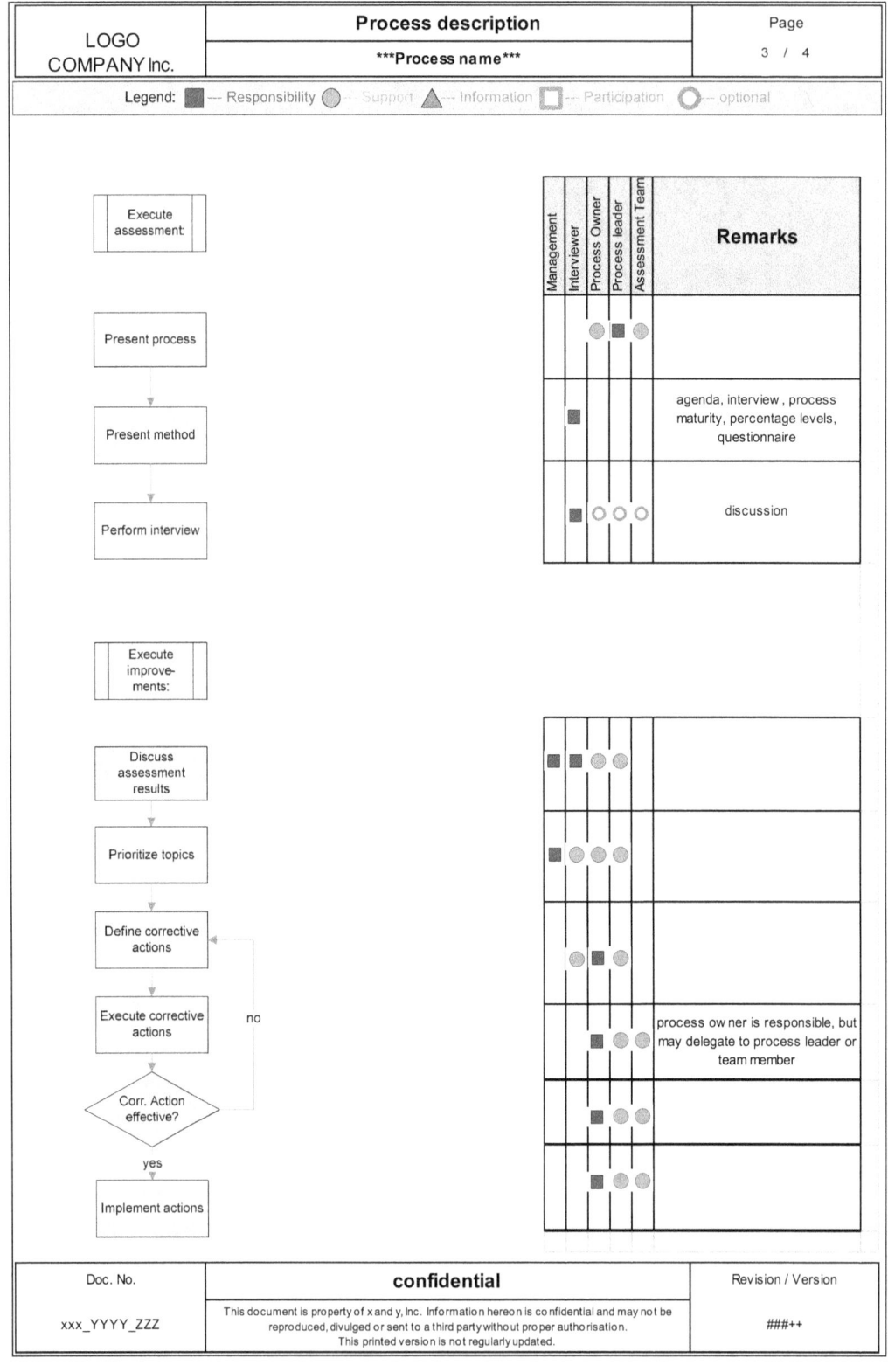

Fig. 11.20 Generic sub processes for the process assessment

# 12. Improvement, Excellence and Self Assessment

## 12.1. Evolution of the quest for quality

In the 20th century, understanding, the methods for handling and management of quality proceeded through a remarkable improvement path. Starting with simple quality control activities like sorting of produced parts, the methodology advanced to management of quality using new methods like statistical process control and total quality management on a company level. Today the top players in quality focus on business excellence with or without respective total quality management models, e.g. EFQM.

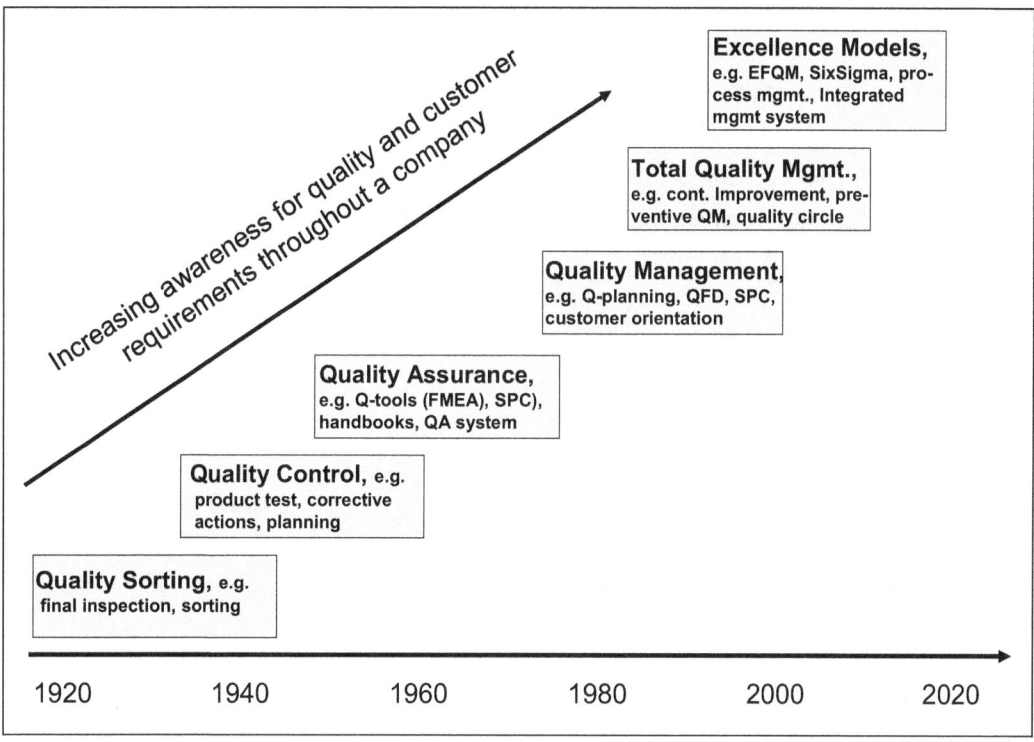

Fig. 12.1 Evolution of quality awareness and methods[1]

The major milestones in the history of quality management also provide an insight into the way quality management developed over time[2]:

    1875         F. W. Taylor starts to break down work into small units ("Scientific Management") in order to cope with more complex products and

---

[1] Zollondz, H.-D., 2011, p. 27
[2] Zollondz, H.-D., 2011, p. 50

| | |
|---|---|
| | processes and to be able to employ unskilled labour for more involved production processes |
| 1900 – 1930 | H. Ford starts car manufacturing in assembly lines and uses process inspection, self-checking etc. |
| 1907 – 1908 | AT&T starts product inspection and test |
| 1908 | Introduction of the t-distribution by W. S. Gosset ("Student") |
| 1924 | First introduction of quality control charts by W. A. Shewart |
| 1931/32 | Publications and lectures on statistical methods by W. A. Shewart |
| 1940 – 1943 | Military standard sampling plans developed by Bell Labs |
| 1946 | Foundation of the International Organization for Standardization (ISO) which later on published many international standards related to quality management |
| 1948 | Initial studies on experimental design by G. Taguchi |
| 1950 | Introduction of the Cause-and-Effect diagram by K. Ishikawa |
| 1950 | Introduction of the PDCA circle by E. Deming |
| 1957 | Publication of the Quality Control Handbook by J. M. Juran and F. M. Gryna |
| 1960 | K. Ishikawa introduces the quality control circle concept |
| 1960 | P. B. Crosby initiates the Zero-Defect approach |
| 1970 | The British Quality Association is formed |
| 1975 – 1978 | Quality Circles in North America are the starting point for later Total Quality Management |
| 1987 | Publication of the first standards on quality systems (ISO9000 family) |
| 1987 | Malcolm Baldrige Award established |
| 1989 | EFQM foundation |
| 1989 | Start of Motorola's SixSigma program |
| 1990 | Start of development of automotive quality standard QS9000 |
| 2000 | Publication of the ISO9001:2000 standard |
| 2002 | ISO/TS16949 replaces QS9000 and it includes ISO9001 verbatim |
| 2015 | Publication of ISO9001:2015 standard with increased focus on process management and risk/opportunity assessment |

## 12.2. Continuous Improvement

One of the most important factors in modern quality management is the idea of continuous improvement within an organization. The continuous improvement approach not only intends to improve the quality of a particular product, but keeps the entire organization in focus, e.g. business processes, customer relationship or the quality management system itself. This means that all employees as well as the management of the organization have to be part of this improvement culture. It is quite obvious that the concept of continuous improvement requires a lot of creativity, motivation and commitment and works best in an open environment with internal stakeholders have to have a high level of awareness for gaps, deviations or weaknesses[1].

The quest for systematic improvement, however, has been in place for many years. Four generations of improvement were identified from the year 1780 up to today[2]:

| | |
|---|---|
| 1780 – 1880 | Engineering improvement (e.g. steam power based mechanization) |
| 1880 – 1980 | Scientific improvement (e.g. introduction of electricity and scientific methods by W. Taylor, H. Ford, W. Shewart) |
| 1980 – 2010 | Program based improvement (e.g. by W. Deming, T. Ohno, J. Juran) |
| 2010 … | Adaptive, systematic management process (e.g. the way of improvement is subject to improvement on a regular base) |

Despite the fact, that there are many other quality management concepts or techniques available, e.g. KAIZEN as the Japanese way of continuous improvement, this section focuses on a few examples only, and does not strive for complete coverage of the field:

- Deming Cycle
- Continuous Improvement Circle
- Employee Suggestion System

An overview of basic techniques and methods for continuous improvement have been summarized by M. Sokovic[3].

## 12.2.1. Deming Cycle[4,5]

The Deming cycle, also known as PDCA cycle, actually developed by Dr. Walter Shewart earlier, was promoted by W. Edwards Deming in the 1950s in Japan as a major contributor to a company's continuous improvement process.

In this context PDCA stands for Plan – Do – Check – Act and represents a four step method to improve business processes within a company (fig. 12.2):

- PLAN

The task here is to examine the current situation in order to find potential for improvement. Examination can be done using an internal / external audit,

---

[1] Zollondz, H.-D. (2011)
[2] Burton, T. T. (2014)
[3] Sokovic, M. (2009)
[4] Singh, V. K. (2013)
[5] Singh, J, Singh, H. (2012)

benchmarking, brainstorming or any other way to identify weaknesses. Subsequently improvement measures are planned.

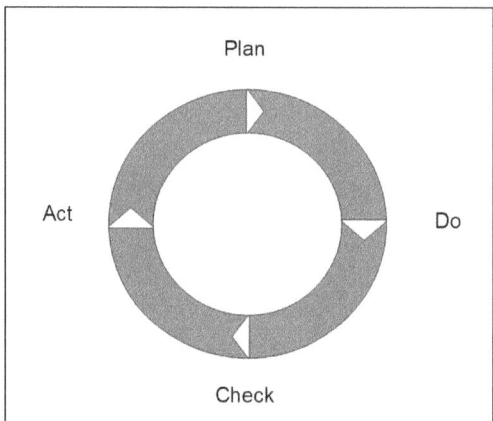

Fig. 12.2 Deming cycle

> DO

Identified improvement measures are implemented on a small scale ("pilot") in order to find out the effectiveness of the measures.

> CHECK

An actual review of the effectiveness of the executed measures has to take place, i.e. the results of the change have to be reviewed.

> ACT

Once it is demonstrated that the change is successful, the measure has to be standardized. For this reason process descriptions and trainings have to adapted and executed and the change has to be implemented throughout the company. If the desired improvement does not materialize or is insufficient, the Deming cycle is repeated.

## 12.2.2. Continuous Improvement Circles

Continuous improvement circles (CIC) are groups of people meeting regularly to discuss product or process quality related topics. The purpose of those meetings is to come up with improvement measures, action items, target dates and responsible persons. If the desired result is not achieved, then a modified plan has to be devised, and the Deming cycle is repeated.

The way the improvement cycles are organized depends on the size of the company. Smaller or medium sized companies often rely on the operating level only, including the following activities:

- Definition of the scope (what needs to be improved?)
- Brainstorming for topics to be improved
- Problem description and assessment (priorities!)
- Problem analysis (root cause analysis, interfaces..)
- Identification of problem solutions or measures to improve the process robustness (including prioritization)
- Definition of measures considering resources required and benefit expected
- Execution of measures (release of resources by management required)
- Check results (see Deming cycle)

The group meeting agenda could be as follows:

(1) Welcome by the team leader                                     5 min
(2) Status review of existing measures and action item list   10 min
(3) Brainstorming for new ideas / measures                     10 min
(4) Definition of responsible persons for further actions       5 min

The size of the group ranges typically from four to seven people. A biweekly meeting is sufficient in most cases. A member of the group should be responsible to track the actions (fig. 12.3). Frequently this is a member of the quality management group or the quality management responsible within the company.

Large companies might have a complete organizational structure for continuous improvement with many continuous improvement (CI) teams with the respective CI team leader and a CI coordinator for major organizational units. The CI coordinators are responsible for the management and administration of the team meetings and report, for instance, to a global head of CI activities (e.g. head of operations). In addition to this operational organization the CI process requires an owner who takes care of CI process improvement, who executes CI training and who may also be responsible for overall reporting of CI results.

## 12.2.3. Employee Improvement Suggestion Scheme[1]

Similar to the previous section "continuous improvement circles" the employee improvement proposal system (EIPS) receives its efficiency from the fact, that for employees participating or being close to a business process or the way things are done, it is easier to observe gaps or problems than for others.

---

[1] Neagoe, L. N., Marascu Klein, V. (2009)

| Continuous Improvement Cycle ||||||||||
|---|---|---|---|---|---|---|---|---|---|
| Team: J. Wittmann, …… |||||||||
| Meeting schedule: Tuesday, biweekly, 10 am |||||||||
| No. | Proposal / measure | priority | Start date | PLAN | DO | CHECK | ACT | final release date |
| 1 | … | 1 | | | | | | |
| 2 | … | 2 | | | | | | |
| 3 | … | 3 | | | | | | |
| 4 | … | 2 | | | | | | |
| 5 | | 2 | | | | | | |

Fig. 12.3 Action item list for the continuous improvement cycle including a visualization of the status of actions

The EIPS intends to improve things stepwise. The fundamental idea of it is based on the Japanese way of continuous improvement ("Kaizen") and is called Kaizen Teian. The effectiveness of the employee improvement proposal system is based on several prerequisites. First of all the employees in the company have to be aware of the need for improvement for the company as well as for themselves, eventually. In addition the people have to be willing to report deficiencies they identified during daily work, i.e. they have to be motivated. The employee's motivation is frequently increased by the implementation of an award system. The amount of money being awarded is typically calculated from the benefit the company has with this proposal or may be a standard payment independent on the benefit for the company. Of course, also non-financial awards are common, e.g. paid time off, a visit to a restaurant, sports event or similar venues. Finally, there has to be a system in place on the company side for evaluation and assessment of employee's improvement proposals. The submitter of new ideas has to receive the result of the assessment within a reasonable amount of time. Otherwise the employee's motivation will severely suffer. The assessment may be performed by a separate committee or by the employee responsible for that particular item to be improved.

It is useful to have a standardized method for the assessment in place, e.g. considering the following assessment criteria which are weighed depending on the company's priorities:

➢ Benefit

Ranges from minor (e.g. < € 1000 as a single event) up to very significant on a company level (e.g. hundreds of thousands of Euros, possibly recurring every year)

- Creativity

Ranges from "well known - is typical in other areas, but has not been implemented for several reasons" up to "absolute new creative idea, never seen before"

- Effort for implementation

Ranges from "basically no effort, less than one hour of work to implement" up to "major efforts required – challenging existing resources, benefit endangered"

- Timeline for implementation

Ranges from "just turn it on, smaller than one hour" up to "extensive long lasting studies have to be executed to provide evidence for feasibility"

- Improvement area

Ranges from "improvement in own area of work" up to "improvement suggestions in an entirely different area of work"

A well-known positive side-effect of a well-managed employee suggestion scheme is an increase of the intrinsic motivation of the employees, since such a scheme gives every individual the capability to be creative and shape his/her work place.

## 12.2.4. Improvement Initiatives

Another pillar of continuous improvement in a company is the company driven improvement initiative, e.g. defect reduction program, yield improvement program, zero defect program or others.

Different from the above described methods in this case the management of the organization identified the need for improvement in a particular area and initiates a program or project within an organizational or even within the entire company.

Improvement initiatives should be set up as a project in a professional way in order to make them successful. This includes e.g.

- A clearly defined organizational set up with a high ranking project sponsor, project leader, project team members
- A clear scope and defined objectives
- Sufficient resources financially as well as regarding human resources, e.g. availability of experts in this field with enough of their work time allocated
- A clear schedule with starting and end date

## 12.3. Six Sigma

### 12.3.1. History and Definition

Beyond temporary improvement initiatives which are limited in time and in the sustainable impact on the employees' attitude for quality the Six Sigma method is a company wide quality initiative, which is integrated in the company organization, with its own infrastructure and long term, clearly not only temporarily, defined roles and responsibilities[1].

The need for this kind of new approach first came up in the United States of America. In the late 1970s Motorola, a microelectronics manufacturer, experienced high cost due to failing products and generally low product quality. So Motorola's management decided to tackle this problem using a systematic approach. When investigating the options for quality improvement it became obvious, that problems during manufacturing (e.g. instabilities, negative trends etc.) lead to quality problems on product level later on. At the same time many quality related statistical methods had proven their effectiveness for decades already. So the idea of a systematic and scientific way of manufacturing process improvement with statistical methods was conceived. The concept of Six Sigma was finally implemented at Motorola in the mid 1980s.

The further development of Six Sigma was twofold. On the one hand Motorola's improved product quality inspired other companies to take over the Six Sigma concept. For instance, General Electric (GE), started to introduce Six Sigma in the mid 1990s and many others followed[2]. On the other hand the application of Six Sigma methods was not limited to manufacturing processes, but was rolled out over the whole organization like development, logistics and services.

Various definitions are in place for Six Sigma, for example:

> "Six Sigma is a systematic methodology aimed at operational excellence through continuous process improvements" [3]

---

[1] Köhler, M., et al. (2014), p. 258
[2] He, Z., Goh, T. N. (2015)
[3] Pande, P. (2003)

> **Six Sigma is**
> "an organized and systematic method for strategic process improvement and new product and service development that relies on statistical methods and the scientific method to make dramatic reductions in customer defined defect rates" [1]

The basic idea of Six Sigma is a process based understanding of cause-and-effects chains:

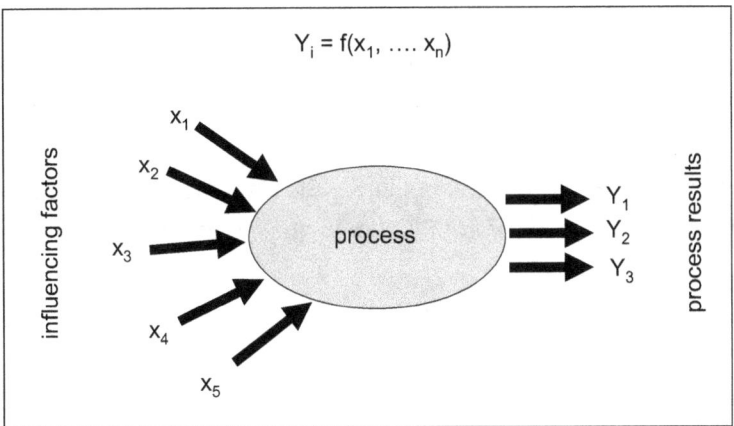

Fig. 12.4 Functional understanding of the Six Sigma concept

The obvious task is now to identify the influencing parameters $x_i$ and their impact on the process results $Y_j$.

For the quantitative description of the process performance "defects per million opportunities" (dpmo) can be calculated [2]

$$dpmo = \frac{c}{n_{parts} \cdot n_{CTQC}} \cdot 1.000.000$$

with

c : number of critical to quality non-conformancies

$n_{parts}$ : number of parts measured

$n_{CTQC}$ : number of critical to quality characteristics per part

A six sigma process is characterized by the mean of the distribution to have a distance of 6 σ to upper and lower specification limit, which, assuming a (typical and realistic) shift of the distribution by ± 1.5 σ, is equivalent to 3.4 defects per million.

---

[1] Linderman, K. (2003)
[2] Koehler M. et al. (2014), p. 258

In addition, six sigma includes financial aspects explicitly in the improvement process.

## 12.3.2. Six Sigma Organization

The Six Sigma organization is characterized by different levels of competence and responsibility regarding contributing to or leading of Six Sigma projects. Those levels are named after Asian martial arts levels of skills like "green belt", "black belt" etc. (fig. 12.5).

Yellow Belts
Yellow Belts have knowledge about the most relevant Six Sigma quality tools of the various Six Sigma project phases. They are team members in smaller Six Sigma projects and are familiar with the way Six Sigma projects are executed.

Green Belts
Green belts are either team members in Six Sigma projects or are leaders of small Six Sigma projects. They are aware of Six Sigma quality tools relevant for the particular Six Sigma project phases.
Being trained project leaders they know the success factors of Six Sigma projects.

Black Belts
Black Belts are experienced with successfully leading major Six Sigma projects by the active use of all the statistical know how. In addition to running Six Sigma projects they typically are also responsible for company internal Six Sigma training in order to further implement Six Sigma in the company.

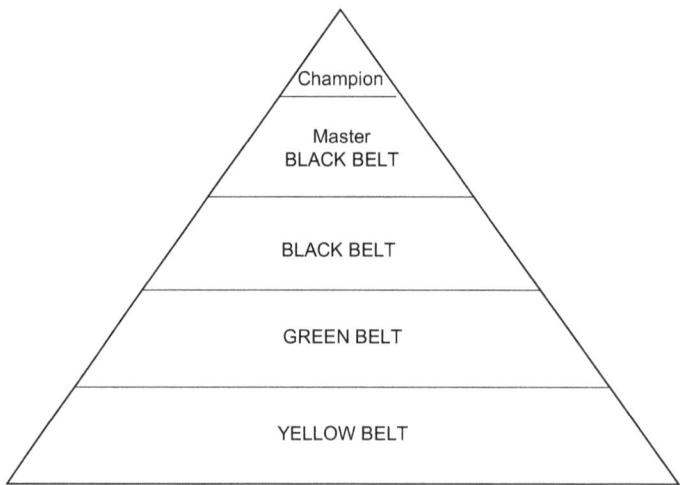

Fig. 12.5 Six Sigma hierarchical organization

Master Black Belt
Master Black Belts both work as Six Sigma project leaders for strategically relevant projects and as coaches for the Black Belts in the various projects (e.g. during project reviews). They manage, coordinate and shape the company wide Six Sigma program and select and prioritize the Six Sigma projects.

Six Sigma Champion
Six Sigma Champions act as godfathers for the particular Six Sigma projects. They set strategic goals and report directly to the general management.

## 12.3.3. Method

Similar to Deming's four step PDCA cycle the Six Sigma method is based on a five step improvement cycle in order to achieve sustainable results rather than straw fire like short term improvements. This cycle, the so called DMAIC cycle, is a structured way of Six Sigma project execution[1] (fig. 12.6).
DMAIC stands for:

- D – Define       : Define the problem as well as the objectives
- M – Measure   : Measure the relevant characteristics and determine the current status
- A – Analyze     : Identify and analyze the root cause of the problem
- I – Improve      : Identify and execute corrective and improvement actions
- C – control      : Ensure sustainability of the executed measures

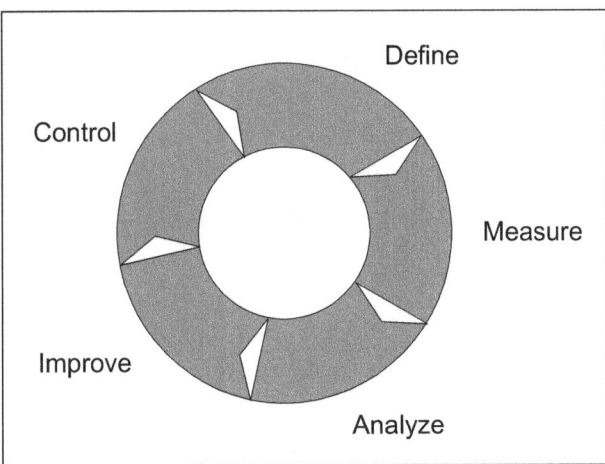

Fig. 12.6 DMAIC cycle

---

[1] Koehler M. et al. (2014), p. 265

The DMAIC steps are further detailed and particular methods and tools are assigned to each step[1,2]. Table 12.1 summarizes the quality tools used in the particular DMAIC phases.

Define

A clear and comprehensible description of the problem is crucial for substantial improvements. It is typically summarized using the so called project charter with the following information:

- Project leader & team members as well as stakeholders
- Objectives (benefit, business case), those should be defined in a "SMART" way (Specifically, Measurable, Actively influenceable, Realistic, Timed)
- Project resources, schedules and scope (including what is not included in the project)
- Mapping of the relevant process (process owner, supplier, customer, process description, inputs, outputs)

Measure

The problem described during the 'Define' phase has to be translated into measurable dimensions. Impact factors on process quality are determined and measured in order to determine the current status. In detail this means:

- Identify the relevant "Critical to Quality" factors (CTQ), the "Voice of the Customer" (VoC), i.e. what are the customer's requirements, and other performance indicators
- Measurement System Analysis in order to ensure the measurement system used is capable
- Define when and how often measurements are executed and collect data
- Based on the previous items determine the performance of the process in a quantitative way (e.g. failure rate, yield, time of execution ...)

Example:

The voice of the customer, e.g. a customer's statement when receiving a product information from the supplier: "this always takes too much time", has to be translated into a metric.

The respective metric could be the number of days between the customer's request and receipt of the information by the customer.

---

[1] Koehler M. et al. (2014), p. 266
[2] De Mast, J., Lokkerbol, J. (2012)

Analyze

The main objective of the Analyze phase is to identify influencing factors and root causes for deviations of the current status from the target status. For this purpose the problem, considered a system, is systematically broken down into single components with their respective function. Subsequently potential component malfunctions and their consequences to the functionality of the system are analyzed, e.g. by applying FTA (failure tree analysis).

Improve

In this phase different solutions might be found which then have to be compared with each other. Finally the most suitable solution is selected and the plan for implementation is created. Then the identified impact factors have to be adjusted in order to improve the process performance.
Afterwards the effectiveness of the measures is tested in a pilot test and control methods are developed.

Control

In this phase improvement measures are standardized (process descriptions, standard operating procedures ...). Previously developed control methods are transferred to control plans and the process performance is monitored using process capability figures.

There is obviously a large overlap between Six Sigma and quality management which conforms with the ISO9001 standard. The difference is that in ISO9001 there is a stringent system in place (which is at least not as explicit in Six Sigma), whereas there is a stronger link to business and financial aspects in Six Sigma.

| Quality tool / method | used in | | | | |
|---|---|---|---|---|---|
| | Define | Measure | Analyze | Improve | Control |
| CTQ tree | X | | | | |
| Arrow diagram | X | | | | |
| KANO model | X | | | | |
| Matrix diagram | X | | | | |
| SWOT analysis | X | | | | |
| Voice of the customer | X | | | | |
| GANTT chart | X | | | | |
| Pareto diagram | X | X | X | X | |
| Flow chart | X | | X | X | |
| Checksheet | X | X | X | X | X |
| Process map | X | | X | X | |
| SIPOC diagram | X | | | X | |
| Critical to quality matrix | X | X | | X | X |
| Quality function deployment (QFD) | X | | X | X | |
| Statistical Process Control (SPC) | X | X | | X | |
| Failure Mode Effect Analysis (FMEA) | X | | | X | |
| Six Sigma Indicators | X | | | X | |
| Process capability analysis | | X | X | | X |
| Control chart | | X | | | |
| Benchmarking | | X | | X | |
| Histogram | | | X | | |
| Bar charts | | | X | | |
| Scatter diagram | | | X | | |
| Ishikawa diagram | | | X | | |
| Affinity diagram | | | X | | |
| Hypothesis test | | | X | X | |
| Analysis of variance | | | X | X | |
| Regression analysis | | | X | X | |
| Correlation analysis | | | X | X | |
| Design of experiments | | | X | X | |
| Brainstorming | | | | X | |
| Poka Yoke | | | | X | |

Table 12.1 Six Sigma Toolbox (examples)[1]

[1] Ismyrlis, V., Moschidis, O. (2013)

## 12.4. Business Excellence & Quality Awards

### 12.4.1. Introduction to TQM & Business Excellence

The term Total Quality Management came up in the mid 1980s[1] and the understanding of TQM as a competitive advantage developed in the 1990s[2].

Due to the fact, however, that Total Quality Management is more a process or a culture within a company than a recipe how to manage an organization, there is no general definition of Total Quality Management (TQM) in place. More and more the term TQM is replaced by "Business Excellence"[3], which was later on defined by the European Foundation for Quality Management.

The concept of TQM or Business Excellence can be characterized by the following facets:

- Strong orientation to the customer and his needs, but without neglecting other stakeholder's interests. In other words, a balance of interests of the various stake holders has to be achieved, which is aligned to an overall company optimum
- Integration, motivation and participation of the employees
- Strong build up and use of know how in the organization
- Continuous improvement on all levels using the right quality tools & methods and employee & management involvement
- Promotion of cooperative relationships to customers and suppliers
- Focus on value add activities and prevention rather than waste and repair
- Balance between short term and long term aspects and systematic strategy development process
- Explicit assessment of the quality of management

Business Excellence (BE) Models are the way to transport the idea, the spirit and the measurability of TQM and the BE level achieved into a company.

More than eighty national and regional quality awards, based on the Malcolm Baldrige National Quality Award and the European Quality Award, allow companies worldwide to challenge and improve their BE model[4].

Despite a clear economic benefit the implementation and use of a Business Excellence Model its practical implementation is sometimes not as straightforward as it could be. The reasons behind are e.g. lack of top management commitment and limited resources[5].

---

[1] Zollondz, H.-D. (2011), p. 324
[2] Becker, M., Meyer, E. (2014)
[3] Friedli, T. (2014)
[4] Dahlgaard, J. J. et al. (2013), p. 2
[5] Dahlgaard, J. J. et al. (2013), p. 3

> "Business Excellence is a strategy that makes quality an integral part of the way business is designed
> Goes beyond the quality of products and services, and takes on a broader meaning of maximizing the effectiveness of the business in meeting or exceeding customer value expectations and using continuous improvement to drive business results
> It is the total quality of how the business operates as a system." [1]

In other words, if top management commitment and/or resources are lacking the introduction or execution of a business excellence approach does not work. Quality, in particular on a business excellence level, is the absolute responsibility of the top management.

## 12.4.2. Deming Prize

### 12.4.2.1. Introduction and History

In the 1950s the quality of Japanese products was not competitive which slowed down reconstruction and development of the Japanese economy after World War II. Together with W. Edwards Deming the Union of Japanese Scientists and Engineers (JUSE) developed a model for the quality oriented management of a company, known as Total Quality Control (TQC)[2]. From there the Japanese national quality award, the Deming Prize, was initiated in 1951.

Initially the Deming prize was awarded to Japanese companies only. In the mean time the Deming Prize is also open to non-japanese company and does have several categories:

- Deming Prize (for companies, institutes, organizations)
- Deming Grand Prize (companies having received the prize earlier with additional significant improvement)
- Deming Prize for individuals (single persons which contributed significantly to the idea of Total Quality Management)

---

[1] Hodlin, S. (2008)
[2] Malorny, C., Dicenta, M. (2014), p. 364

> Deming Distinguished Service Award for Dissemination and Promotion (overseas) (for single persons outside Japan which contributed significantly to disseminating Total Quality Management)
> Nikkei QC Literature Prize (for literature contributing to advancement of Total Quality Management with focus on Japan)

### 12.4.2.2. Assessment Criteria of the Deming Prize

| Contents of the assessment elements | max. number of points |
|---|---|
| **Company policy and its application in quality management** | 20 |
| - clear company policy<br>- guiding quality and customer oriented company objectives<br>- quality and customer oriented company strategy | (10) |
| - company wide implementation of the company policy | (10) |
| **Product development / process innovation** | 20 |
| - new product development is quality oriented<br>- process innovation is quality oriented | (10) |
| - new products meet customer wishes<br>- process innovations improve efficiency | (10) |
| **Maintenance and improvement of product and process quality** | 20 |
| - daily business<br>- standardization, training and education leads to reduction of Failures and stabilization of standard procedures | (10) |
| - continuous improvement<br>- continuous and managed improvement of company activities<br>- this leads to reduction of complaints and defects<br>- this increases customer satisfaction | (10) |
| **Implementation of management systems (quality, quantity, delivery, cost, safety, environment) and their effective execution** | 10 |
| **Collection and analysis of qualitative information and use of Information Technology** | 15 |
| - company collects information<br>- applies statistical methods<br>- and uses this data for development of products and Processes | (15) |
| **Employee development** | 15 |

| | |
|---|---|
| - systematic training and education of employees<br>- this enables the employees to actively support product and process improvement | (15) |

Table 12.2 Deming price assessment criteria and scoring[1]

One of the earliest companies to be awarded the Deming Price is Toyota. The success of the company and the uniqueness of the Toyota Production System (TPS) and the Toyota company culture (the "Toyota Kata") is "living proof" of the effectiveness of TQM.

### 12.4.3. EFQM & EFQM Excellence Award[2,3]

#### 12.4.3.1. Introduction and History

**EFQM**

**European Foundation for Quality Management**

The EFQM was founded in 1989 when 67 heads of companies (CEOs, Presidents) committed themselves to the goals of EFQM[4]. In 1990 the EFQM Excellence Model was developed to measure how effective a company developed and implemented its strategy. Since then the model went through several reviews and changes and consists of three components:

- Eight basic concepts of excellence
- Nine criteria for excellence grouped in nine criteria (fig. 12.7) with 32 subcriteria
- The assessment and scoring method named RADAR

Different kinds of organizations can participate in the EFQM excellence award like large and small companies, public authorities, educational institutions and even divisions of companies.

Meanwhile hundreds of organizations have participated in the EFQM award activities.

#### 12.4.3.2. Eight basic concepts of excellence[5]

The purpose of excellence is to achieve long-term success, i.e. excellent results, for all stakeholders of an organization. Stakeholders are e.g. shareholders, employees, customers the general public and suppliers.

---

[1] Malorny, C., Dicenta, M. (2014), p. 364
[2] EFQM (2013)
[3] DQG (2013)
[4] EFQM (2016)
[5] Uygur, A., Sümerli, S. (2013)

Within the EFQM model eight concepts are defined as the basis of the EFQM excellence model:

➢ Actualizing Balanced Results

Both short-term and long-term requirements of all shareholders have to be met in a planned way. The organization's mission and vision have to be aligned accordingly.

➢ Adding Value for Customers

Customer requirements and expectations have to be understood and estimated in a way that there is a value add on the customer side.

➢ Visionary, Inspiring and total Leadership

It is the management's responsibility to shape the future and to define the direction, the organization is heading to. This includes a clear target setting on the one hand and an open atmosphere for the employees to reach an excellent level.

➢ Management via Processes

Decisions are made based on facts and data in a systematic and comprehensible way. To ensure this the management of organization has to be based on processes. The overall purpose, of course, is to ensure long-term economic success.

➢ Achieving with Employees

Individual targets, common company values & culture and employee appreciation increase the motivation of employees.

➢ Creativity and Maintaining Innovation

Innovation is based on the creativity of the employees and potentially other stakeholders. A culture of permanent improvement, learning and innovation is then the basis for sustained excellent company performance.

➢ Creating Cooperation

A reliable relationship with other organizations, i.e. customers, suppliers, society, is fundamental for a consistent cooperation with them.

➢ Taking Responsibility for a Sustainable Future

In the long run an organization is considered successful when the organization's ethics and values meet and exceed expectations and regulations.

### 12.4.3.3. Nine Elements[1]

The nine elements of the EFQM model, named "criteria", are subdivided in five enabling criteria with 24 subcriteria and four result relevant criteria with 8 subcriteria. The criteria are weighted based on their importance for excellence and for the determination of the overall score assessment later on.

It should be noted that the red lines which connect the boxes in figure 12.7 represent logical connections or cause and effect relationship, e.g. the leadership and employees develop and influence the strategy, and that the strategy has an impact on the processes etc. Also, there is a closed feedback loop from the results criteria back to the 5 enabler criteria (identify need for improvements based on results), and the loop is closed by the way how the changes in the 5 enabler criteria have a hopefully positive impact on the results. It is also noteworthy that the time scale for the criterion processes is days to weeks, months (i.e. it covers operations), whereas the relevant time scale for the other 4 enabler criteria is typically months to years, i.e. medium and long term and mostly strategic.

In other words, the organization is viewed as a system which is represented by the 9 criteria and the connections between them. The ultimate purpose of the system, very much like for living beings is stability, survival, well-being and growth.

Both enablers and results yield a maximum score of 500 points resulting in an overall maximum achievable 1000 points. The following listing provides an overview of the criteria with their respective weighting and a potential questionnaire or basis of information, respectively, for the execution of an assessment.

<u>Enabling Criteria</u>

➤ Leadership (10%)

The company top management is supposed to shape the company's future and keep company values and ethics up to enable success. For this purpose the company's vision, mission and values have to be defined by the management.

Questions, e.g.

- o what are the organization's mission and vision like today and in the coming years; are those known and accepted within the company
- o what are the organization's values and in what way are they implemented, e.g. regarding people, customers, suppliers or partners
- o how does the organization identify the need for changes and in what way does the organization deal with changes (e.g. people information)

---

[1] Herrmann, J., Fritz, H. (2011), p. 258

- what methods are used to identify potentials for improvement of the management system with regard to excellence
- how does the management contribute to increasing the motivation of employees and to support continuous learning

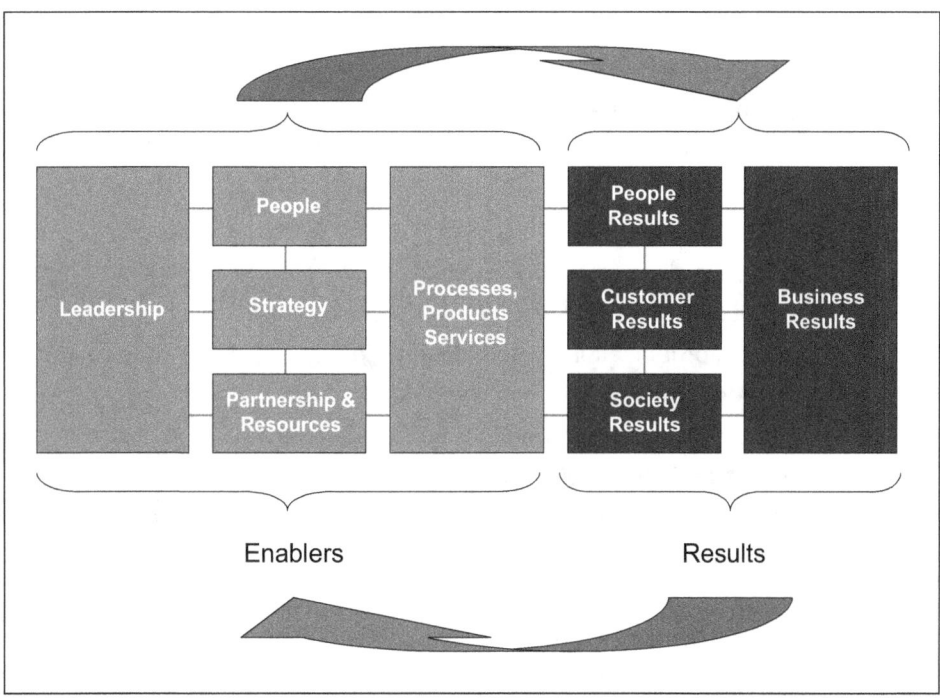

Fig. 12.7 EFQM model criteria[1]

➤ Strategy (10%)

Vision and mission lead to a stakeholder oriented company strategy. Objectives, processes and plans are derived from there to enable the strategy.

Questions, e.g.

- what methods does the organization use to identify market needs and trends as well as competitor's activities
- what way does the organization determine its own performance (e.g. key performance indicators, weaknesses & strengths) and are those benchmarked with external organizations

---

[1] EFQM (2016)

- o what is the organization's strategy, how is it adapted proactively to new market developments and how is it implemented using company objectives, target setting, processes and how is it communicated to the employees

➢ People (10%)

An organization's excellence is expressed by their appreciation for their people and is accompanied by an open, equal and fair culture. People are motivated by open communication and an adequate reward system.

Questions, e.g.

- o what way does the organization plan staffing (number, competences, motivation etc.) to support the strategy
- o does the organization systematically identify and close gaps regarding staff (number, competences, motivation etc.)
- o does the organization encourage independent ways of work and enable open and effective communication
- o what kind of award/assessment system is in place taking care of the employees

➢ Partnerships & Resources (10%)

External partnerships, e.g. with suppliers, have to be actively managed in order to comply with the strategy. This makes the organization effective in operations as well as in respect to social aspects.

Questions, e.g.

- o does the organization actively manage partners and suppliers; in what way does this include common activities to improve the performance along the supply chain
- o does the organization actively manage all financial affairs; is there a financial strategy, planning and controlling supporting the organization's strategy
- o what way does the organization manage handling of buildings, materials, inventory etc.
- o in what way does the organization manage and use technology, know how and information to support the organization's strategy and performance

➢ Processes, Products & Services (10%)

Processes, products and services are subject to planning and management. They have to be improved continuously for the benefit of the stakeholders, in particular of the customers.

Questions, e.g.

- what level of process management is implemented within the organization regarding e.g. definition of processes and key processes, process performance indicators, process owners, process training etc. with regard to the stakeholder's benefit (note the similarity to the process maturity models)
- what methods are used within the organization to identify product and service requirements
- what methods are used within the organization to develop, produce, market and distribute products and services in order to achieve an optimum for the customer
- does the organization actively manage the relationship to the customer (e.g. regular communication, integration in the development process, complaint management)

Results Criteria

➢ Customer Results (15%)

The level of success is measured using key performance indicators and against set targets for key results. Those have to be aligned with the organization's strategy and exceed expectations. Targets and indicators have to be based on the customer's needs, expectations and requirements.

The performance indicators are classified in two categories: subjective and objective indicators. Subjective indicators represent the perception of the customers, objective indicators are based on facts. This is also true for the next two criteria, people and society.

Basis of information, e.g.

- customer surveys
- customer's supplier evaluation
- number of complaints
- supplier awards
- market share
- percent of new customers

The first two are subjective indicators, the last four objective indicators. Subjective and objective indicators are regarded as equally important.

➢ People Results (10%)

An organization's excellence is expressed by achievement or over achievement of results regarding their people's expectations.

Basis of information, e.g.

- employee attitude or satisfaction survey
- number of complaints
- employee turnover rate
- number of staff away sick
- 360° feedback
- number of accidents at work
- programs to improve employee's state of health

➢ Society Results (10%)

An organization's excellence is expressed by achievement or overachievement of results regarding society's expectations.

Basis of information, e.g.

- surveys
- newspaper articles
- public events
- sponsorship (e.g. youth work, sports)
- achievements with respect to environmental or energy management
- positive record of compliance with legal requirements

➢ Business Results (15%)

An organization's excellence is expressed by achievement or overachievement of results regarding the expectations of the business stakeholders. In this criterion the two subcriteria are the financial and the operational performance.

Basis of information, e.g.

- revenue

- profit, EBIT (earnings before interest and taxes)
- cash flow
- market share
- patent portfolio
- cycle time
- improvement of productivity
- time to market for new products
- share of products not older than three years
- rate of decrease of the percentage of defective products

### 12.4.3.4. RADAR[1,2,3]

Within EFQM the PDCA circle (Plan – Do – Check – Act) developed by Deming is transferred to the RADAR control circle (fig. 12.8).

The RADAR assessment system not only looks at the actual results but also includes the way those results were achieved. For the enabling criteria this means that each of the five criteria is subdivided into subcriteria which then are assessed regarding approach, deployment and assessment & refinement and the respective attributes using a percentage number.

> **RADAR**
> Results / Approach / Deployment /
> Assessment / Refinement

Each of these criteria & subcriteria are then assessed from 0% for "no evidence" up to 100% for "comprehensive evidence".

On the other hand, the EFQM models also includes actual results which have to be assessed, too, using a similar approach. Four result criteria are subdivided in a total of eight subcriteria which is then assessed regarding relevance & usability and actual performance and the respective attributes.

---

[1] DGQ (2013)
[2] Uygur, A., Sümerli, S. (2013)
[3] Herrmann, J., Fritz, H. (2011), p. 269

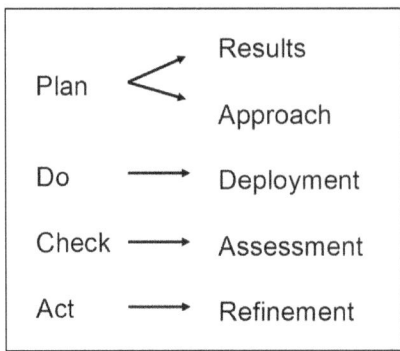

Fig. 12.8 PDCA cycle transformed to the RADAR cycle

|  | RADAR attributes | Explanation |
|---|---|---|
| Approach | Sound | Clearly motivated, using defined processes and based on stakeholders' requirements |
|  | Integrated | Supports the organization's strategy and is in line with other relevant activities |
| Deployment | Implemented | Percentage of areas with approach implemented |
|  | Systematic | Systematic planning and introduction of the measure |
| Assessment & Refinement | Measurement | Deployment, effectiveness and efficiency are monitored on a regular base |
|  | Learning & Creativity | Identification of improvement potentials (e.g. by internal or external best practices) |
|  | Innovation & Improvement | Outputs from other activities (e.g. measurement, learning, creativity) are used to realize improvements and innovations. This includes planning and implementation. |

Table 12.3 Criteria for the RADAR assessment of enablers[1]

---

[1] DGQ (2013)

|  | RADAR attributes | Explanation |
|---|---|---|
| Relevance & usability | Scope and relevance | Relevant and key results are determined based on stakeholder's needs and in line with the organization's strategy and targets. |
|  | Integrity | Those results are significant and accurate and are determined in a timely way. |
|  | Segmentation | An adequate level of segmentation of the results leads to relevant information |
| Performance | Trends | There are positive trends over a significantly large period of time (at least three years). |
|  | Targets | Key results require adequate dedicated targets which are continuously achieved. They are in line with the organization's strategy. |
|  | Comparison | Organization's key results are compared with external organization's achievements (e.g. benchmark) and are competitive relative to the benchmark values. |
|  | Causes | The correlation between enablers and results is well understood. From there it is a valid assumption that the organization's performance continues to be high. |

Table 12.4 Criteria for the RADAR assessment of results[1]

---

[1] DGQ (2013)

### 12.4.3.5. Levels of Excellence

Various levels of excellence are identified in the EFQM model. The purpose of having different levels of excellence is to drive organization stepwise towards higher levels. The following fig. 12.9 provides an overview.

The EFQM model covers the categories large companies, small and medium enterprises and public sector.

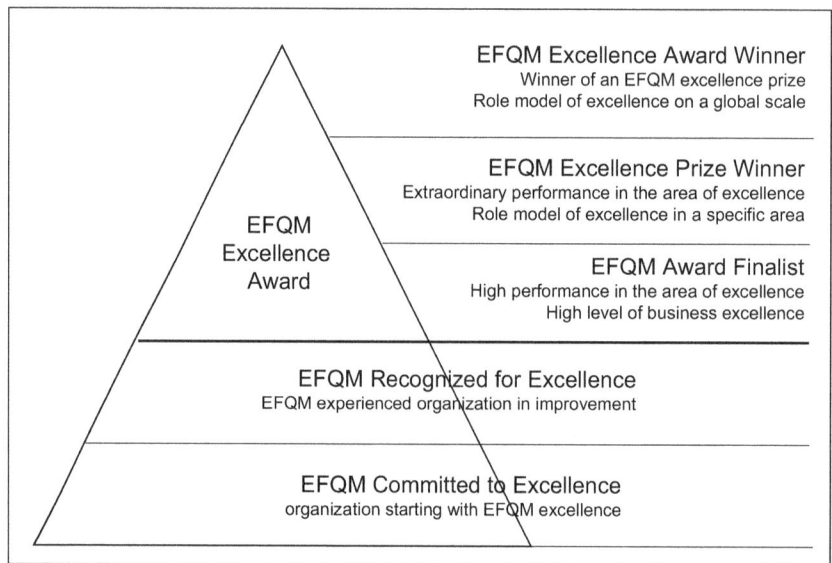

Fig. 12.9 Levels of excellence in the EFQM model[1]

## 12.4.4. Malcolm Baldrige National Award[2]
### 12.4.4.1. Introduction and History[3]

With the implementation of the Malcolm Baldrige National Award in 1987 the US government launched a major quality improvement initiative in order to improve the quality of American products and services as a reaction towards the loss of market share to Japanese competitors due the higher quality of Japanese products. This was especially true for electronics and for motor cars.

Since then the US president in person grants this award to organizations which demonstrate an outstanding performance in the field of quality and productivity improvement based on a TQM model implemented at the organization.

The model consists of three levels with an increasing level of detail starting with:

---

[1] Malorny, C., Dicenta, M. (2007)
[2] NIST (2016)
[3] Malorny, C., Dicenta, M. (2007)

- Leadership
- Strategic Planning
- Customer and Market Focus
- Measurement, Analysis and Knowledge Management
- Human Resources Focus
- Process Management
- Results

There is obviously a strong similarity to the EFQM model. The maximum number of achievable points is 1000, which equals the EFQM system. The MBNQA covers organizations in the following sectors: Education, Health Care, Manufacturing, Nonprofit/Government, Service and Small Business.

### 12.4.4.2. Categories and Items

The Baldrige Excellence Framework is summarized using a systems perspective. It is not surprising, that this framework has certain similarities with the systems representation of an enterprise in the EFQM model. What is missing in the Baldrige framework is the explicit mention of society and partners/suppliers in this top level model visualization.

Each category of the Baldrige Performance Excellence Framework is characterized in more detail by a total of 17 items. The similarity between the models is also apparent in the details of the model.

The following information is aligned to the official Baldrige Award information. To improve clarity, the headlines for the individual items are verbatim from the document[1], without explicit quotation marks. The explanations are our own explanations, i.e. paraphrased from the document.

- Organizational Profile
    - Organizational Description
        - Organizational Environment
        
        Asks for: organization's main product offerings / vision, mission, values / core competencies / organizational culture / workforce profile (age, know how, education) / major assets / regulations (health, safety, financial...) ...
        - Organizational Relationship
        
        Asks for: structure of organization / relationship to stakeholders (customers, suppliers, partners) / main market segments and customers / product requirements ...

---

[1] NIST (2016)

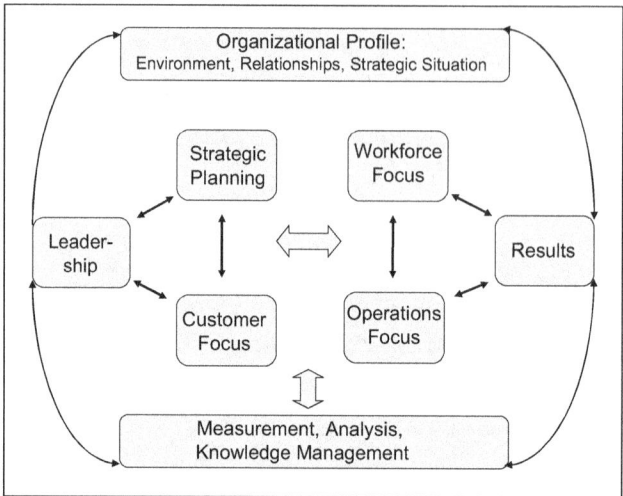

Fig. 12.10 Baldrige Performance Excellence Framework[1]

- Organizational Situation
    - Competitive Environment
    
    Asks for: market position / organizations size and growth trends / competition / key changes / benchmark data…
    - Strategic Context
    
    Asks for: challenges and opportunities for the organization's strategy
    - Performance Improvement System
    
    Asks for: key elements of the performance improvement system
- Leadership
    - Senior Leadership
        - Vision, Values and Mission
        
        Asks for: way of determination and execution of vision/mission / commitment to ethical values / measures to achieve a sustainable organization…
        - Communication and Organizational Performance
        
        Asks for: the way management communicates with staff / the way the organization focuses on key actions to achieve the objectives…
    - Governance and Societal Responsibilities
        - Organizational Governance
        
        Asks for: the set up of the governance system (e.g. fiscal accountability, transparency, independence) / the way senior management's performance is evaluated…

---

[1] NIST (2016)

- Legal and Ethical Behaviour

  Asks for: the way legal compliance is established and respective processes / the way ethical behaviour is supported and executed …

- Societal Responsibilities and Support of Key Communities

  Asks for: how the organization focuses on societal well-being / identification of, involvement in and contributing to communities …

> Strategic Planning
  - Strategy Development
    - Strategy Development Process

      Asks for: the process and execution of strategic planning short and long term / identification of challenges, strengths, weaknesses / reaction to changes (e.g. market, technology, product..) …

    - Strategic Objectives

      Asks for: strategy related key objectives / the way those key objectives are determined

  - Strategic Implementation
    - Action Plan Development and Deployment

      Asks for: development and implementation of an action plan including financial and human resources and performance monitoring

    - Performance Projections

      Asks for: the expected short and long term performance of the above actions

> Customer Focus
  - Voice of the Customer
    - Customer Listening

      Asks for: what way the organization listens to customers (former, actual, other competitor's customers)

    - Determination of Customer Satisfaction and Engagement

      Asks for: what way the organization identifies the level of customer satisfaction in comparison with competitive organization

  - Customer Engagement
    - Product Offerings and Customer Support

      Asks for: the way the organization determines requirements (market, customers) / how does the organization set up and execute customer support activities / how does the organization identify new customers and market segments …

    - Building Customer Relationships

      Asks for: how the organization organizes its customer relationship management / how it deals with customer complaints

- Measurement, Analysis and Knowledge Management
  - Measurement, Analysis and Improvement of Organizational Performance
    - Performance Measurement
    
    Asks for: organizational key performance indicators and measures (short and long term) / collection of relevant data including customer data …
    - Performance Analysis and Review
    
    Asks for: the way the organization reviews its performance …
    - Performance Improvement
    
    Asks for: use of best practice and lessons learned within the organization / how does the organization project future performance / how does the organization execute continuous improvement …
  - Management of Information, Knowledge and Information Technology
    - Data, Information and Knowledge Management
    
    Asks for: handling of organizational data and information / how does the organization manage knowledge …
    - Management of Information Resources and Technology
    
    Asks for: usability of hard- and software / emergency planning and actions …
- Workforce Focus
  - Workforce Environment
    - Workforce Capability and Capacity
    
    Asks for: methods to assess the workforce (capabilities, skills …) / the organization's recruitment and retainment process / what way does the organization maximize the benefit regarding workforce (e.g. performance, work accomplishment..) / how does the organization prepare the workforce for changes …
    - Workforce Climate
    
    Asks for: how does the organization take care of the workforce (e.g. health and safety) / what guidelines and policies are in place …
  - Workforce Engagement
    - Workforce Performance
    
    Asks for: how the organization identifies key factors for workforce engagement and satisfaction / handling of and organizational culture / how does the organization manage workforce performance (e.g. recognition, incentive) …
    - Assessment of Workforce Engagement

Asks for: what way does the organization assess engagement and satisfaction of the workforce / how are business results correlated to workforce engagement …
- Workforce and Leader Development

Asks for: how the learning and development system is set up and evaluated / succession and career planning of the workforce …

- Operations Focus
  - Work Systems
    - Work System Design

    Asks for: work system design and capitalization of core competences / requirements regarding the work system …
    - Work System Management

    Asks for: the way the organization implements, manages and improves the work system …
    - Emergency Readiness

    Asks for: emergency planning for the work system
  - Work Process
    - Work Process Design

    Asks for: what way does the organization design and innovate processes / in what way this is based on requirements or changes in technology or organization etc. / are key requirements and key processes identified …
    - Work Process Management

    Asks for: alignment of key processes with key process requirements / what key performance indicators are in place for those processes / what way does the organization manage the supply chain including the supplier base / what activities are in place regarding process improvement ….

- Results
  - Product and Process Outcomes
    - Customer-Focused Product and Process Results
    - Operational Process Effectiveness Results
    - Strategy Implementation Results
  - Customer-Focused Outcomes
    - Customer-Focused Results
  - Workforce-Focused Outcomes
    - Workforce Results
  - Leadership and Governance Outcomes
    - Leadership, Governance and Societal Responsibility Results

- Financial and Market Outcomes
    - Financial and Market Results

### 12.4.4.3. Scoring[1]

Each item then has to be scored using a two dimensional approach. Dimension 1, "process", addresses category 1 to 6 (see table 12.5), dimension 2, "results" ,covers category 7.

The process dimension considers four aspects for the evaluation: approach, deployment, learning and integration (ADLI), whereas the results dimension evaluation is done looking at level, trend, comparison and integration (LeTCI). Again, a certain similarity to the RADAR method of the EFQM model is visible.

In addition to those prescriptions for the evaluation the model also provides guidelines for the actual scoring as a percentage number. In the process dimension, for instance, "no systematic approach" or "no deployment of any systematic approach" results in a score of 0% or 5 %. The guideline has six levels with a separate description ending at 100% ("effective, systematic approach", "approach is fully deployed").

| Categories and items | Achievable points | |
|---|---|---|
| (1) Leadership | | 120 |
|     Senior Leadership | 70 | |
|     Governance and Social Responsibilities | 50 | |
| (2) Strategic Planning | | 85 |
|     Strategy Development | 40 | |
|     Strategy Implementation | 45 | |
| (3) Customer Focus | | 85 |
|     Voice of the Customer | 45 | |
|     Customer Engagement | 40 | |
| (4) Measurement, Analysis, Knowledge Mgmt. | | 90 |
|     Measurement, Analysis and Improvement of Organizational Performance | 45 | |
|     Management of Information, Knowledge and Information Technology | 45 | |
| (5) Workforce Focus | | 85 |
|     Workforce Environment | 40 | |
|     Workforce Engagement | 45 | |
| (6) Operations Focus | | 85 |

---

[1] NIST (2016)

| | | |
|---|---|---|
| Work System | 45 | |
| Work Progress | 40 | |
| (7) Results | | 450 |
| Product and Process Outcomes | 120 | |
| Customer-Focused Outcomes | 90 | |
| Workforce-Focused Outcomes | 80 | |
| Leadership and Governance Outcomes | 80 | |
| Financial and Market Outcomes | 80 | |

Table 12.5 Baldrige Scoring[1]

A separate guideline exists for the results dimension. The set up is the same as before with different descriptions for the scoring levels. A 0% or 5% scoring level is described as "no organizational scoring results" or "poor results in areas reported" up to 100% with "excellent organizational performance level" or "benchmark leadership". The maximum achievable number of points for each item and category is shown in table 12.5.

Final remark: All TQM models strive to cover all aspects relevant for the success of an enterprise. This naturally includes not only technical, logistic and financial aspects, but to a large degree human aspects. It is remarkable that the QM system standards ISO9001 and ISO/TS16949 almost entirely neglect these aspects. The standard ISO9004 which is NOT for certification, but can be regarded as a guideline, includes many of those human aspects, a strong similarity to the EFQM model is apparent.

## 12.5. Self Assessment

### 12.5.1. Introduction

Effectiveness and sustainability of a management system requires continuous monitoring of activities and results within an organization. This is true both for management systems on an ISO9001 level as well as on the Business Excellence level. At the same time the execution of Self Assessments increases awareness for gaps and achievements significantly.

> "Self-assessment is a comprehensive and systematic review of an organization's activities and results, referenced against a chosen standard."[2]

---

[1] NIST (2016)
[2] DIN EN ISO9004:2009

It is most important to note that the assessment does not only result in a score but also shows up the need for and the way how to improve.

In this context the purpose of a Self Assessment is to identify an organization's strengths and weaknesses on a regular basis in order to determine the status of the organization or of parts of the organization. Since the results of the Self Assessment are used for operational or strategic improvement of the organization it is considered an effective and systematic method of continuous improvement.

The way of execution of a Self Assessment is not prescribed in a specific way. Depending on the culture or experience in the organization various options are possible like self assessments with the help of questionnaires or self assessment workshops.

The structure of the assessment teams needs to be determined before the actual assessment starts and does offer various options, too, e.g.:

- Single persons from the area concerned
- A group of employees from the area concerned
- A group consisting of different levels of hierarchy from the area concerned
- A group consisting of different levels of hierarchy from the area concerned and participants from other areas in addition, preferably the "neighbours" in the internal value chain, i.e. internal customers, suppliers or partners
- One of the options above with an external moderator

### 12.5.2. Execution

The self assessment process typically consists of three phases, namely the preparation phase, the actual execution of the assessment and the improvement phase (fig. 12.11).

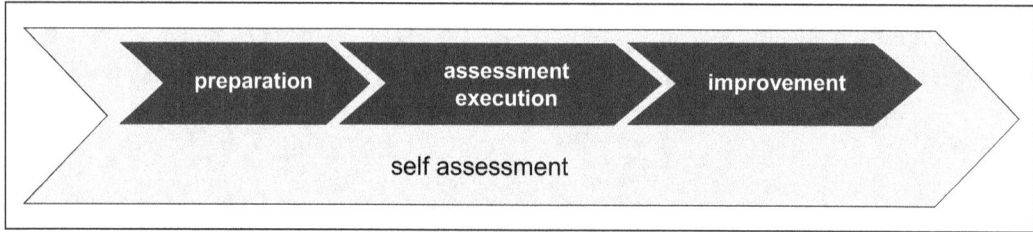

Fig. 12.11 Phases of the self assessment process

Preparation Phase
The preparation phase often starts with an official kick-off meeting, in particular if the self assessment is a periodic event, e.g. an annual or biannual cycle. The purpose of this phase, after discussion and analysis of the current situation, is to

- Determine the scope of the self assessment
- Determine the way of execution (see previous subsection)
- Appoint a self assessment team leader
- Name the team members
- Determine the time line from start to end
- Assess the status of the action items of the previous self assessment cycle

Assessment Execution

The actual execution may be based on a questionnaire or by performing a workshop both resulting in a list of topics or gaps. Weaknesses and strengths have to be determined and documented.

Then, depending on the scope and the focus area, areas of improvement are prioritized based on impact on quality, performance, customer etc. and corrective actions including timelines and responsible persons have to be defined.

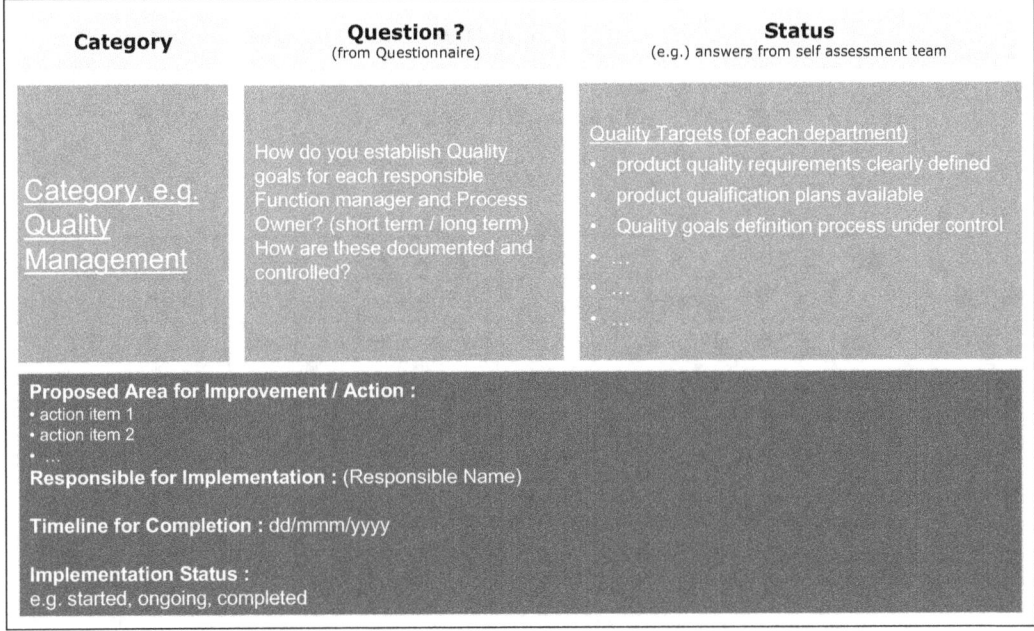

Fig. 12.12 Assessment sheet

Improvement

During the improvement phase corrective actions which were defined earlier are executed. The self assessment team leader or another assigned person is in charge of tracking the status of all actions. Typically, this person provides an update of the status to all participants and to management on a regular basis. Before closure of a corrective action its implementation needs to be confirmed by meaningful evidence.

It is useful to summarize questions, analysis results and status on a sheet of paper in order to have a quick outline of the facts (fig. 12.12).

### 12.5.3. Examples

#### 12.5.3.1. DIN EN ISO9004:2009[1]

In this section the way to execute a self assessment and to score the results is taken from DIN EN ISO9004:2009 (annex A) as an example. The ISO9004 model uses five levels of maturity describing the requirements for the particular levels for key and detailed elements. Table 12.6 shows the generic table.

The level specific requirements are separately defined for each element in ISO9004, annex A. For instance one element is asking for the way activities are organized within the organization. The levels are defined as follows:

Level 1 : Activities are organized in an unsystematic way. Operating procedures or the like are not available or only exist as rudimentary documents

Level 2 : A simple quality management system is implemented and organization of activities is based on organizational functions

| Element | Maturity Level | | | | |
| --- | --- | --- | --- | --- | --- |
| | Level 1 | Level 2 | Level 3 | Level 4 | Level 5 |
| Element 1 | description of level 1 | description of level 2 | description of level 3 | description of level 4 | description of level 5 |
| Element 2 | ... | ... | ... | ... | ... |
| Element 3 | ... | ... | ... | ... | ... |
| Element 4 | ... | ... | ... | ... | ... |

Table 12.6 Illustration of the ISO9004 maturity level evaluation

Level 3 : The quality management system is based on processes. Effectiveness and efficiency of the QM system leads to flexibility regarding organization of activities

---

[1] DIN EN ISO 9004:2009

Level 4 : In addition to the level 3 status interactions between processes are well working. The QM system stimulates activity and improvement, interested parties are identified and their needs are accounted for

Level 5 : In addition to the level 4 the QM system stimulates innovations as well as benchmarking activities. Interested parties include emerging parties and needs are complemented by expectations. All of them are accounted for.

**12.5.3.2. Standard Independent Approach**

In addition to following the guideline ISO9004 provides, of course it may be meaningful, too, to design a questionnaire and scoring system that is aligned with the company's product or specific requirements. Typically those questionnaires are also based on a standard (ISO9001, ISO/TS16949), but include or even have the focus on company specifics. Some examples are shown in fig. 12.13.

| No. | Criteria | comments | strengths | area of improvements | evidence |
|-----|----------|----------|-----------|----------------------|----------|
| ... | ... | ... | ... | ... | ... |
| ... | ... | ... | ... | ... | ... |
| | In what way do you determine and communicate product- and technology roadmaps? | | | | |
| | In what way do you determine and communicate product discontinuation plans? | | | | |
| | ... | | | | |
| | What method do you apply to assess the supplier's quality on a regular base? | | | | |
| | ... | | | | |
| | How do you achieve the target of having all suppliers at least ISO9001:2015 certified? | | | | |
| | ... | | | | |
| | How do you ensure that your subcontractors (e.g. silicon foundry, assembly, test) meet ISO/TS16949 requirements? | | | | |
| | ... | | | | |
| | ... | | | | |
| | What way do you ensure that only customer defined … | | | | |

Fig. 12.13 Example for self assessment questionnaire

Final Remark

In this section various ways and systems have been described to improve the quality of the processes, the product and the way the organization is managed. There can be no general recommendation which method or system to use, since this depends on the concrete circumstances, such as the size, maturity and field of business of an organization.

Clearly the TQM models are the most comprehensive way to develop an organization. In this part 1 of our book, the focus has been to describe various tools and quality improvement systems in a more general way, but always with a focus on HOW to implement quality activities. In part 2 the focus is on specific and typical quality topics in microelectronics and photovoltaics, with frequent hints how similar concrete situations could be handled in other industries.

# 13. Addendum: Statistical Tables

| df | 1-α | | | | | | |
|---|---|---|---|---|---|---|---|
| | 0,8 | 0,85 | 0,9 | 0,95 | 0,975 | 0,99 | 0,995 |
| 1 | 1,377 | 1,964 | 3,078 | 6,314 | 12,706 | 31,821 | 63,657 |
| 2 | 1,001 | 1,386 | 1,886 | 2,92 | 4,303 | 6,965 | 9,925 |
| 3 | 0,978 | 1,25 | 1,638 | 2,353 | 3,182 | 4,541 | 5,841 |
| 4 | 0,941 | 1,19 | 1,533 | 2,132 | 2,776 | 3,747 | 4,604 |
| 5 | 0,92 | 1,156 | 1,476 | 2,015 | 2,571 | 3,365 | 4,032 |
| 6 | 0,906 | 1,134 | 1,44 | 1,943 | 2,447 | 3,143 | 3,707 |
| 7 | 0,896 | 1,119 | 1,415 | 1,895 | 2,305 | 2,998 | 3,5 |
| 8 | 0,889 | 1,108 | 1,397 | 1,86 | 2,306 | 2,896 | 3,355 |
| 9 | 0,883 | 1,1 | 1,383 | 1,833 | 2,262 | 2,821 | 3,25 |
| 10 | 0,879 | 1,093 | 1,372 | 1,813 | 2,228 | 2,764 | 3,169 |
| 11 | 0,876 | 1,088 | 1,363 | 1,796 | 2,201 | 2,718 | 3,106 |
| 12 | 0,873 | 1,083 | 1,356 | 1,782 | 2,179 | 2,681 | 3,055 |
| 13 | 0,87 | 1,079 | 1,35 | 1,771 | 2,16 | 2,651 | 3,012 |
| 14 | 0,868 | 1,076 | 1,345 | 1,761 | 2,145 | 2,625 | 2,977 |
| 15 | 0,866 | 1,074 | 1,341 | 1,753 | 2,131 | 2,602 | 2,947 |
| 16 | 0,865 | 1,071 | 1,337 | 1,746 | 2,12 | 2,584 | 2,921 |
| 17 | 0,863 | 1,069 | 1,333 | 1,74 | 2,11 | 2,567 | 2,898 |
| 18 | 0,862 | 1,067 | 1,33 | 1,734 | 2,101 | 2,552 | 2,878 |
| 19 | 0,861 | 1,066 | 1,328 | 1,729 | 2,093 | 2,54 | 2,861 |
| 20 | 0,86 | 1,064 | 1,325 | 1,725 | 2,086 | 2,528 | 2,845 |
| 21 | 0,859 | 1,063 | 1,323 | 1,721 | 2,08 | 2,518 | 2,831 |
| 22 | 0,858 | 1,061 | 1,321 | 1,717 | 2,074 | 2,508 | 2,819 |
| 23 | 0,858 | 1,06 | 1,319 | 1,714 | 2,069 | 2,5 | 2,807 |
| 24 | 0,857 | 1,059 | 1,318 | 1,711 | 2,064 | 2,492 | 2,797 |
| 25 | 0,856 | 1,058 | 1,316 | 1,708 | 2,06 | 2,485 | 2,787 |
| 26 | 0,856 | 1,058 | 1,315 | 1,706 | 2,056 | 2,479 | 2,779 |
| 27 | 0,855 | 1,057 | 1,314 | 1,703 | 2,052 | 2,473 | 2,771 |
| 28 | 0,855 | 1,056 | 1,313 | 1,701 | 2,048 | 2,467 | 2,763 |
| 29 | 0,854 | 1,055 | 1,311 | 1,699 | 2,045 | 2,462 | 2,756 |
| 30 | 0,854 | 1,055 | 1,31 | 1,697 | 2,042 | 2,459 | 2,75 |
| 40 | 0,851 | 1,05 | 1,303 | 1,684 | 2,021 | 2,423 | 2,705 |
| 60 | 0,848 | 1,046 | 1,296 | 1,071 | 1,997 | 2,39 | 2,86 |
| 120 | 0,845 | 1,041 | 1,289 | 1,658 | 1,98 | 2,358 | 2,617 |
| n large | 0,843 | 1,039 | 1,282 | 1,645 | 1,96 | 2,326 | 2,576 |

Table 13.1. α-quantiles $t_{df,\alpha}$ of the t-distribution with df=n-1 degrees of freedom[1]

---

[1] Linß, G. (2005), p. 615

| G(u) [%] | .0 | .1 | .2 | .3 | .4 | .5 | .6 | .7 | .8 | .9 |
|---|---|---|---|---|---|---|---|---|---|---|
| 50 | 0,000 | 0,003 | 0,005 | 0,008 | 0,010 | 0,013 | 0,015 | 0,018 | 0,020 | 0,023 |
| 51 | 0,025 | 0,028 | 0,030 | 0,033 | 0,035 | 0,038 | 0,040 | 0,043 | 0,045 | 0,048 |
| 52 | 0,050 | 0,053 | 0,055 | 0,058 | 0,060 | 0,063 | 0,065 | 0,068 | 0,070 | 0,073 |
| 53 | 0,075 | 0,078 | 0,080 | 0,083 | 0,085 | 0,088 | 0,090 | 0,093 | 0,095 | 0,098 |
| 54 | 0,100 | 0,103 | 0,105 | 0,108 | 0,111 | 0,113 | 0,116 | 0,118 | 0,121 | 0,123 |
| 55 | 0,126 | 0,128 | 0,131 | 0,133 | 0,136 | 0,138 | 0,141 | 0,143 | 0,146 | 0,148 |
| 56 | 0,151 | 0,154 | 0,156 | 0,159 | 0,161 | 0,164 | 0,166 | 0,169 | 0,171 | 0,174 |
| 57 | 0,176 | 0,179 | 0,181 | 0,184 | 0,187 | 0,189 | 0,192 | 0,194 | 0,197 | 0,199 |
| 58 | 0,202 | 0,204 | 0,207 | 0,210 | 0,212 | 0,215 | 0,217 | 0,220 | 0,222 | 0,225 |
| 59 | 0,228 | 0,230 | 0,233 | 0,235 | 0,238 | 0,240 | 0,243 | 0,246 | 0,248 | 0,251 |
| 60 | 0,253 | 0,256 | 0,259 | 0,261 | 0,264 | 0,266 | 0,269 | 0,272 | 0,274 | 0,277 |
| 61 | 0,279 | 0,282 | 0,285 | 0,287 | 0,290 | 0,292 | 0,295 | 0,298 | 0,300 | 0,303 |
| 62 | 0,305 | 0,308 | 0,311 | 0,313 | 0,316 | 0,319 | 0,321 | 0,324 | 0,327 | 0,329 |
| 63 | 0,332 | 0,335 | 0,337 | 0,340 | 0,342 | 0,345 | 0,348 | 0,350 | 0,353 | 0,356 |
| 64 | 0,358 | 0,361 | 0,364 | 0,366 | 0,369 | 0,372 | 0,375 | 0,377 | 0,377 | 0,383 |
| 65 | 0,385 | 0,388 | 0,391 | 0,393 | 0,396 | 0,399 | 0,402 | 0,404 | 0,407 | 0,410 |
| 66 | 0,412 | 0,415 | 0,418 | 0,421 | 0,423 | 0,426 | 0,429 | 0,432 | 0,434 | 0,437 |
| 67 | 0,440 | 0,443 | 0,445 | 0,448 | 0,451 | 0,454 | 0,457 | 0,459 | 0,462 | 0,485 |
| 68 | 0,468 | 0,470 | 0,473 | 0,476 | 0,479 | 0,482 | 0,485 | 0,487 | 0,490 | 0,493 |
| 69 | 0,496 | 0,499 | 0,502 | 0,504 | 0,507 | 0,510 | 0,513 | 0,516 | 0,519 | 0,522 |
| 70 | 0,524 | 0,527 | 0,530 | 0,533 | 0,536 | 0,539 | 0,542 | 0,545 | 0,548 | 0,550 |
| 71 | 0,553 | 0,556 | 0,559 | 0,562 | 0,565 | 0,568 | 0,571 | 0,574 | 0,577 | 0,580 |
| 72 | 0,583 | 0,586 | 0,589 | 0,592 | 0,595 | 0,598 | 0,601 | 0,604 | 0,607 | 0,610 |
| 73 | 0,613 | 0,616 | 0,619 | 0,622 | 0,625 | 0,628 | 0,631 | 0,634 | 0,637 | 0,640 |
| 74 | 0,643 | 0,646 | 0,650 | 0,653 | 0,656 | 0,659 | 0,662 | 0,665 | 0,668 | 0,671 |
| 75 | 0,674 | 0,678 | 0,681 | 0,684 | 0,687 | 0,690 | 0,693 | 0,697 | 0,700 | 0,703 |
| 76 | 0,706 | 0,710 | 0,713 | 0,716 | 0,719 | 0,722 | 0,726 | 0,729 | 0,732 | 0,736 |
| 77 | 0,739 | 0,742 | 0,745 | 0,749 | 0,752 | 0,755 | 0,759 | 0,762 | 0,765 | 0,769 |
| 78 | 0,772 | 0,776 | 0,779 | 0,782 | 0,786 | 0,789 | 0,793 | 0,796 | 0,800 | 0,803 |
| 79 | 0,806 | 0,810 | 0,813 | 0,817 | 0,820 | 0,824 | 0,827 | 0,831 | 0,834 | 0,838 |
| 80 | 0,842 | 0,845 | 0,849 | 0,852 | 0,856 | 0,860 | 0,863 | 0,867 | 0,871 | 0,874 |
| 81 | 0,878 | 0,882 | 0,885 | 0,889 | 0,893 | 0,896 | 0,900 | 0,904 | 0,908 | 0,912 |
| 82 | 0,915 | 0,919 | 0,923 | 0,927 | 0,931 | 0,935 | 0,938 | 0,942 | 0,946 | 0,950 |
| 83 | 0,954 | 0,958 | 0,926 | 0,966 | 0,970 | 0,974 | 0,978 | 0,982 | 0,986 | 0,990 |
| 84 | 0,954 | 0,999 | 1,003 | 1,007 | 1,011 | 1,015 | 1,019 | 1,024 | 1,028 | 1,032 |
| 85 | 1,036 | 1,041 | 1,045 | 1,049 | 1,054 | 1,058 | 1,063 | 1,067 | 1,071 | 1,076 |
| 86 | 1,080 | 1,085 | 1,089 | 1,094 | 1,098 | 1,103 | 1,108 | 1,112 | 1,117 | 1,122 |
| 87 | 1,126 | 1,131 | 1,136 | 1,141 | 1,146 | 1,150 | 1,155 | 1,160 | 1,165 | 1,170 |
| 88 | 1,175 | 1,180 | 1,185 | 1,190 | 1,195 | 1,200 | 1,206 | 1,211 | 1,216 | 1,221 |
| 89 | 1,227 | 1,232 | 1,237 | 1,243 | 1,248 | 1,254 | 1,259 | 1,265 | 1,270 | 1,276 |
| 90 | 1,282 | 1,287 | 1,293 | 1,299 | 1,305 | 1,311 | 1,317 | 1,323 | 1,329 | 1,335 |
| 91 | 1,341 | 1,347 | 1,353 | 1,359 | 1,366 | 1,372 | 1,379 | 1,385 | 1,392 | 1,396 |
| 92 | 1,405 | 1,412 | 1,419 | 1,426 | 1,433 | 1,440 | 1,447 | 1,454 | 1,461 | 1,468 |
| 93 | 1,476 | 1,483 | 1,491 | 1,499 | 1,500 | 1,514 | 1,522 | 1,530 | 1,538 | 1,546 |
| 94 | 1,555 | 1,583 | 1,572 | 1,580 | 1,589 | 1,598 | 1,607 | 1,616 | 1,626 | 1,635 |
| 95 | 1,645 | 1,655 | 1,665 | 1,675 | 1,685 | 1,695 | 1,706 | 1,717 | 1,728 | 1,738 |
| 96 | 1,751 | 1,762 | 1,774 | 1,787 | 1,799 | 1,812 | 1,825 | 1,838 | 1,852 | 1,866 |
| 97 | 1,881 | 1,896 | 1,911 | 1,927 | 1,943 | 1,960 | 1,977 | 1,995 | 2,014 | 2,034 |
| 98 | 2,054 | 2,075 | 2,097 | 2,120 | 2,144 | 2,170 | 2,197 | 2,226 | 2,257 | 2,290 |
| 99 | 2,326 | 2,366 | 2,409 | 2,457 | 2,512 | 2,576 | 2,652 | 2,748 | 2,878 | 3,090 |

Table 13.2 Quantiles of the standardized normal distribution[1]

---

[1] DGQ 16-32 (1995), p.124

| n | $A_W$ | $A_E$ |
|---|---|---|
| 2 | 1,386 | 1,821 |
| 3 | 1,132 | 1,487 |
| 4 | 0,980 | 1,288 |
| 5 | 0,877 | 1,152 |
| 6 | 0,800 | 1,052 |
| 7 | 0,741 | 0,974 |
| 8 | 0,693 | 0,911 |
| 9 | 0,653 | 0,859 |
| 10 | 0,620 | 0,815 |
| 11 | 0,591 | 0,777 |
| 12 | 0,566 | 0,744 |
| 13 | 0,544 | 0,714 |
| 14 | 0,524 | 0,688 |
| 15 | 0,506 | 0,665 |
| 16 | 0,490 | 0,644 |
| 17 | 0,475 | 0,625 |
| 18 | 0,462 | 0,607 |
| 19 | 0,450 | 0,591 |
| 20 | 0,438 | 0,576 |
| 21 | 0,428 | 0,562 |
| 22 | 0,418 | 0,549 |
| 23 | 0,409 | 0,537 |
| 24 | 0,400 | 0,526 |
| 25 | 0,392 | 0,515 |

Table 13.3 Factors for the calculation of limits in the x bar chart[1,2]

| n | $A_2$ | $A_3$ | $B_3$ | $B_4$ | $D_3$ | $D_4$ | $E_2$ |
|---|---|---|---|---|---|---|---|
| 2 | 1,880 | 2,659 | - | 3,267 | - | 3,267 | 2,660 |
| 3 | 1,023 | 1,954 | - | 2,568 | - | 2,574 | 1,772 |
| 4 | 0,729 | 1,628 | - | 2,266 | - | 2,282 | 1,457 |
| 5 | 0,577 | 1,427 | - | 2,089 | - | 2,114 | 1,290 |
| 6 | 0,483 | 1,287 | 0,030 | 1,970 | - | 2,004 | 1,184 |
| 7 | 0,419 | 1,182 | 0,118 | 1,882 | 0,076 | 1,924 | 1,109 |
| 8 | 0,373 | 1,099 | 0,185 | 1,815 | 0,136 | 1,864 | 1,054 |
| 9 | 0,337 | 1,032 | 0,239 | 1,761 | 0,184 | 1,816 | 1,010 |
| 10 | 0,308 | 0,975 | 0,284 | 1,716 | 0,223 | 1,777 | 0,975 |
| 11 | 0,285 | 0,927 | 0,321 | 1,679 | 0,256 | 1,744 | - |
| 12 | 0,266 | 0,886 | 0,354 | 1,646 | 0,283 | 1,717 | - |
| 13 | 0,249 | 0,850 | 0,382 | 1,618 | 0,307 | 1,693 | - |
| 14 | 0,235 | 0,817 | 0,406 | 1,594 | 0,328 | 1,672 | - |
| 15 | 0,223 | 0,789 | 0,428 | 1,572 | 0,347 | 1,653 | - |
| 16 | 0,212 | 0,763 | 0,448 | 1,552 | 0,363 | 1,637 | - |
| 17 | 0,203 | 0,739 | 0,466 | 1,534 | 0,378 | 1,622 | - |
| 18 | 0,194 | 0,718 | 0,482 | 1,518 | 0,391 | 1,608 | - |
| 19 | 0,187 | 0,698 | 0,497 | 1,503 | 0,403 | 1,597 | - |
| 20 | 0,180 | 0,680 | 0,510 | 1,490 | 0,415 | 1,585 | - |
| 21 | 0,173 | 0,663 | 0,523 | 1,477 | 0,425 | 1,575 | - |
| 22 | 0,167 | 0,647 | 0,534 | 1,466 | 0,434 | 1,566 | - |
| 23 | 0,162 | 0,633 | 0,545 | 1,455 | 0,443 | 1,557 | - |
| 24 | 0,157 | 0,619 | 0,555 | 1,445 | 0,451 | 1,548 | - |
| 25 | 0,153 | 0,606 | 0,565 | 1,435 | 0,459 | 1,541 | - |

Table 13.4. Statistical constants for Control charts (international)[3]

---

[1] Dietrich, E., Schulze, A. (2014), p. 743
[2] Linß, D. (2014), p. 544
[3] Dietrich, E., Schulze, A. (2014), p. 748

| n | $B_{UCL}$ | $B_{UWL}$ | $B_M$ | $B_{LWL}$ | $B_{LCL}$ |
|---|---|---|---|---|---|
| 2 | 2,807 | 2,241 | 0,798 | 0,031 | 0,006 |
| 3 | 2,302 | 1,921 | 0,886 | 0,159 | 0,071 |
| 4 | 2,069 | 1,765 | 0,921 | 0,268 | 0,155 |
| 5 | 1,927 | 1,669 | 0,940 | 0,348 | 0,227 |
| 6 | 1,830 | 1,602 | 0,952 | 0,408 | 0,287 |
| 7 | 1,758 | 1,552 | 0,959 | 0,454 | 0,336 |
| 8 | 1,702 | 1,512 | 0,965 | 0,491 | 0,376 |
| 9 | 1,657 | 1,480 | 0,969 | 0,522 | 0,410 |
| 10 | 1,619 | 1,454 | 0,973 | 0,548 | 0,439 |
| 11 | 1,587 | 1,431 | 0,975 | 0,570 | 0,464 |
| 12 | 1,560 | 1,412 | 0,978 | 0,589 | 0,486 |
| 13 | 1,536 | 1,395 | 0,979 | 0,606 | 0,506 |
| 14 | 1,515 | 1,379 | 0,981 | 0,621 | 0,524 |
| 15 | 1,496 | 1,366 | 0,982 | 0,634 | 0,539 |
| 16 | 1,479 | 1,354 | 0,984 | 0,646 | 0,554 |
| 17 | 1,463 | 1,343 | 0,985 | 0,657 | 0,567 |
| 18 | 1,450 | 1,333 | 0,985 | 0,667 | 0,579 |
| 19 | 1,437 | 1,323 | 0,986 | 0,676 | 0,590 |
| 20 | 1,425 | 1,315 | 0,987 | 0,685 | 0,600 |

Table 13.5. Statistical constants for the calculation of warning and control limits on the s-chart[1]

| n | $D_{UCL}$ | $D_{UWL}$ | $D_M$ | $D_{LWL}$ | $D_{LCL}$ |
|---|---|---|---|---|---|
| 2 | 3,970 | 3,170 | 1,128 | 0,044 | 0,009 |
| 3 | 4,424 | 3,682 | 1,693 | 0,303 | 0,135 |
| 4 | 4,694 | 3,984 | 2,059 | 0,595 | 0,343 |
| 5 | 4,886 | 4,197 | 2,326 | 0,850 | 0,555 |
| 6 | 5,033 | 4,361 | 2,543 | 1,066 | 0,749 |
| 7 | 5,154 | 4,494 | 2,704 | 1,251 | 0,922 |
| 8 | 5,255 | 4,605 | 2,847 | 1,410 | 1,075 |
| 9 | 5,341 | 4,700 | 2,970 | 1,550 | 1,212 |
| 10 | 5,418 | 4,784 | 3,078 | 1,674 | 1,335 |
| 11 | 4,485 | 4,858 | 3,173 | 1,784 | 1,446 |
| 12 | 5,546 | 4,925 | 3,258 | 1,884 | 1,547 |
| 13 | 5,602 | 4,985 | 3,336 | 1,976 | 1,639 |
| 14 | 5,652 | 5,041 | 3,407 | 2,059 | 1,724 |
| 15 | 5,699 | 5,092 | 3,472 | 2,136 | 1,803 |
| 16 | 5,742 | 5,139 | 3,532 | 2,207 | 1,876 |
| 17 | 5,783 | 5,183 | 3,588 | 2,274 | 1,944 |
| 18 | 5,820 | 5,224 | 3,640 | 2,336 | 2,008 |
| 19 | 5,856 | 5,262 | 3,689 | 2,394 | 2,068 |
| 20 | 5,889 | 5,299 | 3,735 | 2,449 | 2,125 |

Table 13.6. Statistical constants for the calculation of warning and control limits on the R-chart[2]

---

[1] DGQ 16-32 (1995), p. 129

| n | $c_n$ | n | $c_n$ | n | $c_n$ | n | $c_n$ |
|---|---|---|---|---|---|---|---|
| 1 | 1,000 | 6 | 1,135 | 11 | 1,228 | 16 | 1,202 |
| 2 | 1,000 | 7 | 1,214 | 12 | 1,187 | 17 | 1,237 |
| 3 | 1,160 | 8 | 1,160 | 13 | 1,232 | 18 | 1,207 |
| 4 | 1,092 | 9 | 1,223 | 14 | 1,196 | 19 | 1,239 |
| 5 | 1,197 | 10 | 1,176 | 15 | 1,235 | 20 | 1,212 |

Table 13.7. Statistical constants for the calculation of warning and control limits on the median chart[1]

| n | $\tilde{A}_2$ | n | $\tilde{A}_2$ | n | $\tilde{A}_2$ | n | $\tilde{A}_2$ |
|---|---|---|---|---|---|---|---|
| 2 | 1,880 | 7 | 0,508 | 13 | 0,328 | 19 | 0,232 |
| 3 | 1,187 | 8 | 0,433 | 14 | 0,281 | 20 | 0,218 |
| 4 | 0,796 | 9 | 0,412 | 15 | 0,275 | 21 | 0,215 |
| 5 | 0,691 | 10 | 0,363 | 16 | 0,255 | 22 | 0,203 |
| 5 | 0,691 | 11 | 0,350 | 17 | 0,251 | 23 | 0,201 |
| 6 | 0,548 | 12 | 0,316 | 18 | 0,234 | 24 | 0,191 |

Table 13.8. Statistical constants for the calculation of warning and control limits on the median chart[2]

# 14. List of References

## 14.1. Books / Reports / Thesis / Lectures

Antonitsin, A., "Statistical methods in reliability testing", Master thesis, Department of Statistics and Actuarial Science, Faculty of Science, Simon Fraser University, 2009

Bergholz, W., Lectures on Microelectrnics, "006 Performance and Productivity", Jacobs University, Spring 2008

Bergholz, W., Wittmann, J., Lectures on Microelectronics, "006 Performance and Productivity", Jacobs University, 2011

Bosch Heft Bd. 9, "Maschinen- und Prozessfähigkeit", Qualitätssicherung in der Bosch-gruppe, Technische Statistik, 3. Auflage, 01.07.2004

Bosch, „Qualitätssicherungsleitlinie (QSL) (Qualitätsicherungsvereinbarung) Industrie / Betriebsmittel / Dienstleistungen (Stand 12/2011)

Boutellier, R., Biedermann, A., „Qualitätsgerechte Produktplanung", in Masing – Handbuch Qualitätsmanagement, Hrsg. Pfeifer, T. und Schmitt, R., Hanser Verlag München, 5. Auflage, 2007

Bronstein, I.N., Semendjajew K.A., „Taschenbuch der Mathematik", Thun, Frankfurt/Main, Verlag Harry Deutsch, 23. Aufl., 1987, ISBN 3-87144-492-8

Brunner, F., Wagner, K., „Qualitätsmanagement – Leitfaden für Studium und Praxis", Hanser Verlag, 5. Auflage, 2011, ISBN 978-3-446-42516-3

---

[1] DGQ 16-32 (1995), p. 128
[2] Dietrich, E., Schulze, A. (2014), p. 748

Bureau Veritas, „Produktionsteil-Freigabeverfahren (PPAP), Vierte Ausgabe, Originaltitel: Production Part Approval Process (PPAP), Deutsche Übersetzung aus dem Englischen, 31.03.2008

CMMI®, "Improving processes for acquiring better products and services", CMMI-ACQ, V1.3 CMMI Product Team, November 2010, TECHNICAL REPORT, CMU/SEI-2010-TR-032, ESC-TR-2010-032
Däubner, T., "Analysis and optimisation of the design control- and design transfer-processes of prefilled pen systems", Master Thesis, Beuth University of Applied Sciences, Berlin, Institute for Distant Learning, 2013

Dietrich E., Schulze, A., "Statistische Verfahren zur Maschinen- und Prozessqualifikation", 7. Auflage, Carl Hanser Verlag, 2014, ISBN 978-3-446-44055-5

DGQ, "SPC – Qualitätskartentechnik", DGQ Band 16-32, Deutsche Gesellschaft für Qualität e.V., Beuth Verlag, 1995, ISBN 3-410-32890-4

Ensthaler, J., „Qualitätsmanagement und Recht", in Masing – Handbuch Qualitätsmanagement, Hrsg. Pfeifer, T. und Schmitt, R., Hanser Verlag München, 5. Auflage, 2007

Evan, K., "ON Semiconductor Quality & Reliability Handbook", HDB 851/D, Rev 8, May 2014

Ford, Measurement System Analysis, MSA, Reference Manual 4$^{th}$ Edition, June 2010, Chrysler Group LLC, Ford Motor Company, General Motors Corporation ISBN#: 978-1-60-534211-5

Friedli, T., Seghezzi, H. D., Mänder, C., Lützner, R., "Konzepte – Modelle – Systeme", in Masing – Handbuch Qualitätsmanagement, Hrsg. Pfeifer, T. und Schmitt, R., Hanser Verlag München, 6. Auflage, 2014

Gellert, W., Küstner, H., Hellwich, M., Kästner, H., "Kleine Enzyklopädie Mathematik", 9. Auflage, VEB Bibliographisches Institut Leipzig, 1974

Gillen, C., "The Efficient Safety Concept of the SpeedE Steer-By-Wire System", 2014

Groeseneken, G., Maes, H. E., Van Houdt, J., Witters, J. S., "Basics of nonvolatile semiconductor memory devices. Nonvolatile semiconductor memory technology", 1-88, 1998

Hammer, M., "The process audit", Harvard Business Review, HARVARD BUSINESS SCHOOL PUBLISHING CORPORATION, 2007

Hartmann, H., "Lieferantenmanagement Gestaltungsfelder, Methoden, Instrumente mit Beispielen aus der Praxis", Praxisreihe Einkauf Materialwirtschaft Band 11, Deutscher Betriebswirte-Verlag GmbH, 2004

Hedderich, J., Sachs, L., "Angewandte Statistik", 14. Auflage, Springer-Verlag, Berlin, Heidelberg, 2012

Herrmann, J., Fritz, H., Qualitätsmanagement Lehrbuch für Studium und Praxis", Carl Hanser Verlag München, 2011

Hofmann, D. „Praktische Ausbildung und Training Optische Koordinatenmesstechnik Absicherung der Qualitätsfähigkeit durch die Ermittlung der Messmittelfähigkeit von optischen Koordinatenmessgeräten am Beispiel eines zylindrischen Arbeitsnormals", University of Applied Scienced, Jena, 2009, Master – Studiengang Laser- & Optotechnologien Praktikum Optische Koordinatenmesstechnik

Juran, J. M., "JURAN'S QUALITY HANDBOOK", 5$^{th}$ Edition, McGraw-Hill, 1999

Kappes, S., „Analyse- und Optimierungsverfahren zur Verbesserung der Vorhersagequalität von multidimensionalen Regressionsanalysen", Bachelor Thesis, Beuth University of Applied Sciences, 2014

Köhler, M.. Frank, D., Schmitt, R., "Six Sigma", in Masing – Handbuch Qualitätsmanagement, Hrsg. Pfeifer, T. und Schmitt, R., Hanser Verlag München, 6. Auflage, 2014

Korkusuz, D., "Process Capability Analysis for non-normal processes with lower specification limits", Master Thesis, Chalmers University of Technology, Gothenburg, Sweden, 2011, Report No. E 2011:076

Krampf, Peter, „Beschaffungsmanagement – Eine praxisorientierte Einführung in Einkauf und Materialwirtschaft", Verlag Franz Vahlen, München, 2012

Kroonder, M., „Qualitätssicherungsvereinbarungen", in Masing – Handbuch Qualitätsmanagement, Hrsg. Pfeifer, T. und Schmitt, R., Hanser Verlag München, 5. Auflage, 2007

Krueger, D. C., "Semiconductor Yield Modeling Using Generalized Linear Models", Dissertation, Arizona State University, May 2011

Linß, G., "Qualitätsmanagement für Ingenieure", Fachbuchverlag Leipzig im Carl Hanser Verlag, 2. Auflage, 2005

Linß, G., "Prüfplanung", in Masing – Handbuch Qualitätsmanagement, Hrsg. Pfeifer, T. und Schmitt, R., Hanser Verlag München, 6. Auflage, 2014

Liu, B., „Product Development Processes and Their Importance to Organizational Capabilities", Master Thesis, Massachusetts Institute of Technology, 2003

Malorny, C., Dicenta, M., „Funktion und Nutzen von Qualitätsauszeichnungen (Awards)", in Masing – Handbuch Qualitätsmanagement, Hrsg. Pfeifer, T. und Schmitt, R., Hanser Verlag München, 5. Auflage, 2007

Malorny, C., Dicenta, M., „Funktion und Nutzen von Qualitätsaufzeichnungen (Awards)", in Masing – Handbuch Qualitätsmanagement, Hrsg. Pfeifer, T. und Schmitt, R., Hanser Verlag München, 6. Auflage, 2014

Masing, W., „Statistik als Basis qualitätsmethodischen Denkens und Handelns", in Masing – Handbuch Qualitätsmanagement, Hrsg. Pfeifer, T. und Schmitt, R., Hanser Verlag München, 6. Auflage, 2014

Melzer-Ridinger, R., „Materialwirtschaft und Einkauf, Band 2 Qualitätsmanagement", R. Oldenbourg Verlag, 1995

Meyna, A., Pauli, B., „Zuverlässigkeitstechnik Quantitative Bewertungsverfahren"
2., überarbeitete und erweiterte Auflage, Carl Hanser Verlag München, 2010

Pande, P., Neuman, R., Cavanagh, R., "The Six Sigma way team field book: an implementation guide for process improvement teams", New Delhi: Tata McGraw-Hill, 2003

Puffer, W., „Technisch-ökonomische Effizienzbetrachtungen für die Halbleiterfertigung", Dissertation Technische Universität München, Lehrstuhl für Technische Elektronik, Fachgebiet Halbleiterproduktionstechnik, 2007

Prefi, T., „Qualitätsmanagement in der Produktentwicklung", in Masing – Handbuch Qualitätsmanagement, Hrsg. Pfeifer, T. und Schmitt, R., Hanser Verlag München, 5. Auflage, 2007

PROSAFE, Product Safety Enforcement Forum of Europe, „Guideline for Businesses to manage Product Recalls & Other Corrective Actions", November 2011

Razak, N., "Correlation between DFMEA and Process Capability at automotive part supplier", Bachelor Thesis, Faculty of Manufacturing Engineering, University Teknikal Malaysia Melaka (UTeM), 2011

Schaefer, U., "Mikroelektronik – Trendanalyse bis 2018 Vorstellung langfristiger Trends 2008 – 2013 – 2018", ZVEI Zentralverband Elektrotechnik- und Elektronikindustrie e.V., Fachverband Electronic Components and Systems, April 2014

Schaller, R. R., "TECHNOLOGICAL INNOVATION IN THE SEMICONDUCTOR INDUSTRY: A CASE STUDY OF THE INTERNATIONAL TECHNOLOGY ROADMAP FOR SEMICONDUCTORS (ITRS)", Dissertation submitted to the Graduate Faculty of George Mason University, Fairfax, VA, 2004

Schaper, M., "Qualitätssicherung – Skript zur Vorlesung", Technische Universität Dresden, Institut für Werkstoffwissenschaft, Fakultät für Maschinenwesen, 2004

Scheid, S. C., "Die verallgemeinerte Lognormalverteilung", Diplomarbeit, Universität Dortmund Fachbereich Statistik, Nov. 2001

Schulze, A., „Statistische Prozessregelung (SPC)", in Masing – Handbuch Qualitätsmanagement, Hrsg. Pfeifer, T. und Schmitt, R., Hanser Verlag München, 6. Auflage, 2014

Sondermann, J.P., „Interne Qualitätsanforderungen und Anforderungsbewertung", in Masing – Handbuch Qualitätsmanagement, Hrsg. Pfeifer, T. und Schmitt, R., Hanser Verlag München, 5. Auflage, 2007

Susto, G. A., „Statistical Methods for Semiconductor Manufacturing", PhD Thesis, XXV Series, Ph.D. School in Information Engineering, Department of Information Engineering, University of Padova, 2013

Van Zant, P., „Microchip Fabrication – Third Edition A practical guide to semiconductor processing", McGraw-Hill, New York, 1997

VDA Verband der Automobilindustrie e.V., Qualitätsmanagement Center (QMC), Berlin, „Qualitätsmanagement in der Automobilindustrie 6, Prozessaudit, Teil 3", Gelbband, 2. vollständig überarbeitete Auflage, Februar 2010

Villacourt, M., "Failure Mode and Effects Analysis (FMEA): A Guide for Continuous
Improvement for the Semiconductor Equipment Industry", SEMATECH Technology Transfer #92020963B-ENG, 1992

Wagner, S. M., Lieferantenmanagement" in Masing – Handbuch Qualitätsmanagement, Hrsg. Pfeifer, T. und Schmitt, R., Hanser Verlag München, 6. Auflage, 2014

Walk, C., "Hand-book on STATISTICAL DISTRIBUTIONS for experimentalists", Particle Physics Group, Fysikum, University of Stockholm, (e-mail: walck@physto.se), Internal Report SUF–PFY/96–01, Stockholm, 11 December 1996, 1st revision, 31 October 1998, last modification 10 September 2007

Warwick Manufacturing Group, University of Warwick, "Weibull analysis" Product Excellence using 6 Sigma (PEUSS), Section 8, 2007

Widmann, D., Mader, H., Friedrich, H., "Technologie hochintegrierter Schaltungen", 2. Auflage, Springer, Berlin, 1996

Wolf, M., "Tool-Entwicklung zur Bestimmung des Reifegrads von Prozessmanagement aus ganzheitlicher Sicht (ready for the process")", Master Thesis, University of Graz, 2008

Zollondz, H.-D., „Grundlagen Qualitätsmanagement Einführung in Geschichte, Begriffe, Systeme und Konzepte", Oldenbourg Verlag München, 3. Auflage, 2011

ZVEI Electronic Components and Systems, "Guideline for General Automotive Quality Agreement for Electronic Components, Rev. 2, January 2006

ZVEI, „Handbook for Robustness Validation of Semiconductor Devices in Automotive Applications", ZVEI - Zentralverband Elektrotechnik- und Elektronikindustrie e. V. (German Electrical and Electronic Manufacturers' Association), Electronic Components and Systems Division, 2007

ZVEI, "Robustness Validation Manual – How to use the Handbook in product engineering", ZVEI – Zentralverband Elektrotechnik- und Elektronikindustrie e.V. (German Electrical and Electronic Manufacturers' Association), Electronic Components and Systems Division, 2010

ZVEI, „Handbook for Robustness Validation of Automotive Electrical/Electronic Modules", ZVEI - Zentralverband Elektrotechnik- und Elektronikindustrie e. V. (German Electrical and Electronic Manufacturers' Association), Electronic Components and Systems Division, Revision: June 2013

## 14.2. Journals & Conferences

Abreu, P., Sousa, S., Lopes, I. D. S. "Using Six Sigma to improve complaints handling", Proceedings of the World Congress on Engineering 2012 Vol III, WCE 2012, July 4 - 6, 2012, London, U.K. 2012

Aguilar-Saven, R. S., "Business process modelling: Review and framework", Int. J. Production Economics 90 (2004) 129–149, 2004

Ahadi, H.R., "An Examination of the Role of Organizational Enablers in Business Process Reengineering and the Impact of Information Technology", IDEA GROUP PUBLISHING, IJT2598, 2004

Ahmad, A., "Testing of complex integrated circuits (ICs) – The bottlenecks and solutions", Asian Journal of Information Technology, 4(9), 816-822, 2005

Akao, Y., "QFD: Past, Present, and Future", International Symposium on QFD '97, Linköping, 1997

Allan A., Edenfeld, D., Joyner, W. H., Kahng, A. B., Rodgers, M., Zorian, Y., "2001 Technology Roadmap for Semiconductors" Allan, Computer 35.1 (2002): 42-53, 2002

Antony, J., Gijo, E. V., Childe, S. J., "Case study in Six Sigma methodology: manufacturing quality improvement and guidance for managers", Production Planning & Control, 23(8), 624-640, 2012

Baleani, M., Ferrari, A., Mangeruca, L., Sangiovanni-Vincentelli, A., Peri, M., Pezzini, S., "Fault-tolerant platforms for automotive safety-critical applications", In Proceedings of the 2003 international conference on Compilers, architecture and synthesis for embedded systems (pp. 170-177). ACM, Oct. 2003

Becker, M., Meyer, E., "TQM excellence model to examine the role of strategic change and organizational", EPISTEMOLOGIA, 11(01), 5-10, 2014

Bossink, B. A., Blauw, J. N., "Strategic ambitions as drivers of improvement at DaimlerChrysler". Measuring Business Excellence, 6(4), 5-11, 2002

Bhuiyan, N., Baghel, A., "An overview of continuous improvement: from the past to the present", Management Decision, 43(5), 761-771, 2005

Capodieci, L., Gupta, P., Kahng, A. B., Sylvester, D., Yang, J., "Toward a methodology for manufacturability-driven design rule exploration", In Proceedings of the 41st annual Design Automation Conference (pp. 311 - 316). ACM, June 2004

Carbone, T., Tippett, D., "Project Risk Management Using the Project Risk FMEA", Engineering Management Journal, Vol. 16, No. 4, Dec. 2004

Catic, D., Arsovski, S., Jeremic, B., Glisovic, J., "FMEA in Product Development Phase", 5th International Quality Conference, May 20th, 2011

Chang, C. L., Wen, C. H. P., Bhadra, J., "Process-variation-aware Iddq Diagnois for nano-scale CMOS designs - the first step", In Design, Automation & Test in Europe Conference & Exhibition (DATE), 2013 (pp. 454-457). IEEE, March 2013

Chesbrough, H. W.,"Why companies should have open business models", In: MIT sloan management review, Yr.. 48, H. 2: 22–28. 2007, Management review, Jg. 48, H. 2: 22–28.

Cho, M., Ban, Y., Pan, D. Z., "Double patterning technology friendly detailed routing", In Computer-Aided Design, ICCAD 2008. IEEE/ACM International Conference on (pp. 506-511). IEEE, November 2008

Dahlgaard, J. J., Chen, C. K., Jang, J. Y., Banegas, L. A., & Dahlgaard-Park, S. M., "Business excellence models: Limitations, reflections and further development", Total Quality Management & Business Excellence, 24(5-6), 519-538, 2013

Da Silva, M. G., Giasolli, R., Cunningham, S., DeRoo, D., "Mems design for manufacturability (dfm)", Sensors Expo, Boston (USA), 2002

Dayan, R., Evans, S., "KM your way to CMMI", Journal of Knowledge Management, VOL. 10 No. 1, 2006, pp. 69-80, Emerald Group Publishing Limited, ISSN 1367-3270

Degraeve, R., Groeseneken, G., Bellens, R., Ogier, J. L., Depas, M., Roussel, P. J., & Maes, H. E., "New insights in the relation between electron trap generation and the statistical properties of oxide breakdown", Electron Devices, IEEE Transactions on, 45(4), 904-911, 1998

De Mast, J., Lokkerbol, J., "An analysis of the Six Sigma DMAIC method from the perspective of problem solving", International Journal of Production Economics 139(2) 604–614, 2012

De Schepper, P., Hansen, T., Altamirano-Sanchez, E., Pret, A. V., El Otell, Z., Boulart, W., & De Gendt, S., "Line edge and width roughness smoothing by plasma treatment", Journal of Micro/Nanolithography, MEMS, and MOEMS,13(2), 023006-023006 (2014)

Dinmohammadi, F., Shafiee, M., "A Fuzzy-FMEA Risk Assessment Approach for Offshore Wind Turbines", International Journal of Prognostics and Health Management, ISSN 2153-2648, 2013

Doggett, A. M., "A statistical comparison of three root cause analysis tools", Journal of Industrial Technology, 20(2), 2-9, 2009

Dwivedi, V., Anas, M., Siraj, M., "Six Sigma: As applied in quality improvement for injection moulding process", International Review of Applied Engineering Research, 4(4), 317-324, 2014

ElMaraghy, H., et. al., "Product variety management", CIRP Annals – Manufacturing Technology (2013), http://dx.doi.org/10.1016/j.cirp.2013.05.007

Emami-Naeini, A., Ebert, J. L., Kosut, R. L., de Roover, D., & Ghosal, S. (2004, June). "Model-based control for semiconductor and advanced materials processing: An overview", In American Control Conference, 2004, Proceedings of the 2004 (Vol. 5, pp. 3902-3909). IEEE.

Emami-Naeini, A., Dick de Roover, "Control in Semiconductor Wafer Manufacturing." Proc. Symposium To Honor Bill Wolovich, 47th IEEE Conference on Decision and Control, Cancun, Mexico, 2008

Evans, R. A., "COMPONENTS VIS-A-VIS SYSTEMS", IEEE Transactions on Reliability 39 (1990) p. 257

Glueck, J., Wagner, R., Bunz, R., Koch, H., & Schmidt, C., "Zero Defect Strategy for Electronic Components in Automotive Applications", In 2006 IEEE International Test Conference, 2006

Grambow, G., Oberhauser, R., Reichert, M., "Towards Automated Process Assessment in Software Engineering", 2012, In: 7th Int'l Conf on Software Engineering Advances (ICSEA'12), November 18-23, 2012, Lisbon, Portugal

Grasser, T., Kaczer, B., Goes, W., Reisinger, H., Aichinger, T., Hehenberger, P., Wagner, P.-J., Schanovsky, F., Franco, J., Luque, M. T., Nelhiebel, M., "The Paradigm Shift in Understanding the Bias Temperature Instability: From Reaction – Diffusion to Switching Oxide Traps", IEEE Transactions on Electron Devices, 2011

Grasser, T., Reisinger, H., Wagner, P.-J., Schanovsky, F., Goes, W. Kaczer, B., „The Time Dependent Defect Spectroscopy (TDDS) for the Characterization of the Bias Temperature Instability", IEEE Transactions on Electron Devices, 2011

Grasser, T., "Fundamentals of RTN, BTI, and Hot Carrier Degradation - A Matter of Timescales", Institute for Microelectronics TU Wien, Vienna, Austria, International Reliability Physics Symposium IRPS 2013

Gunasekaran, A., Kobu, B., "Modelling and analysis of business process reengineering", International Journal of Production Research, 2002, vol. 40, no. 11, 2521-2546

Hallberg, Ö., Peck, D. S., "Recent humidity accelerations, a base for testing standards", Quality and Reliability Engineering International, 1991, 7. Jg., Nr. 3, S. 169-180.

Harloff, T., Salavati, N., "Pannenserie bei General Motors Tödliche Fehlzündung", Süddeutsche Zeitung, Apr 1st, 2014

Hauser, J. R., „How Puritan-Bennett used the House of Quality", Sloan Management Review Reprint Series, Spring 1993, Volume 34, Number 3

He, Z., Goh, T. N., "Enhancing the Future Impact of Six Sigma Management. QUALITY TECHNOLOGY AND QUANTITATIVE MANAGEMENT, 12(1), 83-92, 2015

Hu, C., Tam, S. C., Hsu, F. C., Ko, P. K., Chan, T. Y., & Terrill, K. W., "Hot-electron-induced MOSFET degradation — Model, monitor, and improvement", Electron Devices, IEEE Transactions on, 32(2), 375-385, 1985

Hu, JS, et. al., "Assembly system design and operations for product variety", CIRP Annals – Manufacturing Technology (2011), doi:10.1016/j.cirp.2011.05.004

Ismyrlis, V., Moschidis, O., "Six Sigma's critical success factors and toolbox" International Journal of Lean Six Sigma, 4(2), 108-117, 2013

Kavakli, V., Loucopoulos, P., "Goal-driven business process analysis application in electricity deregulation", Information Systems Vol. 24 No. 3, pp. 187 – 207, Elsevier Science Ltd., 1999

Kmenta, S., "Scenario-Based FMEA: A Life Cycle Cost Perpective"; in Proceedings of DETC 2000 ASME Desing Engineering Technical Conference, Sep 10-14, 2000, Baltimore, Maryland

Kohlbacher, M., "The Perceived Effects of Business Process Management", Science and Technology for Humanity (TIC-STH), 2009 IEEE Toronto International Conference, 26-27 Sept. 2009, Page(s): 399 – 402, ISBN: 978-1-4244-3877-8

Liebmann, L. W., Mansfield, S. M., Wong, A. K., Lavin, M. A., Leipold, W. C., Dunham, T. G., "TCAD development for lithography resolution enhancement", IBM Journal of Research and Development, 45(5), 651-665, 2001

Liebmann, L. W., "Layout impact of resolution enhancement techniques: impediment or opportunity?", In Proceedings of the 2003 international symposium on Physical design (pp. 110-117). ACM, April 2003

Li, N., Carlsson, G., Dubrova, E., Petersen, K. "Logic BIST: State-of-the-Art and Open Problems", arXiv preprint arXiv:1503.04628, 2015

Limpert, E., Stahel, W.A., Abbt, M., "Log-normal Distributions across the Sciences: Keys and Clues May 2001 / Vol. 51 No. 5, BioScience 341

Lin, C. W., Tsai, M. C., Lee, K. Y., Chen, T. C., Wang, T. C., Chang, Y. W., "Recent Research and Emerging Challenges in Physical Design for Manufacturability/Reliability", In Proceedings of the 2007 Asia and South Pacific Design Automation Conference (pp. 238-243). IEEE Computer Society, Jan 2007

Lin, Jun-Shuw., "A novel design of wafer yield model for semiconductor using a GMDH polynomial and principal component analysis." Expert Systems with Applications 39.8 (2012): 6665-6671

Linderman, K., Schroeder, R.G., Zaheer, S., Choo, A.S., "Six Sigma: a goal-theoretic perspective", Journal of Operations Management 21, 193–203, 2003

Locke, E. A., Latham, G. P., "Building a practically useful theory of goal setting and task motivation: A 35-year odyssey", American Psychologist, Vol 57(9), Sep 2002, 705-717

Lu, H., Bailey, C., Yin, C., "Design for reliability of power electronics modules", Microelectronics reliability, 49(9), 1250-1255, 2009

Nahar, A., Butler, K. M., Carulli Jr, J. M., Weinberger, C., "Quality improvement and cost reduction using statistical outlier methods", In Computer Design, 2009. ICCD 2009. IEEE International Conference on (pp. 64-69). IEEE, October 2009

Neagoe, L. N., Klein, V. M., "Quality and Management Tools, An integrated Approach for Quality cost reduction", RECENT, Vol. 11, no. 2(29), July, 2010

Neagoe, L. N., Marascu Klein, V., "Employee Suggestion System (Kaizen Teian) The Bottom-up approach for productivity improvement", Transilvania University of Brasov, Romania, International Conference on Economic Engineering and Manufacturing Systems , Brasov, 26 – 27 November 2009

Mack, C. A., "Fifty Years of Moore's Law", IEEE TRANSACTIONS ON SEMI-CONDUCTOR MANUFACTURING, VOL. 24, NO. 2, MAY 2011

May, G. S., Spanos C. J., "Fundamentals of Semiconductor Manufacturing and Process Control", IEEE, Wiley-Interscience, 2006

Moreno-Lizaranzu, M. J., Cuesta, F., "Improving electronic sensor reliability by robust outlier screening", Sensors, 13(10), 13521-13542, 2013

Ōno, T., "Toyota production system: beyond large-scale production", Productivity press, 1988.

Pan, D. Z., Yu, B., Gao, J. R., "Design for manufacturing with emerging nanolithography. Computer-Aided Design of Integrated Circuits and Systems", IEEE Transactions on, 32(10), 1453-1472, 2013

Pearson, K., "Note on regression and inheritance in the case of two parents", Proceedings of the Royal Society of London, Jan 1895 **58**, p. 240-242

Raharjo, H., Xie, M., Brombacher, A. C., „A systematic methodology to deal with the dynamics of customer needs in Quality Function Deployment", ResearchGate, Expert Systems with Applications 38 (2011) 3653 – 3663, April 2011

Rajsuman, R., "Iddq testing for CMOS VLSI". Proceedings of the IEEE, 88(4), 544-568., 2000

Ren, Y., Liu, L., Yin, S., Han, J., Wu, Q., Wei, S., "A fault tolerant NoC architecture using quad-spare mesh topology and dynamic reconfiguration", Journal of Systems Architecture, 59(7), 482-491, 2013

Ricco, R., "Legal Aspects of product liability and FMEA", Safety and Reliability for Managing Risk, Guedes Soares & Zio (eds), 2006 Taylor & Francis Group, London, ISBN 0-415-41620-5
Singh, J., Singh, H., "Continuous improvement approach: state-of-art review and future implications", International Journal of Lean Six Sigma, 3(2), 88-111, 2012

Rooney, J. J., Heuvel, L. N. V., "Root cause analysis for beginners", Quality progress, 37(7), 45-56, 2004

Schiffauerova, A., Thomson, V. "A review of research on cost of quality models and best practices", International Journal of Quality & Reliability Management, 23(6), 647-669, 2006

Schneiderman, A. M., "Setting quality goals" Quality Progress, 21(4), 51-57, 1988

Shivakumar, P., Keckler, S. W., Moore, C. R., Burger, D., "Exploiting microarchitectural redundancy for defect tolerance", In Computer Design (ICCD), 2012 IEEE 30th International Conference on (pp. 35-42). IEEE, Sep. 2012

Singh, V. K., "PDCA cycle: a quality approach", Utthan – The Journal of Management Sciences, 1(1), 2013

Soković, M., Jovanović, J., Krivokapić, Z., & Vujović, A., "Basic quality tools in continuous improvement process", Strojniški vestnik - Journal of Mechanical Engineering, 55(5), 333-341, 2009

Song,. Y, Wang, B., "Survey on Reliability of Power Electronic Systems", IEEE TRANSACTIONS ON POWER ELECTRONICS, VOL. 28, NO. 1, JANUARY 2013, p. 591

Soni, S., Mohan, R., Bajpai, L., Katare, S. K., "Reduction of welding defects using Six Sigma techniques", International Journal of Mechanical Engineering and Robotics Research, 2(3), 2278-0149, 2013

Storch, M., „Motto-Ziele, S.M.A.R.T.-Ziele und Motivation", in: Birgmeier, Bernd (Hrsg.): Coachingwissen. Denn sie wissen nicht, was sie tun?, VS Verlag für Sozialwissenschaften / GWV Fachverlage GmbH, Wiesbaden 2009, S. 183-205
ISBN 978-3-531-16306-2

Suresh, S., Moe, A. L., Abu, A. B., "Defects Reduction in Manufacturing of Automobile Piston Ring Using Six Sigma", Journal of Industrial and Intelligent Information Vol, 3(1), 2015

Syduzzaman, Rahman, M., Islam, M., Habib, A., Ahmed, S., "Implementing Total Quality Management Approach in Garments Industry", Bangladesh University of Textiles, Bangladesh
European Scientific Journal December 2014 edition vol.10, No.34 ISSN: 1857 – 7881 (Print) e - ISSN 1857- 7431 341

Thaduri, A., Verma, A. K., Gopika, V., Gopinath, R., & Kumar, U., "Reliability prediction of semiconductor devices using modified physics of failure approach", International Journal of System Assurance Engineering and Management, 4(1), 33-47, 2013

Tiederle, V., "Zuverlässigkeitsprognose von Bauelementen", RELNETYX, Reliability Experts Network, Partner of the Mobility Industry for Reliability, Quality, Safety & Validation, 11. Europäisches Elektroniktechnologie-Kolleg, 16. – 20. Apr. 2008

Thümmel, A., „Vorlesungsskript Qualitätsmanagement", Hochschule Darmstadt, University of Applied Sciences, 2010

Uygur, A., Sümerli, S., "EFQM Excellence Model. International Review of Management and Business Research", 2(4), 980, 2013

Van der Aalst, W.M.P., ter Hofstede, A. H. M., Weske, M., "Business Process Management: A Survey", Business Process Management, Lecture Notes in Computer Science Volume 2678, 2003, pp 1-12, 27 May 2003

Van Soestbergen, M., Mavinkurve, A., Rongen, R. T. H., Jansen, K. M. B., Ernst, L. J., & Zhang, G. Q., "Theory of aluminum metallization corrosion in micro-electronics", Electrochimica Acta, 55(19), 5459-5469, 2010

von Tils, V., "Design Requirements for Automotive Reliability", Bosch, ESSDERC 2006, Montreux, Switzerland, Sep 22, 2006

Wang, H., Ma, K., Blaabjerg, F., "Design for reliability of power electronic systems", In IECON 2012-38th Annual Conference on IEEE Industrial Electronics Society (pp. 33-44). IEEE, Oct 2012

Weibull, W., "A statistical distribution function of wide applicability", Journal of Applied Mechanics, Sep 1951

Weis, S., Garbade, A., Fechner, B., Mendelson, A., Giorgi, R., Ungerer, T., " Architectural support for fault tolerance in a teradevice dataflow system. International Journal of Parallel Programming, 1-25., 2014

Wen, L., Ross, R. G., "Comparison of LCC solder joint life predictions with experimental data", Journal of Electronic Packaging, 1995, 117. Jg., Nr. 2, S. 109-115.

Wittmann, J., Bergholz, W., "Wert der Bewertung", Qualität und Zuverlässigkeit, QZ Jahrgang 51 (2006) 11, Carl Hanser Verlag, München

Wittmann, J., "Praxisorientiertes Prozessmanagement", 4cost Kongress, Berlin, Oct 9th, 2014

Xiang, Y., Chantem, T, Dick, R. P., Hu, X. S., & Shang, L., "System-level reliability modeling for MPSoCs" In Proceedings of the eighth IEEE/ACM/IFIP international conference on Hardware/software codesign and system synthesis (pp. 297-306). ACM, Oct. 2010

Xu, Q., Jiang, L., Li, H., & Eklow, B., "Yield enhancement for 3D-stacked ICs: Recent advances and challenges", Design Automation Conference (ASP-DAC), 2012 17th Asia and South Pacific. IEEE, 2012. S. 731-737.

Zimmerman, P., "Double patterning lithography: double the trouble or double the fun?", SPIE Newsroom, 20, 2009

## 14.3. Company internal

Arndt, Hanke, Höhndorf, "Failure Mode and Effects Analysis", Siemens HL, A66500-V30-X-A, released 1993-05-17

Auner, M., "Product Development Manual", Qimonda AG, Doc ID A66417-M1-X-*-7635, released 2008-04-09

Brunner, C., Haßelberg, M., Janker, A., Künne, M., Milferstädt, D., Schneider, W., "Supplier Management", Infineon Technologies AG, Doc ID A66500-P39-X-*-7635, Revision 02, released 2001-08-02

Gerber, K., "Failure Mode & Effects Analysis", Qimonda AG, Doc ID A66500-P0246-X-*-7635, Version 2, released 2009-02-13

Geleng, J., private communication, Munich 1999

Kanert, W., "Training Reliability Testing of Products", Infineon Technologies AG, 07.03.2008, Module 3.3 Mission Profiles

Specht, J., "Product Qualification in Qimonda Product Groups", Qimonda AG, Doc ID A66500-V71-X-*-7635, Version 7, released 2008-05-13

Wagner, T., "Statistical Process Control (SPC)", Infineon Technologies AG, Quality Management Guideline, A66500-M0007-X-06-7635, Issue 2007-05

Wagner, T., "QM Tool Description Statistical Process Control (SPC)", Infineon Technologies AG, A66500-T0002-X-*-7635, Rev. 3, Issue 2008

Wagner, T., "Statistics Handbook", Infineon Technologies AG, Quality Management, A66500-M0011-X-03-7635, Issue 2008-03

Wagner, T., "Measurement System Analysis", Infineon Technologies AG, Quality Management, A66500-M0001-X-07-7635, Issue 2008-09-29

Walter, W., "Tutorial on Metallization Reliability Methodology", Infineon Technologies AG, March 2004

Willer, J., "Flash Technologies Seminar – Memories", Qimonda AG, November 2007

Wittmann, J., "Customer sample quality requirements for Computing and Specialty DRAM", Qimonda AG Quality Management, A66500-S0171-X-*-7635, Version 1, Issue 2008-06-16

## 14.4. International and Consortia Standards

AEC Automotive Electronics Council, Component Technical Committee, "Guideline for part average testing", AEC - Q001 Rev-D, December 9, 2011

AEC Automotive Electronics Council, Component Technical Committee, "Guidelines for statistical yield analysis", AEC - Q002 Rev B, January 12, 2012

AEC Automotive Electronics Council, Component Technical Committee, "ZERO DEFECTS GUIDELINE", Draft version, AEC - Q004, August 31, 2006

AEC Automotive Electronics Council, Component Technical Committee, "FAILURE MECHANISM BASED STRESS TEST QUALIFICATION FOR INTEGRATED CIRCUITS", AEC - Q100 - Rev-H, September 11, 2014

DIN EN ISO 9001:2015, "Quality management systems – Requirements (ISO 9001:2015)"; German and English version EN ISO 9001:2015, DIN Deutsches Institut für Normung e. V. · Jede Art der Vervielfältigung, auch auszugsweise,
nur mit Genehmigung des DIN Deutsches Institut für Normung e. V., Berlin, Beuth Verlag GmbH, 10772 Berlin

DIN EN ISO 9004:2009, "Managing for the sustained success of an organization –
A quality management approach", Trilingual version EN ISO 9004:2009, DIN Deutsches Institut für Normung e.V., Beuth Verlag GmbH, 10772 Berlin

DIN EN ISO 19011:2011, „Guidelines for auditing management systems (ISO19011:2011); German and English version", Beuth Verlag Berlin, Dec 2011

DIN EN ISO 14644-1:1999, "Cleanrooms and associated controlled environments – Part 1: Classification of air cleanliness", International Organization for Standardization, 1999

DIN ISO 21747:2007-03, „Statistische Verfahren – Prozessleistungs- und Prozessfähigkeitskenngrößen für kontinuierliche Qualitätsmerkmale", „Statistical methods – Process performance and capability statistics for measured quality characteristics"

DIN ISO 22514-2:2013, "Statistical methods in process management – Capability and performance – Part 2: Process capability and performance of time-dependent process models", DIN Deutsches Institut für Normung e. V., Beuth-Verlag 2013

DIN ISO 3534-2:2006, „Statistik – Begriffe und Formelzeichen – Teil 2: Angewandte Statistik"

DIN ISO 5479:2004-01, „Statistische Auswertung von Daten - Tests auf Abweichung von der Normalverteilung (ISO 5479:1997)", „Statistical interpretation of data - Tests for departure from the normal distribution (ISO 5479:1997)"

DIN ISO/TS 16949, DIN SPEC 1115, „Qualitätsmanagementsysteme – Besondere Anforderungen bei Anwendung von ISO9001:2008 für die Serien- und Ersatzteil-Produktion in der Automobilindustrie", Normenausschuss Qualitätsmanagement, Statistik und Zertifizierungsgrundlagen (NQSZ) im DIN, Beuth Verlag Berlin, 2009

ISO/TS16949, „Customer Specific Requirements (ISO/TS-16949) Semiconductor Commodity", Jan 12, 2004

IPC/JEDEC J-STD-020D.1, "Moisture/Reflow Sensitivity Classification for Nonhermetic Solid State Surface Mount Devices", March 2008 (Supersedes IPC/JEDEC J-STD-020D
August 2007)

JESD22-A101C, "Steady State Temperature Humidity Bias Life Test", JEDEC Solid State Technology Association, March 2009

JESD22-A102-C, "Accelerated Moisture Resistance - Unbiased Autoclave", JEDEC Solid State Technology Association, October 2000

JESD22-A103D, "High Temperature Storage Life", JEDEC Solid State Technology Association, December 2010

JESD22-A104D, "Temperature Cycling", JEDEC Solid State Technology Association, March 2009

JESD22-A108D, "Temperature, Bias, and Operating Life", JEDEC Solid State Technology Association, November 2010

JESD22-A113F, "Preconditioning of Nonhermetic Surface Mount Devices Prior to Reliability Testing", JEDEC Solid State Technology Association, October 2008

JESD22-A118A, "Accelerated Moisture Resistance - Unbiased HAST", JEDEC Solid State Technology Association, March 2011

JESD47 I, "Stress-Test driven Qualification of Integrated Circuits", JEDEC Solid State Technology Association (2012)

JESD89-3A, "Test Method for Beam Accelerated Soft Error Rate", JEDEC Solid State Technology Association, November 2007

JESD122G, "Failure Mechanisms and Models for Semiconductor Devices", October 2012, JEDEC Solid State Technology Association

JESP146A, „Guidelines for Supplier Performance Rating", January 2008, JEDEC Solid State Technology Association

prEN DIN 9001:2014-08, Deutsches Institut für Normung e.V., Beuth-Verlag, 2014

VDA, „Band 4: Produkt- und Prozess FMEA", 2. überarbeitete Auflage 2006, aktualisiert im Juni 2012

## 14.5. Internet

Aal, A., "Discussion Group (DG) Summary: fast Wafer Level Reliability (fWLR) Monitoring", MELEXIS Microelectronic Integrated Systems, 2010 IIRW FINAL REPORT
http://ieeexplore.ieee.org/stamp/stamp.jsp?arnumber=5706517
[Dec 4th, 2015]

Adams, T. C., "RELIABILITY: Definition & Quantitative Illustration"
http://kscsma.ksc.nasa.gov/Reliability/.../whatReli.pdf
[Mar 21st, 2015]

Alcom, "Supplier Controlled - Safe Launch Plan", September 2013
http://supplier.alpsautomotive.biz/supplyWeb/Doc/SLP.ppt
[Feb 27th, 2016]

Allen, P. E., "LECTURE 010 - INTRODUCTION TO CMOS ANALOG CIRCUIT DESIGN", 2010
http://www.google.de/url?sa=t&rct=j&q=&esrc=s&source=web&cd=6&ved=0CEkQFjAF&url=http%3A%2F%2Fwww.aicdesign.org%2FSCNOTES%2F2010notes%2FLect2UP010_%2528100324%2529.pdf&ei=8DYMVe_MN8TpUreIg9AC&usg=AFQjCNEX8KGthqTNickraC-pb83msPiKng
[Mar 20th, 2015]

Alpha and Omega Semiconductor, "Power Semiconductor Reliability Handbook", 2010 Alpha and Omega Semiconductor, Rev. 1.0
http://www.aosmd.com/media/reliability-handbook.pdf
[Dec 31st, 2015]

Altera, „RELIABILITY REPORT 59 1H 2015", 2015 Altera Corporation
https://www.altera.com/en_US/pdfs/literature/rr/rr.pdf
[Dec 25th, 2015]

Altera (2), „Altera's Automotive Quality Program Building Higher Quality Automotive Electronics", July 2015
https://www.altera.com/content/dam/altera-www/global/en_US/pdfs/literature/br/alteraautomotivequalityprogram.pdf
[Feb 11th, 2016]

Arthur DLittle, "Markt- und Technologiestudie Leistungselektronik Automotive 2015"
http://www.adlittle.de/uploads/tx_extthoughtleadership/ADL_Studie_Leistungselektronik_2015_2006.pdf
[Dec 17th, 2015]

Azizi, N., Yiannacouras, P., "Gate Oxide Breakdown", ECE1768 – Reliability of Integrated Circuits, 2012
http://ambientelectrons.org/wp-content/uploads/2012/02/presentation.pdf
[Jan 2nd, 2016]

Barr, T., "Introduction to economic statistics", Topic 5 Discrete Random Variables, 2010
http://www.tavisbarr.com/classes/intro_stats/topic-05-discrete-rvs.pdf
[Jan 15th, 2015]

Berlin Partner GmbH, "Incoterms 2000", 2007
http://www.businesslocationcenter.de/imperia/md/content/aussenwirtschaft/exportimport/incoterms_2000.pdf
[Apr 24th, 2016]

Boning, D., "Run-by-Run Control Methods", Lecture Notes Spring 2003
http://ocw.mit.edu/courses/electrical-engineering-and-computer-science/6-780-semiconductor-manufacturing-spring-2003/lecture-notes/ln17runbyrun.pdf
[May 27th, 2016]

Bozturk, C., "BURN-IN, RELIABILITY TESTING, AND MANUFACTURING OF SEMICONDUCTORS"
https://www.google.de/url?sa=t&rct=j&q=&esrc=s&source=web&cd=1&cad=rja&uact=8&ved=0ahUKEwj9g_3xtLXJAhVhnXIKHVI1BbQQFggdMAA&url=http%3A%2F%2Fwww.personal.kent.edu%2F~cbozturk%2FBURN-IN%2C%2520RELIABILITY%2520TESTING%2C%2520AND%2520MANUFACTURING%2520OF.ppt&usg=AFQjCNHn8WfhG06rgoqRMoCWFJzVohKHnA
[Nov 29th, 2015]

Burton, T. T., "A History of Lean and Continuous Improvement", Executive White Paper Series, The Center for Excellence in Operations, Inc., 2014
http://ceobreakthrough.com/wp/wp-content/uploads/2015/03/A-History-of-Lean-and-Continuous-Improvement.pdf
[Jan 22nd, 2016]

Business Dictionary, Definition of product development
http://www.businessdictionary.com/definition/product-development.html
[Mar 24th, 2015]

Business Dictionary, Definition of direct material
http://www.businessdictionary.com/definition/direct-material.html
[July 30th, 2015]

Byers, E., "Memory Scaling Challenges Detailed by Micron R&D Director", Micron, LithoVision (2015)
https://nikonereview.com/2015/memory-scaling-challenges-detailed-by-micron-rd-director/
[Dec 17th, 2015]

Cayman Business Systems, Revision N 980815, "Failure Mode and Effects Analysis", 2002
http://elsmar.com/pdf_files/FMEA-N.pdf
[Nov 14th, 2014]

Chang, J. T.-Y., McCluskey, E. J., "Quantitative Analysis of Very-Low-Voltage Testing", Center for Reliable Computing Stanford University, Stanford, CA, 1996
http://www-crc.stanford.edu/crc_papers/changvts96.pdf
[Feb 25th, 2016]

Cirrus, "Cirrus Logic Reliability: Qualification & Monitoring"
http://www.cirrus.com/en/quality/reliability.html
[Mar 27th, 2015]

Continental AG, "Supplier Requirements Manual", Revision Date: 20th of July 2015
http://www.continental-corporation.com/www/download/portal_com_en/themes/global_sourcing/download/supplier_requirements_manuel.pdf
[Feb 25th, 2016]

Continuous probability distribution, www.princeton.edu
https://www.princeton.edu/~achaney/tmve/wiki100k/docs/Continuous_probability_distribution.html
[Jan 15th, 2015]

DGQ Deutsche Gesellschaft für Qualität, „Expertenwissen für DGQ-Mitglieder Das EFQM Excellence Modell 2013"
http://www.dgq.de/dateien/EFQM-Excellence-Modell-2013.pdf
[Jan 7th, 2016]

Dieseldorff, C. G., Tseng, C., SEMI, "Technology Node Transistions Slowing Below 32nm" (June 30, 2014)
http://www.semi.org/en/node/50391
[Nov 12th, 2015]

Dixon, J., Jones, T., "Hype Cycle for Business Process Management, 2011", Gartner Research, ID Number: G00214214, 25 July 2011
http://www.adeptia.com/products/Hype_cycle_BPM_2011.pdf
[Nov 23rd, 2014]

Dovich, R.,"FMEA and you", Perspectives on Quality, Vol. 1, Nr. 7
http://www.fmeainfocentre.com/guides.htm
[Nov 17th, 2014]

EFQM, "Business Excellence Matrix User Guide – EFQM Model 2013 Version"
http://www.efqm.org/sites/default/files/efqm_bem_2013_user_guide.pdf
[Jan 7th, 2016]

EFQM, "Our History EFQM celebrates its 25th anniversary"
http://www.efqm.org/about-us/our-history
[Jan 8th, 2016]

Engineering Statistics Handbook, NIST/SEMATECH e-Handbook of Statistical Methods,
http://www.itl.nist.gov/div898/handbook/
[Dec 31st, 2014]

Fairchild Semiconductor, „Wafer Level Reliability"
http://www.asq-pinetree.org/uploads/6/3/8/7/6387208/wlr_overview_1_for_asq_updated.pptx
[Nov 29th, 2015]

Frankwicz, P. S., Romano, S. E., Moutinho, T., "Process Excursion Detection using Statistical Analysis Methodologies in High Volume Semiconductor Production", Operations Engineering, National Semiconductor, South Portland, Maine, NESUG 2009 Posters
http://www.lexjansen.com/nesug/nesug09/po/PO11.pdf
[Dec 4th, 2015]

Freidinger, R., „Geschäftsprozesse im Unternehmen", Manuskript. zur Vorlesung. Dr. GP0303282.DOC, 2003
www.freidinger.de/Skript/GP0303283.doc
[Nov 30th, 2014]

Furtaw, R., "Texas Instruments General Quality Guidelines", Copyright © 2015, Texas Instruments Incorporated
http://www.ti.com/lit/ml/szzq076h/szzq076h.pdf
[Dec 5th, 2015]

Fujitsu Semiconductor Limited, "FUJITSU SEMICONDUCTOR Quality and Reliability Assurance", ©2002-2012 FUJITSU SEMICONDUCTOR LIMITED, AD00-00003-13E October 2012
http://www.fujitsu.com/downloads/EDG/binary/pdf/catalogs/a000000313e.pdf
[Dec 5th, 2015]

General Motors Company, "General Motors Customer Specific Requirements – ISO/TS16949, Including GM Specific Instructions for PPAP 4th Ed. (see Section 5)", 2015
http://www.iatfglobaloversight.org/docs/REVISION%20Master_GM%20Customer%20Specifics_rev141212_FINAL.pdf
[Nov 11th, 2015]

Germany Trade & Invest, "INDUSTRY OVERVIEW The Automotive Electronics Industry in Germany", Issue 2014/2015
https://www.gtai.de/GTAI/Content/EN/Invest/_SharedDocs/Downloads/GTAI/Industry-overviews/industry-overview-automotive-electronics-industry-en.pdf
[Feb 11th, 2016]

Gruber, Jürgen, „Robustness Validation / Mission Profile Compared to AEC-Q100 Standard Qualification Flow", AESIN Conference, October 2015
http://aesin.org.uk/wp-content/uploads/2015/02/RoodMicrotec_NMI-Robustness-Validation.pdf
[Jan 21st, 2016]

Fangaria, P., "Automotive IC Market to Display Strongest Growth Through 2018 - From luxury to base models, IC content on all new cars is increasing", IC Insights Research Bulletin, 2014 IC Insights, Inc.
http://us1.campaign-archive2.com/?u=77cdbf15a0a91de4d98102b0a&id=e6164a26bd&e=7d246f3eb2
[Dec 15th, 2015]

Hartsell, M., "Best practices drive up basic automotive semiconductor component reliability", Texas Instruments, EETimes, designlines Automotive, Aug 8th, 2008
http://www.eetimes.com/document.asp?doc_id=1272891
[Nov 13th, 2015]

Heist, W., "Quality First – Zero Defect Strategy", 55th EOQ Congress World Quality Congress, Budapest, Hungary, June 20-23, 2011, Knorr-Bremse Systeme für Nutzfahrzeuge GmbH, Germany
http://www.eoq.org/fileadmin/user_upload/Documents/Congress_proceedings/Budapest_June_2011/Proceedings/4_1_heist_s.pdf
[Feb 25th, 2016]

Hodlin, S., "The Road to Business Excellence: The Integrated Quality System Way", DST Output, 2008
http://www.asq-qm.org/resourcesmodule/download_resource/id/432/src/@random4baa4bfa00315/
[Jan 7th, 2016]

Infineon Technologies, "Angle Sensor GMR-Based Dual Die Angle Sensor TLE5012", Data Sheet, Rev. 1.1, 2015-03-12
http://pdf1.alldatasheet.com/datasheet-pdf/view/752237/INFINEON/TLE5012BD.html
[Fab 14th, 2016]

Infineon Technologies, Living Automotive Excellence On the way to Zero Defect products and services, September 2010
http://www.infineon.com/dgdl/Living+Automotive+Excellence.pdf?folderId=db3a30431ce5fb52011d2dd52b231e7d&fileId=db3a30431ce5fb52011d2dd5ca751e7e
[Feb 25th, 2016]

Institute of Quality & Reliability, "Tables of constants for Control charts", 2008
http://www.world-class-quality.com/images/download/20081104091341_Control%20Chart%20Constants%20and%20Formulae.pdf
[Feb 1st, 2015]

Intel, „Intel Quality System Handbook", April 2014
http://www.intel.com/content/www/us/en/manufacturing/quality-system-handbook.html
[Dec 4th, 2015]

Intel, "Intel forges ahead to 10nm, will move away from silicon at 7nm", arstechnica, Gears&Gadgets / Product News & Reviews, 2015
http://arstechnica.com/gadgets/2015/02/intel-forges-ahead-to-10nm-will-move-away-from-silicon-at-7nm/
[Nov 12th, 2015]

Intel (2), "Intel Microprocessor Transistor Count Chart"
http://www.intel.com/pressroom/kits/events/moores_law_40th/

[Dec 16th, 2015]

Intel (3), „Intel Chips Timeline"
http://www.intel.com/content/www/us/en/history/history-intel-chips-timeline-poster.html
[Dec 16th, 2015]

International Automotive Task Force, "MINIMUM AUTOMOTIVE QUALITY MANAGEMENT SYSTEM REQUIREMENTS FOR SUB-TIER SUPPLIERS", Reference to Chrysler Group LLC and Ford Motor Company ISO/TS 16949 customer-specific requirements, August 2014
http://www.iatfglobaloversight.org/docs/Minimum%20Automotive%20Quality%20Management%20System%20Requirements%20for%20Sub-tier%20suppliers%20-%20AUG%2014.pdf
[Feb 24th, 2016]

International Rectifier, "Quality requirements for IC products" (2012)
http://www.irf.com/quality-and-reliability
[Mar 22nd, 2015]

Kallenbach, R., „Trends in Automotive Electronics", Robert Bosch GmbH, Automotive Electronics Division, Reutlingen, Germany, Journal of Electrical Engineering
http://www.jee.ro/covers/art.php?issue=WO1175852866W461617421995c
[Dec 9th, 2015]

Kenol, J., "The basics of FMEA"
http://asqnorthjersey.org/Basics_of_FMEA(ASQ_304.pdf
[Nov 14th, 2014]

Klenke, A., „Tabellenwerk Statistik", Institut für Mathematik, Johannes Gutenberg-Universität Mainz, 2011
http://www.mathematik.uni-mainz.de/~klenke/vorlesungen/vorl_ss14-3/tabellenwerk.pdf
[Jan 5th, 2015]

Klotz, S., University of Tübingen
http://homepages.uni-tuebingen.de/stefan.klotz/seiten/Statistik/StatistikKorrelation.pdf
[Nov11th, 2014]

Kmenta, S., "Scenario-based FMEA Using Expected Cost", IIE Workshop, January 22, 2002
http://www.fmeainfocentre.com/presentations/SFMEA-IIE.pdf
[Nov 18th, 2014]

Lean Academy, Six Sigma Basics V7.6, © 2012 Massachusetts Institute of Technology
http://ocw.mit.edu/courses/aeronautics-and-astronautics/16-660j-introduction-to-lean-six-sigma-methods-january-iap-2012/lecture-videos/MIT16_660JIAP12_3-6.pdf
[Jan 21st, 2015]

Ludwig-Mayerhofer, W., „Verteilungen stetiger Zufallszahlen", Vorlesung Statistik (Master), Universität Siegen, Philosophische Fakultät, Seminar für Sozialwissenschaften
https://www.uni-siegen.de/phil/sozialwissenschaften/soziologie/mitarbeiter/ludwig-mayerhofer/statistik/ludwigm_down_stat2.html?lang=de
[Jan 15th, 2015]

Magna Electronics, "Supplier Quality Requirements Manual", Rev. 2
http://www.magna.com/docs/default-source/suppliers/magna_supplier_quality_requirements_manual_rev2.pdf?sfvrsn=2
[Feb 27th, 2016]

Markgraf, B., Chron
http://smallbusiness.chron.com/controlled-document-per-iso-66514.html
[Nov 2nd, 2014)

Marshall, J., „An Introduction to Reliability testing", The University of Warwick, PEUSS 2011/2012
http://www2.warwick.ac.uk/fac/sci/wmg/ftmsc/modules/modulelist/peuss/slides/section_7b_reliability_testing_slides_compatibility_mode.pdf

[Nov 29th, 2015]

Martin, A., " SPECIAL INTEREST GROUP (SIG): FAST WAFER LEVEL RELIABILITY (FWLR) MONITORING", Infineon Technologies AG, 2007 IIRW FINAL REPORT
http://ieeexplore.ieee.org/stamp/stamp.jsp?arnumber=4469246
[Dec 4th, 2015]

Müller, U., „Physikalisches Grundpraktikum", Mathematisch-Naturwissenschaftliche, Fakultät I, Institut für Physik
http://gpr.physik.hu-berlin.de/Skripten/Einfuehrung/PDF-Datei/Einfuehrung.pdf
[Dec 31st, 2014]

NIST, "2011 – 2012 Criteria for Performance Excellence", Baldrige Performance Excellence Program, National Institute of Standards and Technology, US department of commerce
http://www.nist.gov/baldrige/publications/upload/2011_2012_Business_Nonprofit_Criteria.pdf
[Jan 13th, 2016]

ON Semiconductor, "Effective Automotive Quality"
http://www.onsemi.com/pub_link/Collateral/TND387-D.PDF
[Nov 27th, 2015]

Pericom Semiconductor Corporation, „PERICOM PRODUCT CHANGE NOTIFICATION (PCN)", SPECIFICATION NO.: QA-1420 Rev: F, 2007
https://www.pericom.com/assets/PCN-Files/PCNprocedure.pdf
[Dec 5th, 2015]

PwC, "Spotline on Automotive PwC Semiconductor Report", Technology Institute
Interim Update Global Semiconductor Trends – Special Focus Automotive Industry,
September 2013
https://www.pwc.com/gx/en/technology/publications/assets/pwc-semiconductor-survey-interactive.pdf
[Feb 11th, 2016]

Quality-One, FMEA – Quick Reference Guide
http://www.fmeainfocentre.com/guides/DesignPktNewRating.pdf
[Nov 14th, 2014]

Rabaey, J. M., "Design for Test", EE241 Class Notes, UC Berkeley EE241
http://www.ohio.edu/people/starzykj/network/Class/ee516/Slides/test.pdf
[Feb 19th, 2016]

Raytheon Company, "Design Failure Modes and Effects Analysis"
http://www.raytheon.com/connections/rtnwcm/groups/public/documents/content/rtn_connect_dfmea_pdf.pdf
[Nov 14th, 2014]

Renesas, "Semiconductor Reliability Handbook", Renasas Electronics, Rev. 1, March 2013
http://www.renesas.eu/media/products/common_info/reliability/reliability_handbook/pdf/r51zz0001ej0100.pdf
[Nov 29th, 2015]

Reusch, W., "Poisson-Verteilung",
http://www.physik.uni-wuerzburg.de/~reusch/fehler/wisem0102/vorlesung7.pdf
Universität Würzburg, Physikalisches Institut, 2002
[Jan 12th, 2015]

Rubin, D., Zhao, Y., "Wafer Level Reliability Testing – A Critical Device and Process Development Step", Keithley Instruments, Inc., May 2005
http://www.tek.com/sites/tek.com/files/media/document/resources/2622%20WLR.pdf
[Dec 4th, 2015]

Rudden, J., "Making the Case for BPM: A Benefits Checklist", BPTrends, January 2007

http://www.bptrends.com-www.bptrends.com/publicationfiles/01-07-ART-MakingtheCaseforBPM-BenefitsChecklist-Rudden.pdf
[Nov 23rd, 2014]

Sandhu, J. S., "Quality Manual", Pyramid Semiconductor Corporation, Specification No.: Q-70051, Revision 07, Page 1 of 14
http://www.pyramidsemiconductor.com/download/QualityManual.pdf
[Dec 4th, 2015]

Seiko Epson Corporation, "Seiko Epson Semiconductor Product Quality Assurance Guidebook", ©SEIKO EPSON CORPORATION 2009
http://global.epson.com/products/semicon/technology/pdf/assurance_handbook.pdf
[Dec 3rd, 2015]

SemeLab, "A Short Guide to Quality and Reliability Issues In Semiconductors For High Rel. Applications", Issue 1.0 Jan 2002
http://products.semelab-tt.com/pdf/q_and_r_in_hi_rel.pdf
[Nov 29th, 2015]

Song-Moon, Suh, Applied Materials, "A Design Guideline for Zero Defect Process Tools", Jul 10th, 2013
http://www.semiconwest.org/sites/semiconwest.org/files/docs/SW2013_Song-Moon%20Suh_Applied%20Materials.pdf
[Nov 19th, 2015]

Spanos, C. J., "Special Issues in Semiconductor Manufacturing, Lecture 1: Introduction & IC Yield", Department of Electrical Engineering and Computer Sciences, EE290H, Fall 2005
https://inst.eecs.berkeley.edu/~ee290h/fa05/Lectures/PDF/lecture%201%20intro%20IC%20Yield.pdf
[Nov 27th, 2015]

Steinmann, C., "ISO 15504 (SPiCE) Assessment", HM&S GmbH, 2000
http://www.hms.org/pub/synprove/spice121/spicemotivation.pdf
[Dec 7th, 2014]

ST Microelectronics, "Quality Handbook Our path to excellence", ST Microelectronics, May 2013, Reference : BRHBOOKQ0313
http://www.st.com/web/en/resource/quality_and_reliability/quality_report/qualification_report/ST-Quality-Handbook-V1-1.pdf?sc=quality-handbook
[Dec 5th, 2015]

Turner, A., McMenamin, D., "Using the Built-in Self-Test (BIST) on the MPC5777M", Freescale Semiconductor, Inc., Document Number: AN5131, Application Note, Rev. 0, 10/2015
http://cache.nxp.com/files/microcontrollers/doc/app_note/AN5131.pdf?fpsp=1&WT_TYPE=Application%20Notes&WT_VENDOR=FREESCALE&WT_FILE_FORMAT=pdf&WT_ASSET=Documentation&fileExt=.pdf
[Feb 19th, 2016]

Turner, T., "Fast WLR tests quickly identify process variations", Keithley Instrumens Inc., 2001
http://www.testandmeasurement.com/doc/fast-wlr-tests-quickly-identify-process-varia-0001
[Nov 29th, 2015]

Viga, R., "Grundlagen ingenieurwissenschaftlichen Arbeitens (GIA)", „Zuverlässigkeit und Produktlebenszyklus"
https://www.uni-due.de/~hl271st/Lehre/GIA/gia_kap6_tei2_ss10.pdf
[Jan 4th, 2015]

Vishay Semiconductors, "Quality and Reliability", Document Number: 82501, Rev. 1.3, 26-Aug-05, Vishay Semiconductors
http://www.vishay.com/docs/82501/82501.pdf
[Dec 10th, 2015]

Webber, A., "Calculating Useful Lifetimes of Embedded Processors", Texas Instruments Incorporated, Application Report SPRABX4–November 2014

http://www.ti.com/lit/an/sprabx4/sprabx4.pdf
[Dec 11th, 2015]

Webber, A., "Calculating FIT for a Mission Profile", Texas Instruments Incorporated, Application Report, SPRABY3–March 2015
http://www.ti.com/lit/an/spraby3/spraby3.pdf
[Dec 10th, 2015]

Weiß, M., Österreicher, F., UNTERLAGEN zum STOCHASTIKSEMINAR, Teil 2, IFFB Fachdidaktik und LehrerInnenbildung, FB Mathematik der Universität Salzburg, Salzburg, November 2007
www.uni-salzburg.at/fileadmin/oracle_file_imports/554102.PDF
[Nov 12th, 2014]

WEKA, Verfasst von kringf am Do, 12/27/2012 - 15:02, "Professionelles Rückrufmanagement"
http://www.maschinenrichtlinie-2006-42-eg.de/professionelles-r%C3%BCckrufmanagement
[Oct 14th, 2015]

Winter, R., „ZVEI Fact Sheet Key facts for reliable electronics", ZVEI – Zentralverband Elektrotechnik- und Elektronikindustrie e.V, Division: 'Electronic Components and Systems' and 'PCB and Electronic Systems', Version: 15.6.2012
http://www.zvei.org/Downloads/Fact%20Sheet%20Compendium%20English.pdf
[Dec 9th, 2015]

ZF, „Quality Assurance Directive for Purchased Items", OR83, Edition 2011
http://www.zf.com/media/media/document/corporate_2/company_4/purchasing_and_logistics/purchasing_strategy/quality_guidelines/2011_4/german/qr83ausgabe_2011_offizielle_version_20110809.pdf
[Nov 10th, 2015]

www.ingramcontent.com/pod-product-compliance
Lightning Source LLC
Chambersburg PA
CBHW062346220526
45472CB00008B/1722